Four week loan
Benthyciad pedair wythnos

BASIC AND APPLIED
BONE BIOLOGY

ELSEVIER
science & technology books

Companion Web Site:

http://booksite.elsevier.com/9780124160156

Basic and Applied Bone Biology
David B. Burr and Matthew R. Allen, Editors

Resources for Professors:

- All figures from this book available as both Power Point slides and JPEG files

TOOLS FOR ALL YOUR TEACHING NEEDS
textbooks.elsevier.com

ACADEMIC PRESS

BASIC AND APPLIED BONE BIOLOGY

Edited by

DAVID B. BURR

*Department of Anatomy and Cell Biology, Indiana University School of Medicine, and
Department of Biomedical Engineering,
Indiana University-Purdue University Indianapolis (IUPUI),
Indianapolis, Indiana, USA*

MATTHEW R. ALLEN

*Department of Anatomy and Cell Biology, Indiana University School of Medicine,
Indianapolis, Indiana, USA*

AMSTERDAM • BOSTON • HEIDELBERG • LONDON • NEW YORK • OXFORD
PARIS • SAN DIEGO • SAN FRANCISCO • SINGAPORE • SYDNEY • TOKYO

Academic Press is an imprint of Elsevier

Academic Press is an imprint of Elsevier
32 Jamestown Road, London NW1 7BY, UK
225 Wyman Street, Waltham, MA 02451, USA
525 B Street, Suite 1800, San Diego, CA 92101-4495, USA

Cover credit lines:
Main image: Photograph of human humerus. Dr. Matthew R. Allen
Upper left: Photomicrograph of fracture healing in a rat bone. Dr. David B. Burr and Dr. Keith W. Condon
Left panel two: Photomicrograph of fluorochrome labels in a mouse ulna. Dr. Alexander G. Robling
Left panel three: Scanning electron micrograph of osteocyte-lacunar network. Image by Drs. Lilian Plotkin, Teresita Bellido, and Lynda Bonewald and pseudocolor by Dr. Alexander G. Robling
Left lower panel: Atomic force microscopy image of collagen fiber network in bone. Dr. Joseph Wallace.

British Library Cataloguing-in-Publication Data
A catalogue record for this book is available from the British Library

Library of Congress Cataloging-in-Publication Data
A catalog record for this book is available from the Library of Congress

ISBN: 978-0-12-416015-6

For information on all Academic Press publications
visit our website at elsevierdirect.com

Typeset by MPS Limited, Chennai, India
www.adi-mps.com

Printed and bound in China
14 15 16 17   10 9 8 7 6 5 4 3 2 1

Dedications

This book is dedicated to the wonderful teachers I have had over many years—Denny, Bruce, Harold, Eric, Mitch, and Charles, among many others—who have so patiently taught me about the beauties and intricacies of our skeleton.
And to my wife, Lisa, and son, Erik, who have tolerated and supported my obsession with bone.

—David B. Burr

This book is dedicated to my late father, who was, and always will be, the most influential teacher in my life.
And to Kristine, Sophie, and Gus who provide me with a daily reminder that there is so much more to life than bone biology.

—Matthew R. Allen

Contents

5

SKELETAL DISEASE AND TREATMENT

Preface

More than 10 years ago when we began to teach our graduate-level Basic Bone Biology course at Indiana University, there were several excellent reference works available, primarily targeted to researchers working in a wide range of areas in skeletal biology. These included (and still include) *Principles of Bone Biology*, edited by John Bilezikian, Lawrence Raisz, and John Martin, which has since been expanded to two volumes; "Big Red" (*Osteoporosis*), the excellent and very complete reference edited by Bob Marcus, David Feldman, and Jennifer Kelsey; and the more succinct *Primer of Metabolic Bone Diseases*, updated and republished every few years by the American Society of Bone and Mineral Research. These are still available, and still excellent, but they do not serve well as textbooks for a bone biology course either because they are too extensive, too expensive, or do not cover relevant topics in sufficient depth. Therefore, we have chosen over the years to use primary reference materials—mostly, recent papers published in the peer reviewed literature—for our course. From a didactic standpoint, this is an acceptable approach, and even a desirable one, especially for a graduate course in which the goal is to teach the student how to read and evaluate the literature. However, it became clear over time that this was not a sufficient surrogate for a true textbook.

As the skeletal biology group at Indiana University grew over the years, we incorporated topic experts to deliver lectures in their area of expertise. We soon realized that the course and content experts provided the foundation for building a textbook on basic and applied skeletal biology. As we discussed this idea with our colleagues here at Indiana University and up the road at Purdue University, there was universal support and enthusiasm. Discussion with colleagues outside of our group made it clear that, beyond our own requirements, there was a need and a desire by the academic community for such a text. Writing this textbook began as something of a selfish idea—we needed it for our course—but we truly hope that it will be welcomed and used by others who find it appropriate for their own courses, or as a more modest reference than existing books on a wide range of topics in skeletal biology.

Basic and Applied Bone Biology covers those topics that we feel are relevant to a modern course in skeletal biology. The book is organized, like bone, in a somewhat hierarchical manner. The first section begins with the basic construction of bone, including its cellular structure and dynamics and the basic physiological processes that bone uses to grow and adapt itself over a lifetime. This is succeeded by several chapters related to the technical aspects used to assess bone in health and disease—various imaging modalities; biomechanical measurements useful for assessing bone properties; histomorphometric techniques to evaluate the dynamics of bone modeling and remodeling; and genetic approaches used to tease out the roles of specific genes and proteins, and epigenetic influences in the basic metabolic functions of bone. These early chapters provide the foundation for the next several chapters on skeletal adaptation, highlighting mechanically-induced adaption of bone, fracture healing, and adaptation of the oral cavity associated with orthodontics and implants. Following this, the text transitions (gradually we hope) into areas that are more clinically related, the applied aspects referred to in the title. These chapters address growth and development, metabolic and hormonal processes, and how these are related to health and disease. The text ends with a chapter on pharmaceutical treatments for osteoporosis, which we hope incorporates both the clinical elements of treatment and the biological reasons for, and effects of, these treatments.

Skeletal biology is, by its nature, interdisciplinary. The course that we teach at Indiana University typically includes students in the basic medical sciences, general biology, the dental sciences, several engineering subspecialties, foods and nutrition, kinesiology, and rehabilitation sciences. We have written this textbook to cover a range of topics that we feel would be relevant to these groups, and have attempted to write various chapters in a way that will be understandable to those students whose particular expertise and interest may not be in the area covered by a given chapter. We have also attempted to write the chapters so that they will be suitable for students at various levels of study, including undergraduate, graduate, and even postgraduate. We realize that the danger of this is that some chapters may be too superficial for students who are more expert in the area covered by that chapter. However, the textbook is meant to be supplemented by additional readings that delve into specific topics in greater depth for those who wish to specialize in that area. To this end, we have included a list of 10–15 suggested readings at the end of each chapter that can serve as a starting point for supplementary reading and discussion. Further, we have incorporated study questions at the end of each chapter. We have resisted the temptation to include answers to these questions. They are intended to be used for discussion (although they could also be used for testing), and there may not be a single "correct" answer. We hope that they will permit further exploration of the chapter topic, at the level appropriate for the student.

Finally, we have not only had a lot of fun putting this text together but have also learned a lot in areas that are not within our own expertise. We sincerely hope that it serves the same purpose for you.

David B. Burr, PhD

Matthew R. Allen, PhD

16 February 2013

Acknowledgements

We would like to thank the students/post-doctoral fellows who reviewed and provided critical critiques of each chapter:

Rachel Dirks
Paul Childress
Chris Newman
Lisa Cole
Dr. Jun Sun
Joshua Gargac
Dr. Heather Coan
Amy Sato
Dr. Emily Farrow
Perla Reyes-Fernandez
Dennis Joseph
Rebecca McCreedy
Maxime Gallant

We would also like to thank Dr. Jason Organ for editing the questions at the end of each chapter.

We are grateful to the following individuals for contributing to original figures:

Dr. Keith W. Condon for several histological images in chapters 1, 2, and 7

Mr. Drew Brown for numerous illustrations in Chapter 4 as well as figures 1.4, 9.5, 9.6, 17.6, 17.9, and 17.11

Dr. Joseph Wallace for figures 1.1 and 1.3

Dr. Mitch Schaffler for figures 1.9 and 1.11

Dr. Nicoletta Bivi for figure 2.2

Dr. Alex Robling for figure 4.6

Dr. Stuart Warden for figure 5.15 and 9.17

Rafael Pacheco-Costa and Dr. Lynn Neff for figure 2.3

Drs. Lilian I. Plotkin, Teresita Bellido, and Lynda Bonewald for the original SEM in 9.11

Dr. Hau Zhou for the clinical photomicrographs in figures 7.16, 7.17, 7.18 and 7.21

Dr. Steven Doty for the EM images in 2.9

We would also like to thank Dr. James Fleet for contributing to the early drafts of Chapter 13.

List of Contributors

Ozan Akkus, PhD Department of Mechanical and Aerospace Engineering, Biomedical Engineering, and Orthopaedic Surgery, Case Western Reserve University, Cleveland, Ohio, USA

Matthew R. Allen, PhD Department of Anatomy and Cell Biology, Indiana University School of Medicine, Indianapolis, Indiana, USA

William J Babler, PhD Department of Oral Biology, Indiana University School of Dentistry, Indianapolis, Indiana, USA

Teresita Bellido, PhD Department of Anatomy and Cell Biology, Indiana University School of Medicine, Indianapolis, Indiana, USA

Nicoletta Bivi, PhD Department of Anatomy and Cell Biology, Indiana University School of Medicine, Indianapolis, Indiana, USA

Angela Bruzzaniti, PhD Department of Oral Biology, Indiana University School of Dentistry and Department of Anatomy & Cell Biology, Indiana University School of Medicine, Indianapolis, Indiana, USA

David B. Burr, PhD Department of Anatomy and Cell Biology, Indiana University School of Medicine, Indianapolis, Indiana, USA

Tim Corbin, MS Department of Therapeutic Discovery, Transgenic Division, Amgen Inc., Thousand Oaks, California, USA

Linda A. DiMeglio, MD, MPH Department of Pediatrics, Indiana University School of Medicine, Indianapolis, Indiana, USA

Robyn K. Fuchs, PhD Department of Health and Rehabilitation Sciences, Indiana University School of Medicine, Department of Physical Therapy, School of Health and Rehabilitation Sciences, Indiana University, Indianapolis, USA

Tien-Min Gabriel Chu, DDS, PhD Department of Restorative Dentistry, Indiana University School of Dentistry, Indianapolis, Indiana, USA

Kathleen M. Hill Gallant, PhD, RD Department of Anatomy and Cell Biology, Indiana University School of Medicine, Indianapolis, Indiana, USA

Erik A. Imel, MD Departments of Medicine and pediatrics, Indiana University School of Medicine, Indianapolis, Indiana, USA

Daniel L. Koller, PhD Department of Medical and Molecular Genetics, Indiana University School of Medicine, Indianapolis, Indiana, USA

Kelly Krohn, MD Senior Medical Advisor, Eli Lilly Laboratories, Indianapolis, Indiana, USA

Jiliang Li, PhD, MD Department of Biology and Center for Developmental and Regenerative Biology, Indiana University-Purdue University Indianapolis, Indianapolis, Indiana, USA

Bruce H. Mitlak, MD Distinguished Medical Fellow, Lilly Research Laboratories, Indianapolis, Indiana, USA

Lilian I. Plotkin, PhD Department of Anatomy and Cell Biology, Indiana University School of Medicine, Indianapolis, Indiana, USA

Alexander G. Robling, PhD Department of Anatomy and Cell Biology, Indiana University School of Medicine, Indianapolis, Indiana, USA

Sean Shih-Yao Liu, DDS, MS, PhD Department of Orthodontics, Indiana University School of Dentistry, Indianapolis, Indiana, USA

David L. Stocum, PhD Department of Biology and Center for Developmental and Regenerative Biology, Indiana University-Purdue University Indianapolis, Indianapolis, Indiana, USA

Joseph M. Wallace, PhD Department of Biomedical Engineering, Indiana University-Purdue University Indianapolis, Indianapolis, Indiana, USA

Connie M. Weaver, PhD Nutrition Science, Purdue University, West Lafayette, Indianapolis, USA

Kenneth E. White, PhD Department of Medical and Molecular Genetics, Indiana University School of Medicine, Indianapolis, Indiana, USA

BASIC BONE BIOLOGY AND PHYSIOLOGY

Bone Morphology and Organization

David B. Burr[1] and Ozan Akkus[2]

[1]Department of Anatomy and Cell Biology, Indiana University School of Medicine, Indianapolis, Indiana, USA
[2]Department of Mechanical and Aerospace Engineering, Biomedical Engineering, and Orthopaedic Surgery, Case Western Reserve University, Cleveland, Ohio, USA

THE FUNCTIONS OF BONE

Bone is multifunctional, playing roles in mechanical support and protection, mineral homeostasis, and hematopoiesis. In recent years, it has become clear that bone also serves an important endocrine function.

The mechanical functions of bone are by far the most widely recognized and studied. Both trabecular and cortical bone serve this function, although the nature of this function is partly specific to each. The dense cortical bone comprises most of the bone mass and takes on most of the role for load bearing. Although the more porous cancellous bone also supports load, one of its important functions is to redirect stresses to the stronger cortical shell. The mechanical function of bone extends beyond simple load bearing, which requires a certain degree of strength and stiffness. Because of its organization as a multiscale material, it is also highly adapted to avoid fractures caused by repetitive loading at physiologic levels, i.e. failure in fatigue.

Bone also serves a protective function, especially in those vital areas such as the torso and head where injury can be fatal. In these locations, the bone microstructure is not different from that of bone in other locations, but it is organized in a manner that can absorb maximum energy with minimum trauma to the bone itself. For instance, the cranial vault is constructed of two thin plates of dense bone that sandwich porous cancellous bone (the porous appearance of this bone is why it is sometimes called *spongy bone*). Ribs are also constructed in this way, but with less dense cancellous bone. In the case of the ribs, the inherent curvature of the bone also increases its ability to absorb impact energy. Developmentally, bones that serve a protective function (e.g. calvarium and ribs) are formed, at least in part, through intramembranous ossification rather than through endochondral ossification, (see Chapter 4) and are part of the axial skeleton.

It is not widely realized that bone is a blood-forming (hematopoietic) organ, but regions composed largely of *spongy* bone such as the iliac crest, vertebrae, and proximal femur are good sources of red blood cells throughout life. The marrow cavity within the bone is an important site of red marrow, indicative of hematopoiesis, during growth and development, but is largely composed of yellow fat in adults. White fat and brown fat are also found in the human body, and while these are acted upon by osteocalcin, which is produced by osteoblasts, these types of fat are not actually found within the bone marrow itself. Yellow marrow fat originates from the same precursor cells that differentiate to become bone-forming osteoblasts. It provides an energy store and may contribute to lipid metabolism by regulating triglycerides. Because of the large surface area, regions with a lot of cancellous bone are also responsible for rapid turnover of bone tissue and play an important role in the long-term control of calcium balance.

Bone turnover can be sensitive to changes in energy metabolism that occur as a function of aging, hormone deficiency, or the production of skeletal hormones, and this provides the means for long-term exchange of calcium and phosphate (as well as other minerals such as iron and magnesium). Although calcium provides bone with its stiffness and much of its strength, calcium ions in the mineral phase are also important for enzyme reactions, blood clotting, muscle contraction,

Basic and Applied Bone Biology.
DOI: http://dx.doi.org/10.1016/B978-0-12-416015-6.00001-0

and the transmission of nerve impulses. Both cortical and cancellous bones are sites of long-term storage and of the rapid exchange of ions within the mineral phase. The release of minerals through bone remodeling is a relatively lengthy process requiring timescales of days to weeks, and replacing these minerals completely takes even longer. However, the extensive surface area represented by the many osteocyte lacunae and canalicular channels provides a significant opportunity for short-term exchange to meet immediate demands. This is the source of the halo that can sometimes be observed around osteocyte lacunae, and which can appear either hypermineralized or hypomineralized depending on the content of labile calcium at this surface.

Bone has been identified as an endocrine organ that helps to mediate phosphate metabolism and energy metabolism by secreting two hormones, fibroblast growth factor 23 (FGF-23), and osteocalcin. Most of the body's FGF-23 is produced by osteocytes, the most abundant cell type in bone. It causes a reduction in renal reabsorption of phosphate and a decrease in serum levels of 1,25-dihydroxyvitamin D_3. Working with other hormones, bone helps to coordinate processes in the kidney and intestine that regulate its own mineralization. There is now also some evidence that the undercarboxylated form of osteocalcin, which is released from the bone matrix during resorption, helps to regulate pancreatic beta-cell proliferation and enhance insulin secretion. It also acts on adipocytes outside of bone to produce adiponectin, which can reduce insulin resistance. This has the dual effect of increasing glucose utilization and reducing fat in the body cavities. Regulation of the body's energy stores by bone may also be mediated by leptin, acting through the autonomic nervous system and hypothalamus. Thus bone, through its several hormones, helps to coordinate processes in the bone marrow, brain, kidney, and pancreas that affect skeletal tissue mineralization, fat deposition, and glucose metabolism.

BONE IS ORGANIZED AS A MULTISCALE MATERIAL

To achieve these functional goals, bone is organized in a hierarchical manner, from nanometer- to millimeter-sized structures (Fig. 1.1). This contributes not only to its mechanical role in support and movement of the body, but also to its other roles. At the nanostructural level, bone is composed of organic and mineral components, mainly consisting of a matrix of cross-linked type I collagen mineralized with nanocrystalline, carbonated apatite. The collagen and mineral

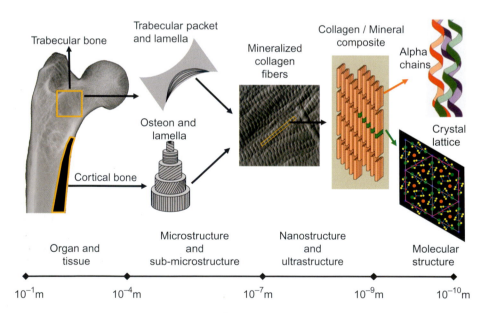

FIGURE 1.1   **The hierarchical organization of bone.** At the macroscopic level, bone is seen as a composite, with dense cortical bone forming an outside shell and cancellous (spongy, trabecular) bone within the marrow cavity. The cancellous bone serves to attenuate loads and to direct forces to the stronger cortical bone. At the microscopic level, cortical bone is composed of many secondary haversian systems, or osteons, that are the product of bone resorption and replacement with human new bone. These osteons are composed of a central canal carrying a blood vessel, nerves, and lymphatics surrounded by layers of concentric lamellae. Trabecular bone is also lamellar, but its structure comprises a combination of lamellae that run approximately parallel to the trabecular surface and the remnants of older remodeled bone that can appear osteonal in some cases. At the ultra- and nanostructural levels, bone is a composite of collagen fibers with plates of mineral interspersed within the collagen fibrils (intrafibrillar) and between the collagen fibers themselves (interfibrillar). The collagen fibrils are composed of molecules forming a triple helix composed of two α_1 chains and a single α_2 chain. *Part of figure courtesy of Beck et al., 1998, J. Struct. Biol. 122, 17–29.*

combine to form a composite material, with mineral providing stiffness to the structure and collagen providing resilience and ductility. At a microscopic [micrometer (10^{-6} m)] level, the individual collagen fibers with interspersed mineral are organized in different ways, depending specifically on the rate, location, and substrate (if any) on which it is formed. At this microstructural level, the organization of bone tissue is very much related to its functional needs: rapid formation for stabilization (in fracture healing) or rapid growth during development; or slower formation to adapt to changing mechanical needs or to replace preexisting bone to provide repair of damaged regions and maintain their unique mechanical properties. At this level, bone can be denser (cortical or compact bone) or rather porous (cancellous, trabecular or spongy bone), depending on the specific mechanical or biological needs and its location.

BONE COMPOSITION

Approximately 65% of bone by weight is composed of mineral (primarily carbonated apatite), but as a living tissue, its organic component, mostly type I collagen, contributes about 20–25% to its composition (Fig. 1.2). The remainder (approximately 10%) is composed of water that is bound to the collagen-mineral composite, and unbound water that is free to flow through canalicular and vascular channels in bone. Unbound water can be redistributed as the bone undergoes loading and probably contributes to the signals detected by cells, informing them of loading conditions (see Chapter 9). Water is exchanged on a nearly 1:1 basis with mineral, so as bone becomes

more mineralized, water content declines, and vice versa. This is important to its mechanical behavior; more highly mineralized bone is stiffer because it has more mineral, but also because it has less water. In addition, although it is stiffer, drier bone tends to be more brittle, and therefore break more easily.

About 90% of the organic portion of bone is type I collagen, with smaller amounts of types III and V collagen also found in the zone surrounding the bone cells. The remaining 10% is made up of noncollagenous proteins (NCPs) which play a vital role in regulating collagen formation and fibril size, mineralization, cell attachment, and microcrack resistance. Of this small amount of NCPs, about 85% is extracellular and the remainder is found within bone cells.

THE NANOSCALE ORGANIZATION OF BONE

Collagen

At a fundamental level, bone is composed of collagen fibers interspersed with plates of mineral both within and between the fibrils. Individual collagen molecules are formed from two α1 chains and a single α2 chain that assemble into a triple helix (Fig. 1.3A). Each chain is about 1000 amino acids in length and the helical center portion of collagen molecules comprise repeating units of a Gly-X-Y triplet. The periodic repetition of glycine residues is essential to the formation of the triple helix structure. While almost all amino acids are present in collagen, X and Y groups are often occupied by proline and hydroxyproline residues. Both these amino acids form a ring with the main backbone of the chain, resulting in improved helical rigidity. Hydroxyproline is particularly critical, and unique, to collagen as its hydroxyl group is essential for hydrogen bonding with water molecules. This interaction is so critical because the stability of the triple helix is maintained by a sheath of water molecules attracted by hydroxyproline. During the intracellular production stage, the nonhelical registration peptides (N-propeptide and C-propeptide) at both ends of the molecule secure the chains together by sulfur cross-links. The triple helix with its terminal propeptides is known as the procollagen molecule. Following the exocytosis of molecules, these propeptide regions are cleaved enzymatically, leaving nonhelical domains at both ends of the molecule, termed the *N-* or *C-telopeptides* (at the N-terminus or C-terminus, respectively). Cleavage of the registration peptides forms the mature collagen molecule, composed of the helical triple helix region and the nonhelical terminal N- and C-telopeptides.

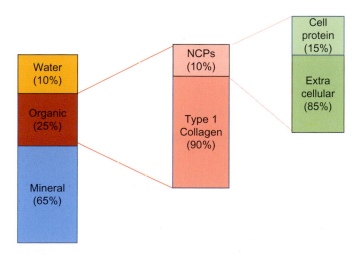

FIGURE 1.2 **Bone is composed of an organic matrix, mineral, and water.** Most of the organic matrix is type I collagen, but noncollagenous proteins (NCPs) that contribute to mineralization and adhesion are also present. Most of this is extracellular, although a small amount of protein is also found within the cells.

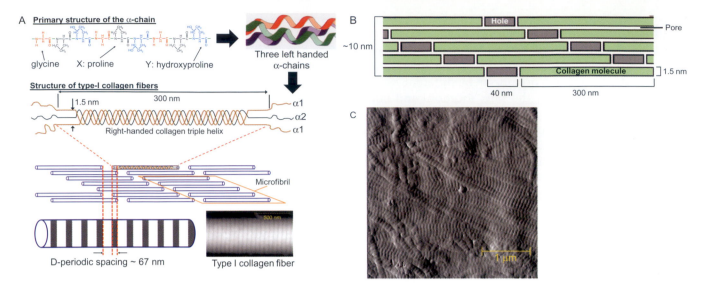

FIGURE 1.3 **Collagen fibrils are constructed from many collagen molecules.** (A and B) Each collagen molecule is approx. 300 nm long and 1.5 nm thick. The molecules are stacked in a quarter-staggered array such that there are 67 nm hole zones between the ends of the molecules and spaces between the laterally contiguous molecules known as pores. (C) The hole and overlap zones give the collagen fibril its characteristic banded appearance when viewed with atomic force microscopy. Both holes and pores enable the deposition of mineral during primary and secondary mineralization processes. The diameter of the entire collagen fiber can vary; its thickness is regulated by the action of noncollagenous proteins such as decorin. *Part of figure courtesy of Beck et al., 1998, J. Struct. Biol. 122, 17—29.*

Lateral and longitudinal aggregation of collagen molecules is essential to extend from the nanoscale to the microscale. In this assembly, five molecules form a microfibril in a semihexagonal arrangement. Microfibrils aggregate laterally and longitudinally to make up fibers that are eventually about 150 nm in diameter and 10 μm in length. Electron microscopy images of collagen fibers reveal an approximately 67 nm banding pattern, called the D-banding pattern, which represents the space between the ends of contiguous collagen molecules, and the overlap between the end regions of laterally contiguous molecules (Fig. 1.3B). The mean diameter of collagen fibrils and their spacing is less in osteoporotic bone than in healthy bone, which may increase the bone's fragility.

Collagen fibrils are connected by different kinds of cross-links that can have profound effects on the material properties of the tissue, and ultimately on the mechanical behavior of the whole bone (Fig. 1.4). These can be broadly grouped into those formed through enzymatic processes and those formed through processes of nonenzymatic glycation, which create advanced glycation end products (AGEs).

Enzymatically Mediated Collagen Cross-Linking

Pyridinoline and deoxypyridinoline are two mature cross-links of collagen that are derived from an enzymatic pathway initiated by the enzyme lysyl oxidase. Pyridinoline is the maturation product of two hydroxylysyl (Hyl) residues from the telopeptide with a

Hyl from the α-helix, whereas the deoxypyridinoline analog contains a lysine residue from the α-helix. These trivalent cross-links are very stable. The content of these *mature* cross-links in human bone collagen increases sharply up to the age of about 10—15 years and thereafter remains constant or possibly declines slightly, although the number of pyridinoline and deoxypyridinoline cross-links can change with treatments that alter bone turnover. An increased pyridinoline:deoxypyridinoline ratio has been related to increased compressive strength and stiffness in bone, but probably has no effect on toughness or ductility.

Nonenzymatically Mediated Collagen Cross-Linking

There are several cross-links of collagen that result from the nonenzymatic condensation of arginine, lysine, and ribose. Pentosidine, an AGE, constitutes the smallest fraction of these nonenzymatically glycated cross-links, but is often used as a marker for their total content because it is one of the only AGEs that can be accurately quantified. Other AGEs include Nε-carboxymethyllysine, furosine, imidazolone, and vesperlysine. As AGE formation occurs over a period of years, proteins with long-half lives, such as collagen, can accumulate substantial amounts of AGEs with age; for example, pentosidine accumulation triples over the last three decades of life. In addition, AGEs have been shown to reduce collagen fibril diameter. Because they

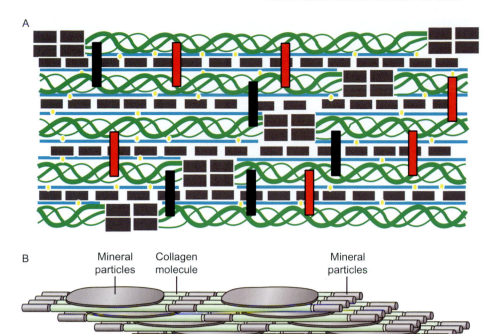

B

Mineral particles   Collagen molecule   Mineral particles

40 nm

FIGURE 1.4 (A) Collagen molecules (green helix) are cross-linked within the fibril by bonds that are formed through enzymatic processes, and by those formed without the need for an enzymatic reaction. Enzymatically formed cross-links (black bars), such as pyridinoline or deoxypyridinoline, form near the ends of the molecules, the C and N-termini. Nonenzymatically formed bonds (red bars), such as pentosidine, are randomly located between the molecules. Mineral (gray blocks) is deposited in the hole and pore zones between the collagen fibrils. Water (blue lines) and hydrogen bonds contribute to the bonding of mineral and collagen within the fibril. (B) Hole and pore zones between the molecules contain plates of bone mineral (hydroxyapatite). Water is bound to collagen in these spaces, and this alters the distribution of load sharing between the collagen and bone mineral deposited in this location.

can be formed in the presence of sugars (such as glucose or ribose), they are accumulated by individuals with diabetes mellitus; this is one contributor to the increased bone fragility found in these people.

Accumulation of AGEs in the extracellular matrix of bone also regulates the proliferation and differentiation of bone-forming cells, the osteoblasts, through interaction with the AGE-specific receptor (RAGE). Binding to RAGE activates NF-κB in osteoblasts and stimulates the production of cytokines. Accumulation of AGEs in collagen can impair osteoblast proliferation and differentiation, reduce osteocalcin secretion, and cause disruptions in cell-matrix interaction and cell adhesion that ultimately affect bone formation.

AGEs also may regulate both osteoclastogenesis and osteoclast activity. Osteoclastic resorption is slowed in the presence of AGEs, in part, perhaps, because the solubility of collagen is reduced. These pathways—AGE regulation of osteoclast differentiation and activity, as well as effects on matrix solubility—may contribute to normal or even elevated bone mass that is found in people with type 2 diabetes who have high concentrations of AGEs in their bone. However, the presence of AGEs makes the bone material (tissue) brittle, and thus more susceptible to fracture even though there is more bone mass.

Collagen Orientation

Historically, collagen in bone has been reported to be regularly organized, with collagen fiber bundles arranged parallel to each other in adjacent sheets (lamellae), either perpendicular to each other or arranged alternately in adjacent lamellae. This stemmed from the different microscopic appearance of bone under polarized light (Fig. 1.5). It is likely that this is partly a function of the optics and plane of section, rather than the way that collagen bundles are arranged. Under cross polarized light, collagen fiber bundles that are oriented transverse to the plane of viewing appear light, or birefringent, whereas those that are oriented parallel to the plane (i.e. longitudinally) are dark. This is because transversely oriented fiber bundles rotate the plane of polarized light with respect to the viewing plane, whereas those that run longitudinally do not. Alternately arranged, or intermediate, collagen fiber bundles are interpreted as representing a combination of fiber arrangements in successive lamellae. In reality, there are many variations on this theme, and it is likely that collagen fiber bundles even within a lamella are arranged in many different orientations, with the predominant orientation being responsible for what is observed either microscopically or by X-ray diffraction.

The collagen in these layers has been shown to be preferentially oriented with respect to the predominant stress in the bone. Longitudinally oriented fibers are predominantly found in portions of bone that are under tension (i.e. being pulled further apart), whereas transversely oriented fibers are more abundant in regions that are usually under compression (i.e. being pushed closer together). This has been shown by mapping the numbers of light or dark osteons across sections of bone in which the primary loading directions are known. This has also been shown experimentally, by altering the direction of loading and

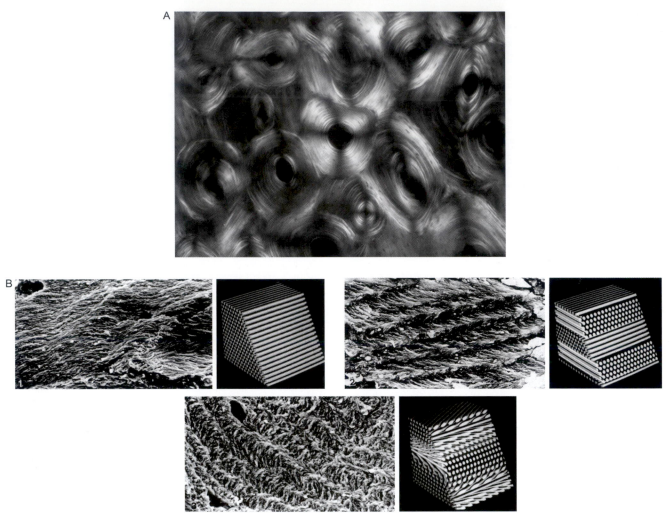

FIGURE 1.5   (A) Variations in collagen orientation can be viewed using polarized light. Fibers that are oriented transversely to the direction of light appear bright, whereas those oriented along the path of the light are dark. In a cross section of cortical bone, dark osteons are composed of longitudinally oriented fibers. Some osteons on the right side of this image would also be characterized as alternately fibered. (B) Electron microscopy images of parallel and alternately fibered collagen bundles in bone are shown along with a schematic representation (top two images). Collagen in bone may in reality be organized in a helicoidal arrangement in which collagen orientation changes only slightly from lamella to lamella. The bottom image shows this effect in three dimensions. *Panel (B) reproduced with permission from Ruggeri et al., 1997.*

observing the collagen fiber direction in the newly formed bone. Both approaches suggest that there is a relationship between collagen fiber orientation and the predominant direction of loading.

For reasons to do with optics and light transmission, bone collagen may only appear in polarized light to be oriented preferentially in these directions. In reality, bone collagen may be organized in a *twisted plywood* configuration that is continuously rotated through 180° cycles (Fig. 1.5B). In this scheme, the collagen fibers gradually, rather than abruptly, change orientation from one successive lamella to another. Under polarized light this would make the bone appear lamellar, with differing light and dark areas. It is as if one took

a piece of plywood in which the fibers in successive plies were perpendicular to each other, twisted it, and then examined the orientation of the individual plies *end-on*. The fibers would then appear as arches, rather than as discrete and oriented fibers. Because the fiber orientation repeats in this model, the tissue-level structure appears lamellar, thus giving bone its characteristic laminar appearance under the microscope.

Whether the collagen is twisted or not, how it becomes oriented in the directions it does is something of a mystery. It has been suggested that the orientation of the osteoblast depositing it determines the collagen orientation, and that the mineral is simply deposited in the spaces between collagen fibrils. An alternative suggestion is that collagen is deposited without any preferred orientation, and that the deposition of the mineral, which is charged, causes both the collagen and the mineral to become oriented in directions that are dependent on the mechanical environment. Which of these ideas is correct is still under debate.

Bone Mineral

Bone mineral is composed of highly substituted, poorly crystalline carbonated apatite mineral, which nucleates within the gap regions between the ends of the collagen fibrils, also known as hole zones, as well as along the pores that run longitudinally between the fibrils (Fig. 1.4). Mineral is initially deposited as an amorphous calcium phosphate, along with large amounts of calcium carbonate. As bone tissue matures, the carbonate content is reduced and mineral crystals grow, becoming more plate-like, and orient themselves parallel to one another and to the collagen fibrils. The long axis of the mineral plate, or c-axis, aligns with the longitudinal axis of the bone. The average size of the mineral crystals in bone tends to span a wide range, but the majority (98%) have a thickness less than 10 nm. As bone ages, the mineral crystals become larger and more crystalline due to changes in ion substitutions and mineral stoichiometry. Therefore, the measured average size of mineral crystals is highly dependent on tissue age. However, it can be difficult to distinguish between smaller crystals with many imperfections and larger crystals with a few imperfections; both may demonstrate similar crystalline properties.

More soluble carbonate can exist in a labile form on the surface of the crystallites, but can also substitute for the phosphate and hydroxyl groups in carbonated apatite. These substitutions within the mineral allow it to be more easily resorbed. Such substitutions distort the shape and size of the crystals, and reduce the stability of the mineral lattice. During episodes of acid load, systemic bicarbonate (HCO_3^-) is consumed to buffer the blood pH. The HCO_3^- deficiency is offset by carbonate and phosphate ions present in bone's mineral reservoir. Thus, in chronic acidosis, the bone's mineral reservoir helps to maintain acid-base homeostasis but also results in bone loss. In addition, different cations (e.g. magnesium, sodium, and strontium) can substitute for the calcium ions, and fluoride can substitute for the hydroxyl group in the apatite lattice; in some cases the mechanical properties of bone are altered and in others the activity of osteoblasts and osteoclasts is affected. At one time, sodium fluoride (NaF) was considered a promising anabolic therapy

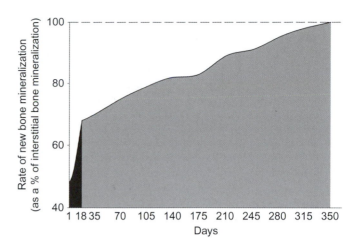

FIGURE 1.6 Primary mineralization of bone occurs within the first 3 weeks after the osteoid is deposited (black colored region). Secondary mineralization occurs in part through a slower growth and maturation of the crystals, and can require a year or more to complete (gray colored region).

for osteoporosis. There is evidence that NaF stimulates osteoprogenitor cells and preosteoblasts, promoting direct bone formation without the need for prior resorption. Moreover, fluoroapatite (the mineral with fluoride substituted for the hydroxyl groups) is more resistant to resorption than is the carbonated apatite. However, the substitution of fluoride into the mineral crystal increases the brittleness of bone and may therefore hasten rather than delay fracture. Although this may or may not occur with other ion substitutions in the mineral crystal, it is instructive in showing that hydroxyapatite is finely adapted to the bone's specific mechanical needs.

There are two sequential and continuous phases to the deposition of mineral in bone: an initial and rapid increase in the number of mineral crystals due to heterogeneous nucleation (primary mineralization), and a slower growth and maturation of those crystals to an eventual size of about $40 \times 3 \times 7.5$ nm. During primary mineralization, mineral is rapidly deposited within the collagen framework, achieving 65—70% of its total mineralization within about 3 weeks after the initial deposition of collagen. During the secondary phase of mineralization, the bone matrix continues to accumulate mineral at a slower, more progressive rate until the amount of mineral reaches a physiologic limit (Fig. 1.6). Estimates for the completion of secondary mineralization range from a few months to many years.

Noncollagenous Extracellular Matrix Proteins

There are numerous NCPs in bone that regulate and direct the construction and maintenance of the extracellular matrix. Although these proteins only account for about 2% of the bone by weight, they play vital roles in embryogenesis and development, regulate the formation and size of collagen fibrils, control mineralization, and provide conduits for cellular signaling and attachment. They can be divided into several large classes (Table 1.1), including:

1. Proteoglycans [heparin sulfate, hyaluronan, small leucine-rich proteoglycans (SLRPs), and versican];
2. Glycoproteins [alkaline phosphatase (ALP), fibronectin, thrombospondin (TSP1 and 2), and vitronectin];
3. Proteins of the small integrin-binding ligand N-linked glycoprotein (SIBLING) family that are associated with bone mineralization [dentin matrix acidic phosphoprotein 1 (DMP-1), matrix extracellular phosphoglycoprotein (MEPE), osteopontin, sialoproteins];
4. Osteocalcin (or bone Gla protein); and
5. Osteonectin (also known as secreted protein acidic and rich in cysteine (SPARC)).

Proteoglycans and Glycosaminoglycans

Proteoglycans are a broad class of molecules defined by a core protein covalently bonded to a variable number of sulfated glycosaminoglycan side chains. Proteoglycans range widely in size, although those in bone tend to be in the smaller range. Proteoglycans in bone help to regulate mineralization by affecting apatite nucleation and growth. Hyaluronan and its receptor, CD44, work together to direct skeletal development. Hyaluronan is a long chain of non-sulfated glycosaminoglycan. It is found mostly in the periosteum and along the endocortical surfaces of bone, but is also present around all major cell types, including osteocytes, within the bone matrix. Versican is a chondroitin sulfate-containing proteoglycan that is important for cartilage formation and is therefore found in the developing skeleton. However, it is also found in osteoid in adult bone, where it may inhibit or regulate mineralization. Heparan sulfate is produced by osteoblasts and osteoclasts, and plays a role in cell-cell communication. It binds FGF and can act as a coreceptor on this protein. It also binds and modulates the activities of transforming growth factor beta (TGF-β) and osteoprotegerin/tumor necrosis factor receptor superfamily member 11B (OPG), both important signaling molecules during the process of bone remodeling and repair.

SLRPs are small proteoglycan molecules that are involved in constructing the collagenous matrix, controlling the aggregation and size of collagen fibrils, and possibly assisting in collagen-mineral interactions.

Perhaps the most important SLRPs are decorin and biglycan: both maintain osteoblast numbers, but they perform that function at different stages of osteoblast development. Decorin is expressed early during cell differentiation, by preosteoblasts, and is downregulated during terminal differentiation of osteoblasts. Biglycan, on the other hand, can induce apoptosis in osteoblast progenitor cells and is upregulated in mature osteoblasts. It is also found in osteocytes and in the pericellular regions of the matrix, and may act as a sensor of shear stress in this location. Both decorin and biglycan function in a complementary manner to maintain osteoblast number. In addition, both bind to collagen and to TGF-β, and can therefore modulate growth factor activity.

Glycoproteins

There are a number of glycoproteins in bone, and some of their functions are not completely understood. However, several are critical to the regulation of bone mineralization. ALP is used as a biomarker for bone formation because it hydrolyzes pyrophosphates, which inhibit mineral deposition by binding to mineral crystals. Neutralizing the pyrophosphates in bone allows normal crystal growth and leads to normal mineralization. ALP is produced by many different organs in addition to bone (e.g. kidney and liver). Therefore, alterations in ALP levels are not necessarily an accurate indicator of the activity of mineralization processes in bone. However, bone-specific alkaline phosphatase (BSAP) can also be measured and represents a widely used and beneficial marker of bone formation and mineralization. Low levels or loss of function of ALP results in a condition known as hypophosphatasia, which causes hypercalcemia and can lead to death in children. TSP1 and 2 are anti-angiogenic NCPs that are important during the early stages of bone formation and are found in mesenchymal stem cells and chondrocytes during cartilage development. TSP2 is a promoter of the mineralization process and increases in osteoid undergoing mineralization. Fibronectin and vitronectin are two other glycoproteins that bind to cells. The former may be important in the early stages of bone formation and cell proliferation. In contrast, vitronectin regulates cell attachment and spreading. It is found in the osteoclast plasma membrane, and may collaborate with osteopontin in attaching osteoclasts to the mineral matrix.

SIBLING Proteins

The SIBLING family of phosphoproteins includes bone sialoprotein (BSP), DMP-1, MEPE, and osteopontin. All of these phosphoproteins play a role in bone

TABLE 1.1   Noncollagenous Proteins in Bone

PROTEOGLYCANS AND GLYCOSAMINOGLYCANS		
Heparan sulfate	Produced by osteoclasts and osteoblasts Plays important roles in cell-cell interactions	
Hyaluronan	Nonsulfated glycosaminoglycan Hyaluronan in periosteum, endosteum, and around cells CD44 is the cell surface hyaluronan receptor and plays a role in development	
Small leucine-rich proteoglycans	Provide structural organization in bone	
	Biglycan	Found in pericellular location undergoing morphological delineation Upregulated in osteoblasts and found in osteocytes May act as shear sensors Binds collagen and TGF-β
	Decorin	First appears in preosteoblasts and is downregulated in more terminal osteoblastic cells Binds to collagen and TGF-β and may regulate fibril diameter Inhibits cell attachment to fibronectin
	Fibromodulin	Binds to distinct regions of collagen fibers Binds TGF-β
	Osteoadherin	Contains RGD sequence Function unknown
Versican	CS-containing PG found in osteoid May capture space destined to become bone	
GLYCOPROTEINS		
Alkaline phosphatase	Potential Ca^{2+} carrier Hydrolyzes inhibitors of mineral deposition such as pyrophosphates Loss of function leads to hypophosphatasia Bone formation marker Nonspecific and bone-specific forms (BSAP)	
Fibronectin	Produced during early stages of bone formation Binds cells in an RGD-independent manner May be involved in proliferation	
Thrombospondin	Role in development—found in early stages of bone formation (MSCs and chondrocytes during cartilage development) Anti-angiogenic	
Vitronectin	Involved in cell attachment and spreading; shows specificity for osteopontin	
SIBLING FAMILY OF GLYCOPROTEINS		
Bone sialoprotein	Limited pattern of expression Marks late stage of differentiation and early stage of mineralization	
Dentin matrix acidic phosphoprotein 1 (DMP-1)	Expressed by osteocytes and osteoblasts Has affinity for hydroxyapatite and the N-terminus of type I collagen Regulates mineralization	
Matrix extracellular phosphoglycoprotein	Expressed by osteocytes and osteoblasts Regulates mineralization Negative regulator of osteoblast activity	
Osteopontin	Secreted by bone cells in early stages of osteogenesis Promotes adhesion of different tissues (cement line and periodontal ligament) Inhibits mineral formation and crystal growth	

(Continued)

TABLE 1.1 (Continued)

OTHER IMPORTANT NONCOLLAGENOUS PROTEINS	
Osteocalcin	Enhances calcium binding, controls mineral deposition
	Expressed by osteoblasts and osteocytes
	Bone remodeling marker
	Overexpressed in cancer and some autoimmune diseases
Osteonectin	Binds to collagen, HA, and vitronectin
	Located at sites of mineral deposition (possible nucleator)
	May play role in osteoblast proliferation

Ca^{2+}, calcium ion; CS, chondroitin sulfate; HA, hyaluronic acid; MSC, mesenchymal stem cell; OPG, osteoprotegerin; PG, proteoglycan; RGD, Arg-Gly-Asp.

mineralization. Osteopontin is secreted by osteoblasts in the early stages of osteogenesis. It inhibits mineral formation and crystal growth, and is found locally in regions of lower mineralization, such as the cement line in bone and the periodontal ligament surrounding the teeth. It may also act to provide a scaffold between tissues with different matrix composition, and to provide cohesion between them. It has been suggested that this is the bone *glue* that provides fiber matrix bonding, as well as crack bridging in the case of microcrack formation. Osteopontin also binds to osteoclasts and promotes the adherence of the osteoclast to the mineral in bone during the resorption process. DMP-1 is expressed by osteocytes and osteoblasts. It has a high affinity for hydroxyapatite and the N-telopeptide region of type I collagen, and functions to locally regulate the mineralization process. It has been implicated in DNA binding, gene regulation, and integrin binding. The absence of DMP-1 causes elevated FGF-23 and results in hypophosphatemic rickets. It is unknown whether DMP-1 plays a role in the differentiation of osteoblasts to osteocytes. MEPE is another protein of the SIBLING family that regulates bone mineralization locally. It is found predominantly in odontoblasts and osteocytes, where it is highly expressed during the mineralization process. MEPE is highly expressed in tissues undergoing rapid mineralization, for example in the woven bone of a fracture callus, as well as in endochondral and intramembranous ossification. Animal studies suggest it is a negative regulator of osteoblast activity; the absence of MEPE results in a high bone mass and resistance to bone loss.

Osteocalcin

Osteocalcin enhances calcium binding and controls mineral deposition. It is expressed by osteoblasts and osteocytes. For this reason, it is used as a marker of bone formation, although it may also function to regulate osteoclasts and their precursors.

Mice in which osteocalcin is absent have severe osteopetrosis. Therefore, osteocalcin can be more accurately viewed as a marker of bone remodeling, and its level increases with the remodeling rate even in those cases, such as postmenopausal osteoporosis, in which there is a severe imbalance between formation and resorption.

Osteonectin

Osteonectin is located at sites of mineral deposition, where it binds to hydroxyapatite, collagen, and vitronectin, and may promote nucleation of new mineral crystals. It may also play a role in osteoblast proliferation and its absence results in osteopenia or low bone mass. It binds to several different growth factors [FGF-2, platelet-derived growth factor (PDGF), and vascular endothelial growth factor (VEGF)] and may regulate their activity.

THE MICROSTRUCTURAL ORGANIZATION OF BONE

At the microstructural level, bone can be organized in a variety of different ways, determined by its function and the manner in which it is deposited. Most, but not all, bone is to some degree lamellar, meaning that collagen and mineral exist in discrete sheets that can be visualized under the microscope. The lamellae create circumferential bands of bone, each $3-7\,\mu m$ thick, that give the appearance of tree rings, each separated by an interlamellar layer approximately $1\,\mu m$ thick. The lamellae may be arranged around the endocortical (wall of the marrow cavity) or the periosteal (outer border of the bone) circumference of the bone (circumferential lamellae), within individual trabeculae, or concentrically around individual vascular channels (concentric lamellae; Fig. 1.7).

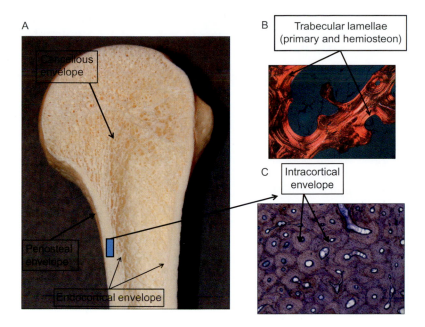

FIGURE 1.7 (A) Macroscopically, bone appears as either porous cancellous bone or denser cortical bone. This structure creates four different kinds of surfaces, called *envelopes*, upon which bone cells can act. (B) Trabeculae in the cancellous bone compartment consist mostly of primary lamellae. However, remodeled areas (areas in which bone has been resorbed and reformed) can also form *hemiosteons*, similar to half osteons. (C) The intracortical envelope in humans is packed with secondary osteons, or haversian systems.

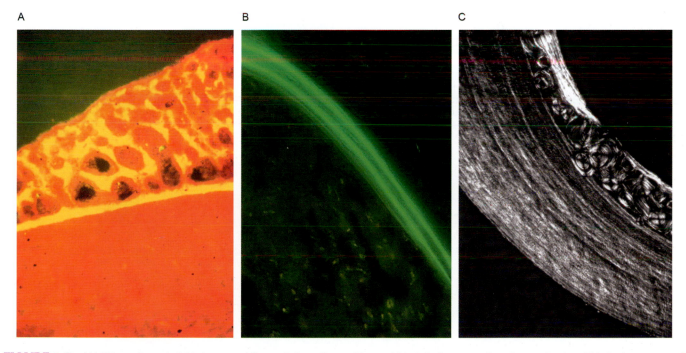

FIGURE 1.8 (A) Woven bone is laid down rapidly, and the collagen fibers within it lack any preferred organization. This is demonstrated by the diffuse tetracycline labeling (yellow) found among the pores within it. (B) Lamellar bone is deposited in sheets in a more organized fashion than woven bone. (C) The lamellar structure can be seen using polarized light microscopy.

Woven Bone

Woven bone is rapidly formed and highly disorganized, and is therefore not arranged in a lamellar pattern (Fig. 1.8). Its rapid formation is the result of a large cell: bone volume ratio. It is usually, but not always, deposited de novo without any previous hard tissue or cartilage model (anlage). It is composed of small and randomly arranged type I collagen fibers that are rapidly mineralized, probably resulting in a tissue that is more highly mineralized than lamellar bone. Because it forms so quickly, it initially presents as a lattice structure, with large pores present within the mineralized structure. This is primarily a repair tissue, forming the callus that bridges the gap during fracture healing to provide stability for the bone during the healing process. It also occurs in response to inflammation, such as in osteomyelitis. However, woven bone is also formed in nonpathologic situations when mechanical loads are much higher than usual or are presented in a way to which the bone is not fully adapted, and is found in the region of the growth plate during endochondral ossification during normal skeletal development.

Primary Bone

There are three types of primary bone that are differentiated by their microscopic organization: primary lamellar bone, plexiform (or laminar) bone, and primary osteons. They are morphologically distinct and impart different mechanical and physiologic properties to satisfy their different functions. They are united by the commonality that they must be deposited directly onto a substrate of either bone or cartilage (or calcified cartilage), without resorption of preexisting bone.

Primary Lamellar Bone

Primary lamellar bone (Fig. 1.8) is the principal type of bone formed on the periosteal surface. It is characterized by a series of parallel laminar sheets. It can become quite dense and has few vascular canals. Therefore, it is very strong, and provides a primarily

mechanical function. However, it is also deposited on the surfaces of the marrow cavity and on trabeculae within the marrow, where it can be quite labile. It may turn over rapidly and be replaced, and may therefore serve to support calcium metabolism.

Plexiform Bone

Plexiform bone (Fig. 1.9) sometimes called fibrolamellar bone, is generally not found in humans (although it has been reported to occur around the time of the major growth spurts), but is found in many animals, especially those that grow rapidly (e.g. cows and sheep). It is a combination of nonlamellar bone, which forms a core substrate, and primary lamellar bone, which is deposited on the surface of the substrate. The nonlamellar portion forms de novo within the fibrous periosteum as buds of fine-fibered bone, composed of small and randomly oriented collagen fibrils (Fig. 1.10). These buds of bone unite with adjacent buds to form a bridge of bone separated from the surface of the preexisting bone by a space that includes vascular elements. Plexiform bone derives its name from this interconnecting vascular plexus. The initial bridging also provides a way to rapidly increase bone strength, as small amounts of bone on the outer surface will contribute significantly to its strength (see Chapter 6). The bridges of bone provide several surfaces upon which lamellae can be deposited, and are one of the reasons that the bone can form so rapidly. As the lamellae form on the surface of the nonlamellar bridges, they gradually fill in these vascular

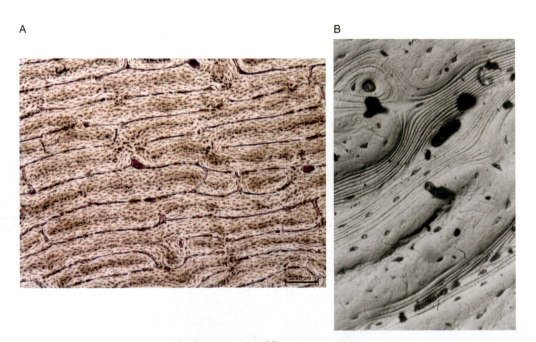

A   B

FIGURE 1.9   Plexiform bone is composed of lamellar bone laid down on a woven bone core. (A) This give it its *bricks and mortar* appearance which can be best visualized using reflected light microscopy. (B) Backscattered electron microscopy. *Courtesy of Dr. Mitchell Schaffler.*

spaces, while retaining smaller spaces for the vessels of approximately 25—50 μm in diameter.

Primary Osteons

Primary osteons are formed by infilling of enlarged vascular channels, usually found within well-organized lamellar bone. The osteonal lamellae are concentrically deposited on the surface of the canal until only a small vascular canal remains. Bone cells are arranged in several circular layers (resembling a solar system) around the vascular canal. Primary osteons are only about 50—100 μm in diameter or smaller, typically have fewer than 10 lamellae, and do not have a well-defined boundary separating them from the rest of the existing matrix. It has been suggested that the existence of primary osteons is related to body size and rapid growth; the presence of primary osteons in rapidly growing deer antler supports this notion.

Secondary Bone

Primary bone is new bone made in a space where bone has not previously existed, although it may be formed on an existing bone surface. When bone is the product of resorption of previously deposited bone followed by deposition of new bone in its place, it is called *secondary bone*. This distinction is important because the apposition of primary bone, which requires only formation, is different from that of secondary bone, which requires a coordinated (coupled) series of processes to resorb and replace the bone that was already there. This is one way of repairing microscopic damage, which is constantly being created. The result of this process of resorption and replacement is a *secondary osteon*, or *haversian system* (Fig. 1.11). Secondary osteons form longitudinally arrayed fibers embedded in a matrix composed of interstitial lamellae, but separated from the matrix by a

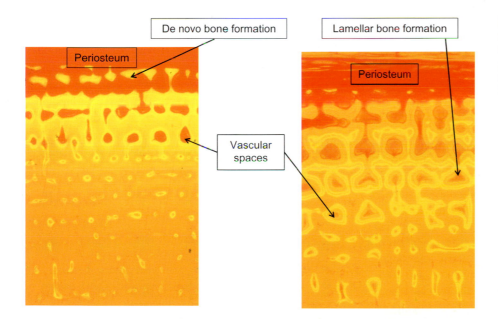

FIGURE 1.10 **The formation of plexi-form bone.** Bone formation begins with intramembranous ossification within the periosteal membrane. Vascular spaces surrounding these bone cores are filled by lamellar apposition.

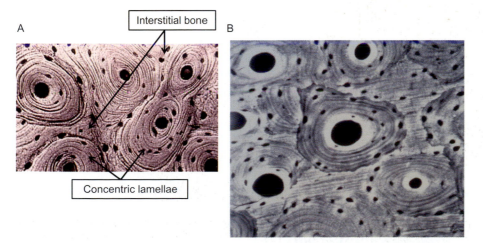

FIGURE 1.11 **Secondary bone is formed by removal of preexisting bone and replacement with new bone.** (A) In the cortex, this results in a secondary osteon, or haversian system, that surrounds a canal transmitting vessels and nerves associated with the vessels. (B) The osteon is bounded by a cement line that separates it from interstitial bone. *Courtesy of Dr. Mitchell Schaffler.*

1. BASIC BONE BIOLOGY AND PHYSIOLOGY

ductile interface, the cement line. The secondary osteon is distinguished from a primary osteon in being larger (100−250 μm in diameter), having more concentric lamellae (approximately 20−25 lamellae), and having a cement line at its outer boundary. The number and size of the osteons varies with age in predictable ways, becoming more numerous but smaller as we grow older. As with primary osteons, the lamellae surround a central haversian canal (approximately 50 μm in diameter) that carries a neurovascular bundle. Secondary osteons are about 1−10 mm long, running at an average angle of 11−17° with respect to the long axis of the bone. However, the orientation of any individual osteon may be quite variable, as the vascular spaces they contain branch extensively. Moreover, haversian vessels are connected in a vascular plexus by other vessels that run between them in a more or less transverse direction. The canals in which these vessels run are called Volkmann canals. The vessels in these canals also connect the haversian capillaries with the marrow vasculature, and with the vascular plexus in the periosteal membrane.

The cement line, or reversal line, represents a remnant of the reversal phase of bone remodeling, i.e. the point at which osteoclastic bone resorption stops and bone formation begins. It clearly demarcates the secondary osteon from its surrounding bone matrix (Fig. 1.11). Cement lines are mechanically important structures that serve as fiber reinforcements to the bone tissue. It is well established using histologic, birefringence, and electron microscopic techniques that the cement line is collagen deficient. There is some debate about whether cement lines are highly mineralized or deficient in mineral, but in either case their mechanical function in preventing or deflecting crack growth would be the same. In addition to mineral and collagen, the cement line also contains high levels of certain kinds of NCPs, such as glycosaminoglycans and osteopontin. This makes a great deal of sense, as osteopontin plays a role in osteoclast adhesion during resorption and the cement line is where the osteoclasts stop resorbing. Both proteoglycan and osteopontin inhibit mineralization, and it also makes sense that the cement line would be mineral deficient.

Osteons have poor shear strength and a weak fiber-matrix interface at the cement line. The functional importance of cement lines lies in their ability to control fatigue and fracture processes, absorb energy by stopping crack propagation, and provide viscous damping in compact bone. They create significant point-specific stiffness variations in bone and increase the static toughness of bone by preventing deleterious crack growth, thereby improving fatigue properties. A viscous cement line may relieve locally high shear stresses by allowing deformation at this interface; the low shear stiffness makes it difficult to transmit energy to a growing crack.

Interstitial Bone

Inevitably, bone that is remodeled will leave behind traces of old bone that is preexisting but not remodeled. This bone may appear to be lamellar but disorganized because its lamellae, in cross section, are incomplete. This interstitial bone represents the remains of either primary or secondary lamellar bone, and fills the gaps between adjacent secondary osteons (Fig. 1.11). Because it has not been recently remodeled, its mean tissue age is older than osteonal bone (although it may formerly have been a complete osteon), and it is more highly mineralized and more susceptible to accumulation of microcracks.

THE MACROSCOPIC ORGANIZATION OF BONE

At the macroscopic level, bone can be divided into dense cortical (or compact) bone and more porous cancellous bone, which is composed of trabecular struts.

These types of bone are distinguished not only by their porosity, but also by their location and function.

Cortical Bone

Cortical bone is found as the primary component of the shafts, or diaphyses of the long and short bones of the extremities (Fig. 1.7). The haversian canals in cortical bone create a porosity of about 3−5%, although this increases with age and with osteoporotic changes to the skeleton. Compact bone is also found surrounding the cancellous bone of the vertebral body, at the ends (or metaphyses) of the long bones, in the iliac crest, and in the skull. It provides both support and protection.

Cancellous Bone[1]

Cancellous bone is found primarily in the metaphyses of the long bones, as well as in the vertebrae,

---

[1]Cancellous bone is sometimes also called spongy bone or trabecular bone. Cancellous is preferred over trabecular when referring to this bone as a structure, because cancellous imparts the idea of a porous structure. When referring to individual struts, the term *trabeculae* can be used. *Spongy bone* is not preferred, because this bone is not spongy from a mechanical standpoint, even though its appearance resembles the structure of a sponge.

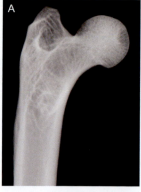

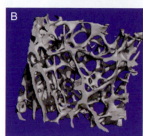

FIGURE 1.12   Architecturally, cancellous bone is composed of plates and rods running at various angles that are thought to reflect the orientation of the major stresses within the bone (A,B). These trabeculae act as struts to support the outer cortical structure, much as the struts of the Eiffel Tower act to support its outer framework (C). *C reproduced with permission from Gustaf Östberg, Materials & Design, 23, 7, 2002, 633—640.*

ribs, and iliac crest. It is composed of plates and rods of bone, each about 200 μm thick, but comprising only about 25—30% of the total tissue volume, with the remainder being marrow space (Fig. 1.7). In cancellous bone, the lamellae are arranged more or less parallel to the trabecular surface, but one finds what appear to be *half osteons*, or hemiosteons, that represent previous periods of surface resorption and subsequent formation (i.e. remodeling events) (Fig. 1.7). One surface of the hemiosteon borders the marrow cavity, rather than a canal, and is separated from the rest of the trabecula by a cement line. These hemiosteons are the product of the same type of bone remodeling found in cortical bone but, because they start on a longer surface rather than on the surface of a haversian canal, they do not have the circular appearance of whole secondary osteons. As they are adjacent to the marrow cavity, and can derive their blood supply from it, they do not need, nor do they contain, a central vascular channel. Occasionally, complete osteons can be found within a trabecula, although because osteonal diameter is similar to or larger than the thickness of a trabecula (approximately 150—180 μm), this is somewhat rare; the osteons found in trabeculae tend to be smaller than those found in cortical bone. If the trabecular strut becomes very thick, bone resorption can occur parallel to the primary orientation of the trabecula and through its approximate center. This initially creates a tunnel in the trabecula (trabecular tunneling), which is

essentially a longitudinal section of an entire bone remodeling unit. This process eventually divides the single trabecula into two trabecular plates.

Cancellous bone derives its primary mechanical benefit from its architecture, which provides structural support without increasing the weight of the entire bone. Because of its location within the marrow cavity, by itself it is not an efficient weight-bearing structure. One important mechanical function is to provide a means for the bone to funnel the stresses imposed on it to the stronger, more massive cortical bone. Because it is highly interconnected, it provides a series of struts that can strengthen bone, similar to the way that the struts in the Eiffel Tower make it a strong structure (Fig. 1.12). The importance of a connected architecture can be demonstrated by a simple analogy to a bench with cross-struts between the primary load-bearing legs (Fig. 1.13). When these struts are connected, the stool can hold weight effectively. However, when those struts are not connected or not connected efficiently, even if they are still present the stool will not serve as well as a load bearing structure.

The architecture of cancellous bone can be characterized by the number of trabeculae (Tb.N), how thick they are (Tb.Th), and how far apart they are (trabecular separation, Tb.Sp). Each of these factors contributes to the overall cancellous bone volume, but the same bone volume can contain trabeculae that are organized in different ways, i.e. more

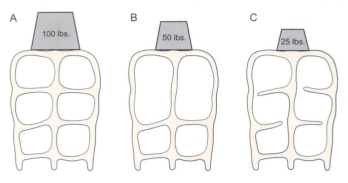

FIGURE 1.13　**The importance of bone architecture, independent of the amount of material (i.e. mass).** Cancellous bone in which the cross struts are disconnected (C) is not very capable of supporting a load, even though there is as much mass as in (A). A smaller mass that is buttressed properly (B) may be able to support more load than a greater mass in which connectivity is compromised. Figures indicate maximum load.

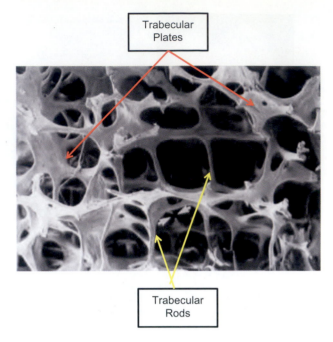

FIGURE 1.15　**Cancellous bone is composed of broader plates and thinner rods.** When bone is lost, as in osteoporosis, more of the trabecular plates convert to a rod-like configuration. *Reproduced with permission from Mosekilde L. Bone and Mineral 1990, 13—35.*

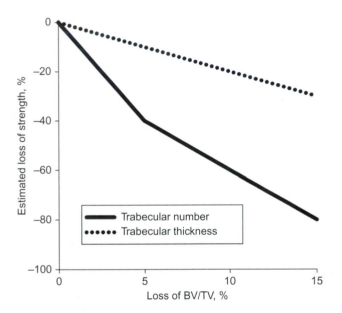

FIGURE 1.14　The thickness of the trabecular plates and their number do not contribute equally to the mechanical strength of the structure. A decrease in trabecular number has a much greater negative impact on bone strength than does the loss of an equal amount of bone through trabecular thinning. This again demonstrates the importance of connectivity in cancellous architecture. BV/TV = bone volume fraction, or the proportion of the tissue area that contains bone (rather than marrow or other spaces). BV, bone volume; TV, trabecular volume.

trabeculae that are thinner or fewer, thicker trabeculae. There is likely to be an ideal relationship between Tb.Th and Tb.N, dependent on the location and primary direction of loading. However, it has been shown several times that the loss of complete trabeculae (reducing Tb.N) reduces the strength and stiffness of bone by two to three times more than does losing the same amount of bone via trabecular

thinning (Fig. 1.14). The reason for this is that the loss of whole trabeculae reduces connectivity within the entire structure, which makes the structure much less capable of bearing weight and less able to direct stresses to the cortex than does maintaining the connections but making them thinner. In healthy human bone, the trabeculae tend to be shaped as plates of bone rather than as circular or elliptical rods (the extent to which the trabeculae are plate-like or rod-like is sometimes called the Structure Model Index). This provides greater strength in the load-bearing direction. It is more difficult to bend a plate in the direction of its greatest width than it is to bend a rod. When people begin to lose bone, as in osteoporosis, their bone plates become more rod-like and lose their connectivity (Fig. 1.15). In some locations, such as the vertebrae, trabeculae will be preferentially lost off-axis from the primary direction of loading so that the remaining struts run in a preferred direction aligned to the loading direction. This anisotropy (directional dependence) enables the structure to maintain its strength in the primary direction of loading, but puts it at risk if the axis of loading is changed.

Cancellous bone beneath cartilage of the articular surface of the joints may function to *cushion* the joint by attenuating forces generated during movement. Immediately beneath the joint, the cancellous bone condenses to form a thin plate of compact bone, called

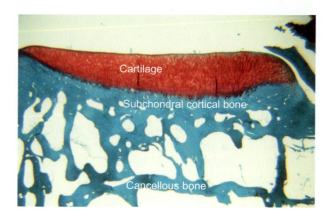

FIGURE 1.16 Beneath the cartilage in joints, there is a dense cortical bone plate called subchondral bone. This may serve to support the joint and regulate stresses. This becomes more porous cancellous bone as one moves distally toward the bony diaphysis.

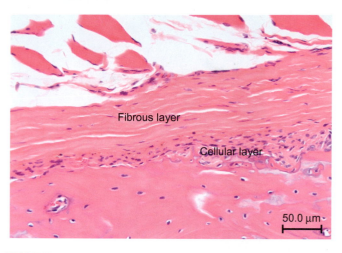

FIGURE 1.17 The periosteal sheath covers the outer surface of bone. It is composed of two layers, an outer fibrous layer, and an inner layer that is more highly cellular. These cells become active osteoblasts and contribute to bone apposition along this surface.

the subchondral plate (Fig. 1.16). It has been suggested that the composite of subchondral plate and cancellous bone beneath it play a role in the development of age-related arthritis, or osteoarthritis. The precise role of subchondral cortical and cancellous bone in the joint is still under debate.

The spaces between the trabecular struts are regions in which blood cells are formed (i.e. red marrow). The differentiation of cells in the bone lineage can be partly diverted toward forming adipocytes, and the marrow within the diaphyses then becomes more fatty (i.e. yellow marrow) with age. However, some red marrow is present throughout life in the ends of the bones, as well as in the vertebrae, iliac crests, and ribs, thus making bone a primary blood-forming organ.

Skeletal Envelopes

At the macroscopic level, there are four distinct surfaces in bone and these are defined by their location (Fig. 1.7): periosteal, endocortical, cancellous, and intracortical. It is convenient to divide the bone into these *skeletal envelopes* because the different surfaces play different roles in the health of the bone, and differ in the manner in which they respond to mechanical loads (see Chapter 9). They also are morphologically distinct, although several of them are interconnected, which provides pathways for communication among them.

The outer surface of the diaphyses is called the *periosteal surface*, and it is covered by a thin fibrocellular membrane (periosteum) that contributes to bone formation along this surface. It is composed of two parts: an outer fibrous portion with a few fibroblast-like cells, and a deep, or cambium, layer that is populated with highly osteogenic cells (Fig. 1.17). The cambium layer is the source of cells responsible for the growth, development,

modeling/remodeling, and fracture repair of our bones. This accounts for the highly sensitive response of the periosteum to mechanical stimulation, infection, tumors, or mild injury. The cells of the periosteum are proliferative and capable of forming either highly organized lamellar bone under these conditions or disorganized woven bone in pathologic situations. The periosteum is well vascularized and innervated by both sympathetic and pain-sensitive fibers. Because mesenchymal cells are also present, cells in the deep layer of the periosteum can also differentiate into chondrocytes and form cartilage, most notably in adults during the fracture healing process.

The surface surrounding the marrow cavity is the *endocortical* (or *endosteal*) surface. This surface is not covered by a membrane, but is instead lined with a discontinuous (fenestrated) layer of osteoprogenitor cells (lining cells; Fig. 1.18). A space containing extracellular fluid exists between these cells and the surface of the bone matrix, and may provide a barrier between the bone fluid within the canalicular and lacunar spaces and the extracellular fluid found in the marrow cavity and vessels. These lining cells are associated with capillaries near the bone surface and sinusoids in the marrow. Because diffusion is the principal mechanism for exchange between the extravascular fluid compartment and the endothelial layer of bone capillaries, the fenestrated nature of the cells lining this surface may act as a membrane, thus controlling fluxes of hydrophilic ions between the vascular and extravascular fluids. This *membrane* is important for the regulation of rapid calcium exchange between bone and the extracellular fluid compartment.

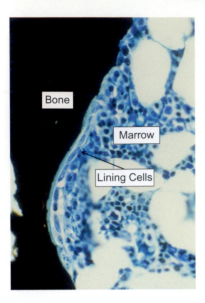

FIGURE 1.18 **Quiescent osteoblasts, or lining cells, cover the endocortical surface of bone.** Notice that there is a space between the cells and the bone surface. This space contains fluid and may provide a barrier between the marrow cavity and the bone.

Trabecular surfaces are morphologically and functionally similar to the endocortical envelope: cells line both surfaces and they are exposed to all the elements of the marrow cavity. Because vessels within the trabeculae are relatively uncommon, this fenestrated layer of cells may regulate nutrients that enter or leave the cellular canalicular system. Bone formation and resorption on trabecular surfaces respond to mechanical and biological signals in a manner parallel to that of endocortical surfaces.

The surfaces of haversian canals represent a fourth skeletal envelope, the *intracortical envelope.* These surfaces are also covered by a fenestrated layer of resting osteoprogenitor cells and, because there is a neurovascular bundle that includes one or two vessels in the canal, can likewise serve to regulate nutrient fluxes between the vascular system and the extracellular fluid compartments that provide nutrients for tissue within the dense cortical bone.

THE BLOOD SUPPLY AND INNERVATION OF BONE

Blood Supply

Calcified tissues present special problems of organization because material cannot diffuse rapidly through the dense tissue to nourish the cells; bone cells must therefore be within about 250 μm of their blood supply. This may be one reason that human bone (and that of other larger mammals, e.g. rabbits, dogs,

and horses) is internally structured with primary and secondary osteons, as otherwise bone with a cortex thicker than 0.5 mm would not be able to support healthy osteocytes in the center of the cortex. It may also be the reason why the average diameter of a trabecula is about 200 μm. Moreover, the circular arrangement of cells around a central haversian canal in compact bone provides a geometrically efficient system for supplying the maximum amount of bone tissue from the minimum number of vessels. This means that secondary osteonal bone has a smaller surface area of canals per unit volume of bone than found in primary bone, and the connections between the vessels, the Volkmann canals, are further apart.

Vessels in bone—and in haversian canals—have the structural characteristics of capillaries, and are often paired within a canal. Venous sinusoids and lymphatic vessels are not found in haversian canals, although prelymphatic vessels may be present. The absence of venous sinusoids is one of the characteristics of a true capillary bed. The vessels are about 15 μm in diameter, but tend to be wider near to the endosteal surface than the periosteal surface. The vessel walls contain no smooth muscle, but are fenestrated capillaries lined by an incomplete layer of endothelial cells. They are similar in this way to vessels in other blood-forming organs like the spleen and bone marrow. The endothelial cells of an osteonal vessel contain numerous pinocytic vesicles that may facilitate transport of water and nutrients across the capillary wall. A continuous 40−60 nm basement membrane surrounds the vessel and limits the rate of ion transport across the capillary wall. Endothelial extensions into the lumen of the capillary may be present and perhaps provide a greater surface area for exchange.

In long bones, blood enters the marrow cavity through the nutrient artery, which then fans out to run longitudinally within the marrow cavity. These arteries will supply and form sinuses, and may provide branches back toward the endosteum. From the marrow cavity, arterial blood enters the intracortical capillaries and flows outward. Also, the sinuses in the marrow cavity form a confluence from which blood is collected and distributed via venous branches back out of the marrow through nutrient and metaphyseal veins. About two-thirds of the blood supply to cortical bone enters through the endocortical surface, with the remainder entering through vessels within the canals and along the surface of the periosteum, which is highly vascularized. The vessels in the bone cortex and in the marrow are sensitive to increased medullary hemostatic pressure, and alterations in pressure can cause pain. The flow in the medullary cavity is regulated by sympathetic nerve fibers that either reduce or increase flow as necessary.

Innervation of Bone

Very little is known about the innervation of the bone, or about the receptors and neurotransmitters that are expressed by nerves in bone. Although at one time it was thought that the bone itself did not contain sensory nerve fibers, it is now known that the marrow, periosteum, and even mineralized bone are supplied by primary afferent nerves as well as by postganglionic autonomic fibers from the sympathetic nervous system. The fibers of the primary sensory nerves are thin (although they can assemble to form bundles), poorly myelinated or unmyelinated, and have low conduction velocities. Sensory fibers express calcitonin gene-related peptide (CGRP) in peptide-rich C- or Aδ fibers, as well as substance P, which is predominantly associated with unmyelinated nerve fibers. These peptides are most highly expressed in the periosteum, which is why periosteal tissues are so sensitive to pain. Neurofilament heavy polypeptide (NF200 or NF-H) is expressed by thinly myelinated primary afferent nerves (Aδ fibers) and at concentrations about equal to those of CGRP and substance P within bone. Because these fiber types are poorly myelinated, or unmyelinated, they have low conduction velocities: <2 m/s for unmyelinated C-fibers and 2−30 m/s for thinly myelinated Aδ fibers. No thickly myelinated nerve fibers with higher conduction velocities (typical of Aβ fibers) have been identified in bone.

Interestingly, most of the afferent nerve fibers in bone express neuromodulin/GAP-43, a protein that is associated with axonal growth and regeneration. This may be necessary because of the continuous remodeling that occurs within bone, which requires vessel growth and reinnervation of the newly formed bone. It may also provide protection in the case of injury, in which case reinnervation would be critical. Studies have also observed neural sprouting of CGRP+ nerve fibers following fracture or in developing arthritis following reinnervation. High levels of nerve regeneration, possibly driven by nerve growth factor (NGF), have been noted in both malignant and nonmalignant bone diseases. The majority of nerve fibers in bone (approximately 80%) express the high affinity nerve growth factor receptor/tyrosine kinase receptor A (Trk-A), which is most highly expressed in the periosteum. Because NGF sensitizes nociceptors, the interaction of Trk-A with NGF may hypersensitize bone to pain.

Sympathetic nerve fibers generally accompany blood vessels within the bone to supply the marrow cavity and control blood flow in and out of it. They are represented by both adrenergic and cholinergic populations. The adrenergic fibers accompany blood vessels and have been associated with skeletal regions undergoing high rates of remodeling. Expression of vasoactive intestinal peptide (VIP) and neuropeptide Y (NPY) has been shown to influence osteoclast and osteoblast function, and may also play a role in regulating bone remodeling.

BONE FLUID COMPARTMENTS

There are two fluid compartments in bone that allow metabolic exchange and ion transport between different regions of bone, as well as between cells. The first is water that is bound within the collagen fiber and is theoretically exchangeable on a 1:1 basis with bone mineral. This bound water is found in the interstitial fluid compartment and represents about 40% of the total water present in bone. The second compartment comprises water that is not bound and can therefore flow through the canalicular system within the bone in response to mechanical loading. This extracellular bone fluid probably not only nourishes the cells but may also transmit signals from one cell to another. The fluid compartment outside the bone tissue itself, i.e. within the marrow and within the haversian canal, can be considered part of this extracellular fluid compartment. It communicates freely with the bone fluid around the cells, but exchange between the cells and the marrow cavity (or the contents of the haversian canal) is regulated by the fenestrated cellular boundary on the bone surface.

Nutrients may pass from the blood into the extracellular fluid compartment surrounding the vessels in the haversian canal, and from there through the canaliculi to be absorbed interstitially, especially during bone formation and mineralization. The extracellular fluid compartment provides a site for the exchange of materials between the vascular system and the bone, and may be important in calcium exchange or in calcium signaling to the bone cells. The extracellular fluid space is large because of the large number of canaliculi connecting the cells in bone, and all of this space is available for ion exchange. This is therefore likely to be a prime site for the rapid exchange of ions in response to short-term increases in the need for mineral.

BONE MASS AND BONE QUALITY

The strength of bone is derived from a combination of bone mass [i.e. the amount of bone, often measured as bone mineral content (BMC)] and everything else that contributes to its strength, now termed *bone quality*. Bone quality is an ambiguous term, and must be defined in order to be meaningful. The generally accepted definition of bone quality includes four primary physiologic and structural qualities: (1) rate of bone turnover; (2) bone architecture and geometry; (3) properties of the collagen-mineral matrix; and (4) microdamage accumulation. These are not independent, as the rate of turnover has a potent effect on each of the other characteristics. Osteocyte density is sometimes included as an additional independent quality factor unrelated to bone density.

Rate of Bone Turnover

Bone is constantly being resorbed and reformed (see Chapter 4). Each time this occurs, there is a temporary deficit in the amount of bone in the skeleton; the faster the rate of turnover, the greater the deficit. Beyond this effect on bone mass, even a cursory examination of a partly resorbed trabecular strut will indicate that small losses of bone that are currently undetectable by any noninvasive measurement of BMC may have profound effects on the strength of trabecular bone (Fig. 1.19). In addition to creating weakness that goes well beyond the small loss of tissue mass, resorption cavities can create stress concentrations that may provide sites for the nucleation of new microcracks, thus further weakening the strut. Resorption cavities cause larger decreases in stiffness than does trabecular thinning, even when there is an identical loss of bone mass (Fig. 1.19). Therefore, BMC measurements as surrogates of bone strength must be evaluated very carefully.

Bone Trabecular Architecture

One reason that resorption cavities may increase bone fragility more than would be expected based on the amount of bone loss alone is that the greater number of resorption sites increases the probability of trabecular perforation, which contributes to fragility independent of BMC by reducing the connectivity

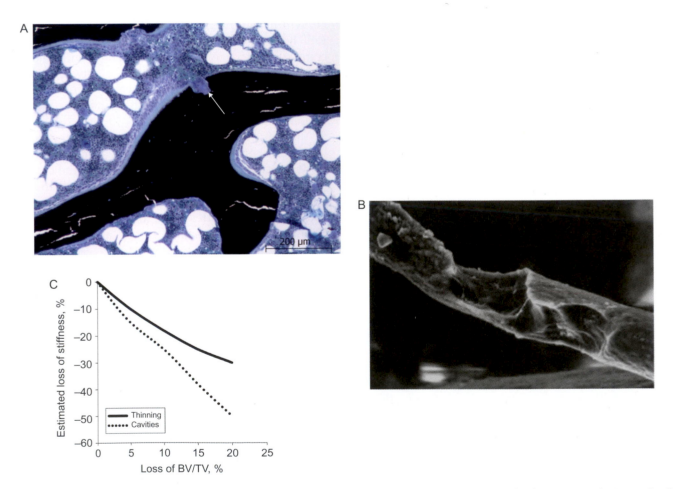

FIGURE 1.19 (A) Resorption of bone creates stress concentrations (at tip of arrow) that can form a nidus for the initiation of microcracks. (B) Excessive resorption can also weaken the trabecula, leading to a loss of connectivity. (C) A small increase in the rate of bone turnover can cause a disproportionate reduction in strength as the number of perforations increases exponentially with increased bone remodeling. Likewise, even without a complete loss of the trabecula, resorption cavities cause 2× more stiffness loss than does trabecular thinning, even with an equivalent loss of bone volume. BV, bone volume; TV, trabecular volume. *Part (B) reproduced with permission from Mosekilde L. Bone and Mineral 1990, 13–35.*

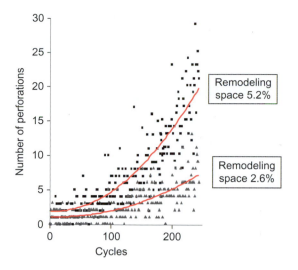

FIGURE 1.20 There is an exponential relationship between the rate of remodeling which creates remodeling spaces and the number of trabecular perforations that are created by remodeling. Doubling the rate of bone turnover can lead to a fourfold increase in trabecular perforations over a period of many years of remodeling. *Data courtesy of Dr. Harrie Weinans.*

between the trabecular plates. This disconnection among supporting elements contributes to weakness independent of the amount of material forming the structure. The number of trabecular perforations is dependent on the rate of bone turnover (Fig. 1.20), with a small increase in turnover resulting in an exponential increase in the number of trabecular perforations. This means that as the remodeling rate increases, the strength of the trabecular bone declines much more rapidly. As indicated above, this has a larger effect on strength than would an equivalent loss of bone from all trabecular surfaces equally.

Tissue Material (Intrinsic) Properties

Increased bone mineralization contributes to its strength and stiffness, and is generally a positive aspect of bone quality. However, bone that is too highly mineralized, such as is found in some bone diseases (e.g. osteopetrosis), can become brittle and increase the risk of fracture. Increasing amounts of mineralization—even within the normal physiologic range—can reduce the energy required to fracture a bone either in impact (e.g. a fall on the hip) or in less traumatic overloading of the skeleton. If very many areas of bone become highly mineralized, or poorly mineralized, the compositional heterogeneity of the tissue can be reduced, which may affect its mechanical integrity. Point-to-point variation in the amount of mineral has an important mechanical effect that

will delay the fracture of bone—just as it does in other heterogeneous composite materials (indeed, this is why some composite materials are constructed in this way).

The amount and maturity of type I collagen in bone, and the manner in which it is cross-linked, can have a profound effect on its strength, stiffness, and energy required to fracture. Lower collagen content reduces the energy required to fracture a bone. An increase in trivalent cross-links may increase the compressive strength and stiffness of bone. Nonenzymatic cross-links form AGEs that can reduce collagen fibril diameter and make bone more fragile, independent of bone mass. Bone from older individuals (>70 years) has three times more nonenzymatic cross-links than bone from those younger than 50 years.

The interaction of collagen and mineral adds another level of complexity. Although collagen supports most of the matrix strain in tension, coupled deformation between collagen and mineral allows collagen to transfer load to the mineral component. This evenly distributes forces and reduces stresses throughout the matrix. Smaller collagen fibers oriented around osteocytes alter local strains at the nano- and micrometer level, thus diverting cracks away from and preserving living cells. The orientation of mineral crystallites also changes the strain environment at the nanoscale.

None of these effects is dependent on bone mass; they therefore represent a component of bone quality.

Microdamage Accumulation

Microscopic damage occurs in bone as a consequence of daily activities that repeatedly load the bone (Fig. 1.21). The amount of damage present in bone is a function both of the amount of damage that is produced and the amount that is repaired through normal physiologic processes. Either increased production of damage or suppressed repair can elevate the level of microdamage in bone. The accumulation of microdamage may increase the fragility of bone, although under most conditions it is unlikely that sufficient damage can be accumulated in the tissue to cause a measurable change in properties. However, the accumulation of damage that occurs with age, in conjunction with the loss of bone mass, may contribute to the increased bone fragility that occurs in postmenopausal women. Microdamage accumulation in bone can reduce the stiffness of the tissue and decrease its strength, but can also have a positive effect in releasing energy and delaying bone fracture.

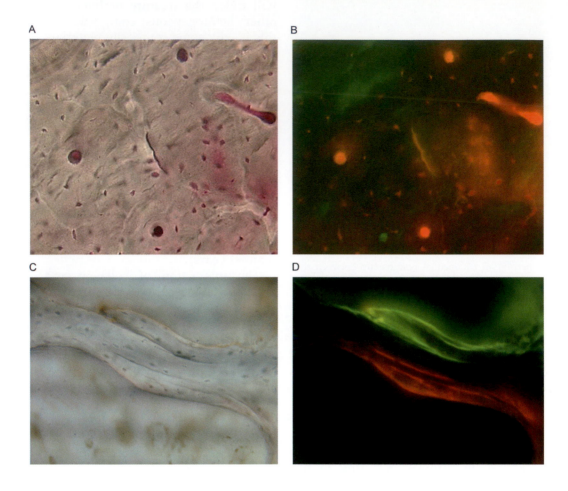

FIGURE 1.21 **Skeletal microdamage is traditionally visualized microscopically following en bloc staining with basic fuchsin.** Microcracks can be observed using brightfield (A) or ultraviolet fluorescence (B) microscopy, the latter taking advantage of the natural fluorescent properties of fuchsin. Alternatively microcracks can be stained en bloc with fluorochromes such as calcein and alizarin (C and D). These stains cannot be visualized using brightfield (C) but can with appropriate filter sets and UV fluorescence (D).

STUDY QUESTIONS

1. Define the three main classes of noncollagenous extracellular matrix proteins, and give examples and functions of proteins from each class.
2. Explain how microcracks are kept from becoming larger at the nanoscalar and microstructural levels.
3. Explain how cancellous and cortical bone work together to provide mechanical support and protection to bone.
4. Compare and contrast woven bone, lamellar bone, primary bone, secondary bone, and interstitial bone.
5. Compare and contrast the four skeletal envelopes: periosteal, endocortical, trabecular, and intracortical.
6. Describe the vasculature that exists within bone and its unique aspects compared to other tissues.
7. Describe the autonomic innervation of bone.
8. Define the four characteristics of bone quality.

Suggested Readings

Bonnucci, E., Motta, P.M., 1990. Ultrastructure of Skeletal Tissues. Bone and Cartilage in Health and Disease. Kluwer Academic Publishers, Boston.

Brookes, M., Revell, W.J., 1998. Blood Supply of Bone: Scientific Aspects. Springer-Verlag, London.

Castañeda-Corral, G., Jimenez-Andrade, J.M., Blook, A.P., Taylor, R. N., Mantyh, W.G., Kaczmarska, M.J., et al., 2011. The majority of myelinated and unmyelinated sensory nerve fibers that innervate bone express the tropomyosin receptor kinase A. Neuroscience. 178, 196—207.

Dempster, D., Felsednberg, D., van der Geest, S., 2006. The Bone Quality Book. Elsevier, Amsterdam.

Enlow, D.H., Brown, S.O., 1957. A comparative histological study of fossil and recent bone tissues. Part III. Mammalian bone tissues. Tex. J. Sci. 10, 187—230.

Foote, J.S., 1916. A contribution to the comparative histology of the femur. Smithsonian Contrib. Knowl. 35, 1—242.

Fuchs, R.K., Allen, M.R., Ruppel, M.E., Diab, T., Phipps, R.J., Miller, L. M., et al., 2008. In situ examination of the time-course for secondary mineralization of Haversian bone using synchrotron Fourier transform infrared microspectroscopy. Matrix Biol. 27, 34—41.

Fukumoto, T.J., 2009. Bone as an endocrine organ. Trends Endocrinol. Metab. 20, 230—236.

Gurkan, U.A., Akkus, O., 2008. The mechanical environment of bone marrow: a review. Ann. Biomed. Engng. 36, 1978—1991.

Jee, W.S.S., 1983. The skeletal tissues. In: Weiss, L. (Ed.), Histology: Cell and Tissue Biology. Elsevier Biomedical, New York.

Kaplan, F.S., Hayes, W.C., Keaveny, T.M., Boskey, A., Einhorn, T.A., Iannotti, J.P., 1994. Form and function of bone. In: Simon, S.R. (Ed.), Orthopaedic Basic Science. American Academy of Orthopaedic Surgeons, Chicago.

Karsenty, G., MacDougald O., Rosen C.J., 2012. —BONE Special Issue: Interactions between bone, adipose tissue and metabolism. vol. 50 429—579.

Martin, R.B., Burr, D.B., Sharkey, N.A., 1998. Skeletal Tissue Mechanics. Springer-Verlag, New York.

Ruppel, M.E., Miller, L.M., Burr, D.B., 2008. The effect of the microscopic and nanoscale structure on bone fragility. Osteoporos Int. 19, 1251—1265.

Urist, M.R., 1980. Fundamental and Clinical Bone Physiology. JB Lippincott Company, Philadelphia.

Bone Cells

Teresita Bellido[1], Lilian I. Plotkin[2] and Angela Bruzzaniti[3]

[1]Roudebush Veterans Administration Medical Center, Indianapolis, Indiana, USA [2]Department of Anatomy and Cell Biology, Indiana University School of Medicine, Indianapolis, Indiana, USA [3]Department of Oral Biology, Indiana University School of Dentistry and Department of Anatomy & Cell Biology, Indiana University School of Medicine, Indiana, USA

OSTEOCLASTS

Osteoclasts are the primary cells involved in bone resorption. Osteoclast activity is essential for bone modeling (Fig. 2.1), which changes the shape of bones during growth, and for bone remodeling (Fig. 2.2), which maintains the integrity of the adult skeleton. The basic multicellular unit (BMU) that remodels bone consists of the coupled activity of osteoclasts that degrade bone and osteoblasts that form bone. Dysregulation of osteoclast activity leads to either increased bone mass (if cells have reduced development or functionality) or reduced bone mass (if cells have increased development or functionality).

For bone remodeling to occur, osteoclast precursors are recruited to the bone surface, where they undergo proliferation followed by differentiation and fusion into mature, multinucleated cells. Cellular polarization results in the formation of specialized functional domains that enable osteoclasts to attach to the bone surface, acidify and degrade the mineralized matrix, and migrate along the bone surface. Upon completion of bone resorption, osteoclasts undergo programmed cell death (apoptosis).

Osteoclast Morphology and Function

Several important structural and functional domains are present in mature osteoclasts that enable them to function as bone-resorbing cells. A characteristic feature of mature osteoclasts is their multinucleation, which occurs as a result of the fusion of mononuclear precursor cells. A key step in the activation of mature osteoclasts is cytoskeletal and membrane reorganization that

results in cellular polarization and the formation of an apical membrane domain in contact with the bone surface and an opposing basolateral membrane domain located away from bone. The membrane domain that plays an important role in the adhesion of osteoclasts to bone is the sealing zone, which is found in mature osteoclasts at the bone-cell interface. The sealing zone delineates the bone-resorbing space beneath the osteoclast and circumscribes another membrane domain known as the ruffled border. The ruffled border membrane and the transition zone between the ruffled border and the sealing zone allow the secretion of hydrolytic enzymes and the internalization of degraded bone matrix products (Figs 2.3 and 2.4).

In vitro, the sealing zone appears as a ring of filamentous actin (F-actin; known as the actin ring or podosome belt), consisting of dynamic, densely packed small actin punctate structures called podosomes (Fig. 2.3). Upon osteoclast attachment to bone, podosomes appear as separate, disorganized, small F-actin dots; within minutes they are reorganized into small rings and after several hours into a peripheral podosome belt/sealing zone. Podosomes are highly dynamic adhesion structures that turn over very rapidly with an apparent half-life of 2–11 min. This turnover depends on the polymerization and depolymerization of the central F-actin core. Although podosomes are a key feature of the osteoclast sealing zone, they are also found in highly migratory cells such as carcinoma cells, endothelial cells, macrophages, and transformed fibroblasts. Podosomes are found at sites of extracellular matrix degradation and the expression of metalloproteinases MMP-14 (or MT1-MMP) and MMP-9 in the same areas strongly supports

Basic and Applied Bone Biology.
DOI: http://dx.doi.org/10.1016/B978-0-12-416015-6.00002-2

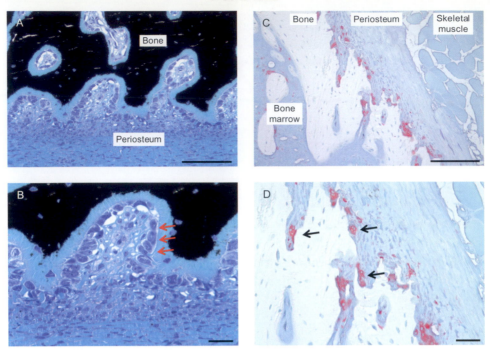

FIGURE 2.1　Bone cells during bone modeling. Bone cells work separately during bone modeling of cortical bone surfaces facing the periosteum. (A and B) Red arrows point to a row of osteoblasts in fetal baboon bone sections. Mineralized bone is stained black with von Kossa and osteoblasts on osteoid (light blue) and osteocytes within bone are stained with McNeal. Scale bar, 100 μm. (C and D) Red arrows point to a row of osteoblasts and black arrows point to osteoclasts in rat bone sections stained with tartrate-resistant acid phosphatase and toluidine blue. Scale bar, 50 μm.

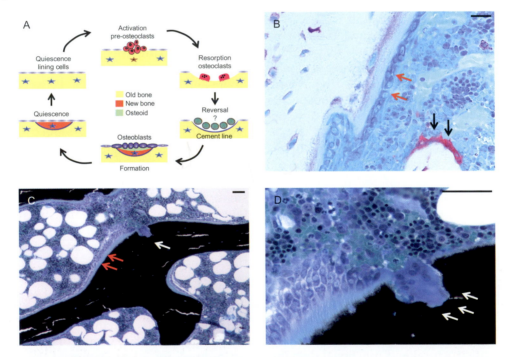

FIGURE 2.2　Bone cells during bone remodeling. Bone cells work coordinately during bone remodeling. (A) Scheme representing the bone remodeling cycle. Bone lining cells cover quiescent surfaces of bone. Upon signals derived from apoptotic osteocytes, preosteoclasts are recruited to particular sites on the bone surface, where they differentiate into mature osteoclasts that resorb mineralized bone. Resorption is followed by secretion of matrix proteins by osteoblasts and subsequent deposition of minerals. Osteocytes regulate each step of the remodeling cycle through molecules that regulate both resorption and formation activity. (B) Osteoclasts (black arrows) and osteoblasts (red arrows) on rat bone surfaces undergoing remodeling stained with tartrate-resistant acid phosphatase (red) and toluidine blue. (C and D) A multinucleated osteoclast resorbing bone (white arrows) and a row of osteoblasts depositing osteoid (red arrows) in a basic multicellular unit (BMU); bear cancellous bone stained with von Kossa and McNeal. Scale bars, 50 μm.

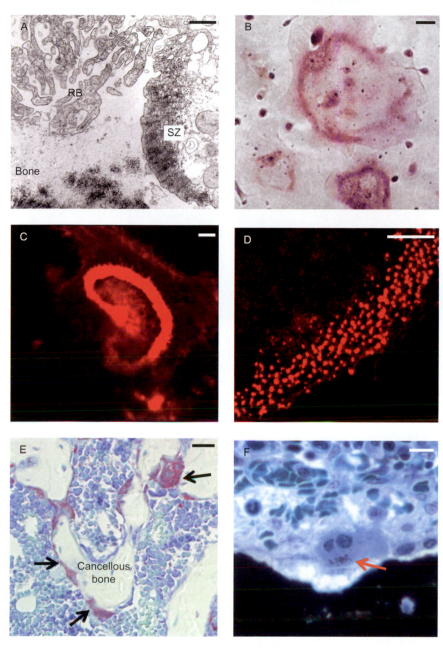

FIGURE 2.3 **Osteoclast morphology.** (A) Electron microscope image of a mature resorbing osteoclast on dentin labeled for cortactin, an F-actin binding protein, which is enriched in the sealing zone. The sealing zone (SZ), ruffled border (RB) and bone are indicated. (B) Tartrate-resistant acid phosphatase-stained multinucleated osteoclast on glass. (C and D) Confocal microscopy image of an osteoclast on dentin, stained with rhodamine-phalloidin to visualize the actin-rich SZ/podosome belt. The higher magnification in (D) reveals individual podosomes. (E) Tartrate-resistant acid phosphatase-stained osteoclasts (arrows) on trabecular surfaces in murine distal femur. (F) Osteoclast on a resorption Howship lacuna with engulfed von Kossa-stained particles in the cytoplasm, indicated by the red arrow. Scale bars, 10 μm (A-D,F) or 50 μm (E).

a role for podosomes in extracellular matrix degradation, as well as in adhesion.

Osteoclast polarization into distinct membrane domains is partly controlled by the engagement of integrin receptors within the sealing zone with Arg-Gly-Asp (RGD)-containing matrix proteins, osteopontin, and bone sialoprotein in bone. Integrins belong to a large family of transmembrane proteins that function as heterodimers of α and β subunits such as $\alpha_V\beta_3$, $\alpha_V\beta_5$, and $\alpha_2\beta_1$. The expression and activity of integrins α_v and β_3 are induced during osteoclast differentiation and regulated by RANK ligand [tumor necrosis factor ligand superfamily member 11/receptor activator of the nuclear factor NF-κB ligand (RANKL)] and macrophage colony-stimulating factor 1 (M-CSF). The importance of integrin $\alpha_V\beta_3$ in bone resorption is underscored by evidence that mice lacking integrin β_3 exhibit a progressive increase in bone mass due to osteoclast dysfunction, and competitive RGD ligands are sufficient to block osteoclast attachment and bone resorption.

In addition to integrins, numerous adaptor and signaling proteins are found in the sealing zone, which implicates these proteins in a variety of signaling cascades important to osteoclast function. For example, the cytoplasmic domains of integrins serve as platforms for signaling proteins such as the proto-oncogene

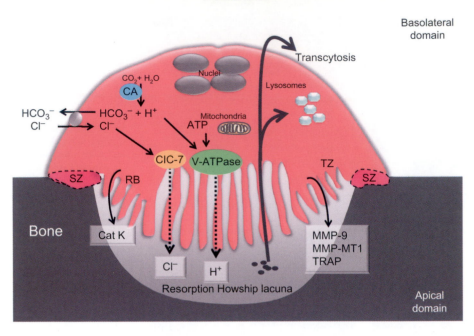

FIGURE 2.4 **Osteoclast morphology and function.** Schematic representation of a resorbing osteoclast showing the sealing zone (SZ), ruffled border (RB), transition zone (TZ), nuclei, and resorption lacuna. Mature osteoclasts are polarized, with the apical domain directed toward the bone surface. Protons generated by carbonic anhydrase (CA) are transported via the V-ATPase to the RB membrane. Secretion of protons (H^+) and chloride ions (Cl^-) through the RB membrane acidifies the resorption lacuna. A coupled basolateral bicarbonate/chloride exchanger maintains electroneutrality, avoiding changes in pH and/or membrane polarization of the cell. Mitochondria generate the necessary ATP. Bone dissolution occurs through the action of secreted cathepsin K (Cat K), matrix metalloproteases (MMPs) and tartrate-resistant acid phosphatase (TRAP). Bone degradation products are released into the bone microenvironment, internalized into the cell, and degraded via the lysosomes or secreted at the basolateral membrane via transcytosis.

tyrosine-protein kinase Src, which is crucial for osteoclast attachment and resorption. Src regulates the disassembly of podosomes and the formation of the ruffled border membrane, at least in part by its ability to interact with protein-tyrosine kinase 2-beta/focal adhesion kinase 2 (Pyk2) and the E3 ubiquitin-protein ligase CBL. The GTP-hydrolyzing enzymes (GTPases) Rho, Ras-related C3 botulinum toxin substrate (Rac), and the guanine nucleotide exchange factors that activate them, also play a central role in modifying the resorptive capacity of osteoclasts by modulating actin cytoskeleton remodeling.

The ruffled border is a highly convoluted membrane domain formed by the fusion of targeted transport vesicles with the apical membrane. The movement of vesicles to and from the ruffled border is controlled by microtubules, microfilaments, and small GTPases such as Rab-7, Rab-3D, and Rac1. The vacuolar proton pump (H^+/ATPase) is also critical for the directional movement of vesicles to the ruffled border. In addition, the proton pump is important for acidifying the matrix within the resorptive lacunae, and for ligand-receptor dissociation and receptor recycling within endosomes. Acidification of the extracellular compartment occurs by the secretion of protons (H^+) across the ruffled border; the protons are generated by

carbonic anhydrase (Fig. 2.4). To prevent intracellular polarization, proton secretion is balanced by the parallel extrusion of chloride ions (Cl^-) through the ruffled border membrane, a process that is controlled by the chloride channel, H^+/Cl^- exchange transporter 7 (CIC-7). Intracellular electron neutrality is maintained by the coupled activity of a bicarbonate HCO_3/Cl^- exchanger localized on the basolateral membrane, which in turn is regulated by carbonic anhydrase. HCl formed in the resorptive space dissolves the hydroxyapatite component of bone matrix. This exposes the organic matrix, largely consisting of type 1 collagen, which is then enzymatically digested by secreted cathepsin K and matrix metalloproteinases such as MMP-14. Osteoclasts also express and secrete tartrate-resistant acid phosphatase (TRAP), which is often used as a histologic marker of osteoclasts (Fig. 2.3). The exact physiologic role of TRAP is unclear, but it is thought to be involved in dephosphorylating osteopontin and bone sialoprotein, which is necessary for integrin binding and for generating reactive oxygen species involved in matrix degradation.

The ruffled border membrane and the sealing zone (actin ring) are essential features of a resorbing osteoclast, and abnormalities in either structure lead to compromised bone resorption. For example, osteoclasts from

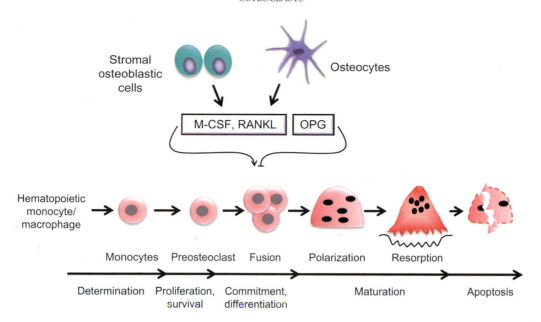

FIGURE 2.5 **Osteoclast generation and fate.** Osteoclast differentiation is governed by RANK ligand [tumor necrosis factor ligand superfamily member 11/receptor activator of the NF-κB ligand (RANKL)] and macrophage colony-stimulating factor 1 (M-CSF) as well as other cytokines secreted by osteoblasts and osteocytes that control various steps of the osteoclast differentiation process, including precursor proliferation, commitment, differentiation, and maturation. Osteoprotegerin (OPG), which is also secreted by osteoblasts and osteocytes, acts as a decoy receptor for RANKL and reduces osteoclast differentiation.

mice lacking integrin β_3, protein-tyrosine kinase 2-beta/focal adhesion kinase 2 (Pyk2), or Src, exhibit abnormal actin rings, fail to spread, and generate fewer and shallower resorptive lacunae on dentin than do wild-type osteoclasts. In addition, abnormalities of various components of the acidification process, such as mutations in H^+/ATPase or ClC-7, cause osteopetrosis, a rare inherited disease in which bones become very dense because of defective osteoclast activity. Mutations in the *TCIRG1* gene, which encodes the a3 subunit of the proton pump, account for more than 50% of osteopetrosis cases in humans. Furthermore, inactivating mutations of the human *CTSK* gene (encoding cathepsin K) result in pyknodysostosis, a disorder characterized by increased bone mass, dwarfism, and facial dysmorphism. Thus, polarization, acidification, and enzymatic dissolution of bone are all critical for osteoclast function.

Once bone dissolution occurs, the products of resorption can be internalized by osteoclasts and travel inside vesicles to lysosomes (Fig. 2.3), where they are degraded or are transported to the basolateral surface and discharged to the extracellular milieu. Alternatively, degraded bone products can be directly released to the bone microenvironment after osteoclast retraction from the resorptive lacunae.

Osteoclast Differentiation and Fusion

Osteoclasts are derived from the hematopoietic monocyte-macrophage lineage. The earliest recognized precursor of the osteoclast resides within the granulocyte-macrophage colony-forming unit (CFU-GM) in bone marrow. CFU-GMs also give rise to granulocytes and monocytes. The hematopoietic-derived mononuclear precursors fuse to form the mature osteoclast, which is multinucleated (Fig. 2.5). Multinucleation is a key morphological feature of osteoclasts and the presence of three or more nuclei, together with TRAP expression, distinguishes mature osteoclasts from immature cells. Although mature osteoclasts can accumulate more than 20 nuclei, not all are active and the mechanism leading to selective activation/inactivation of nuclei is currently unclear. Nevertheless, the number of nuclei and the overall size of the osteoclast appear to be important for resorption, with the largest osteoclasts exhibiting decreased bone resorbing activity.

At the initiation phase of bone remodeling, hematopoietic precursors in the circulation or bone marrow are recruited to the BMU where they subsequently undergo osteoclastogenesis (Fig. 2.5). The recruitment of osteoclast precursor to bone modeling/remodeling sites is controlled by several factors, including calcium gradients, osteocyte- and osteoblast-derived cytokines, and matrix metalloproteinases. Osteoclast precursors proliferate in response to growth factors, such as interleukin-3 (IL-3) and colony-stimulating factors like granulocyte-macrophage colony-stimulating factor (GM-CSF) and M-CSF, to form committed osteoclast precursors. These committed mononucleated preosteoclasts differentiate and fuse to form multinucleated osteoclasts under the influence of RANKL, a member of the tumor necrosis factor (TNF) family of ligands (Figs 2.4 and 2.5). The fusion of osteoclast mononuclear precursors to form the mature polykaryon also requires transmembrane proteins such as dendritic-cell specific transmembrane protein (DC-STAMP); V-ATPase subunit D2, a subunit of the vacuolar ATPase; and DAP, the receptor for the signaling adaptor protein, TYRO protein tyrosine kinase-binding protein/DNAX-activation protein 12 (DAP12).

M-CSF and RANKL are critical for osteoclastogenesis, and deletion of genes encoding M-CSF, RANKL, or tumor necrosis factor receptor superfamily member 11 A (RANK), the receptor for RANKL expressed by osteoclasts and their precursors, inhibits osteoclast differentiation leading to osteopetrosis in mice. Both M-CSF and RANKL are expressed by bone marrow stromal cells (BMSCs) and osteoblastic cells. Importantly, osteocytes have now been identified as a major source of M-CSF and RANKL, as well as osteoprotegerin/tumor necrosis factor receptor superfamily member 11B (OPG), the RANKL decoy receptor, suggesting that osteocytes are central regulators of osteoclastogenesis in vivo (Fig. 2.5). M-CSF contributes to osteoclast differentiation, migration, and survival by binding to its receptor, macrophage colony-stimulating factor 1 receptor (CSF-1 R/c-Fms), on osteoclast precursors, whereas RANKL facilitates osteoclast formation via direct binding to the tumor necrosis factor receptor superfamily member 11 A (RANK) receptor. RANKL is expressed on the surface of osteoblastic cells and osteocytes and is also secreted as a soluble form. Although the soluble form of RANKL is found in the circulation and its presence is sufficient to induce differentiation of osteoclast precursors in vitro, its actual role in osteoclast formation in vivo remains unproven.

M-CSF interaction with its receptor tyrosine kinase c-Fms induces recruitment of signaling molecules Src, CBL, phosphoinositide 3-kinase (PI3-K), extracellular signal-regulated kinases (ERK), 1-phosphatidylinositol 4,5-bisphosphate phosphodiesterase gamma/phospholipase C-gamma (PLC-γ), signal transducer and activator of transcription 1 (STAT1), and growth factor receptor-bound protein 2 (GRB2) to the phosphorylated tyrosine residues on the activated receptor (Fig. 2.6). Osteoclasts stimulated with M-CSF undergo Src tyrosine kinase-dependent cytoskeletal reorganization, spreading and migration, resulting in a transient decrease in bone resorption while cells are reattaching to an adjacent bone resorption area. Signaling cascades downstream of activated c-Fms are affected by the binding of CBL, which acts as a scaffold protein to link c-Fms with other molecules such as PI3-K, RAC-alpha serine/threonine-protein kinase (Akt/PKB), or ERK.

However, CBL also acts as E3 ubiquitin ligase and simultaneously targets activated c-Fms for degradation, thereby attenuating signaling and macrophage proliferation. High doses of M-CSF have also been shown to rescue signaling, spreading and differentiation in integrin β3 knockout osteoclasts but do not rescue the resorptive activity of these osteoclasts, indicating that c-Fms alone is insufficient to support bone resorption by osteoclasts. Moreover, M-CSF knockout mice recover over time from a decrease in osteoclast number and activity, whereas RANKL knockout mice do not, suggesting that RANKL plays a more dominant role in osteoclastogenesis than does M-CSF in vivo. Furthermore, RANKL can stimulate osteoclast formation and resorption in mice even in the absence of functional M-CSF. M-CSF however, augments the effect of RANKL on osteoclastogenesis by inducing the expression of RANK in osteoclast precursors, thereby priming these cells to differentiate in response to RANKL.

RANKL mediates several aspects of osteoclast differentiation, including the fusion of mononucleated precursors into multinucleated cells, acquisition of osteoclast-specific markers, attachment of osteoclasts to bone surfaces, stimulation of resorption, and promotion of osteoclast survival. RANKL expression and secretion by osteoblasts is upregulated by hormones and cytokines such as vitamin D, parathyroid hormone (PTH), and IL-1, IL-6 and IL-11, which promote osteoclastogenesis. Binding of RANKL to the trimeric RANK receptor complex expressed in osteoclast precursors leads to the activation of several signal transduction pathways, involving the recruitment of the adapter protein TNF receptor-associated factor 6 (TRAF6) to the intracellular domain of RANK. TRAF6 activates kinase-dependent signaling as well as transcription factors. Among them, nuclear factor-κB (NF-κB) has been shown to undergo nuclear translocation leading to upregulation of c-Fos/AP-1. In turn, c-Fos binds to nuclear factor of activated T-cells (NFATc1) and upregulates genes crucial for osteoclast differentiation and function (Fig. 2.6). Although several signaling pathways are activated by RANKL in osteoclasts, deletion of genes encoding NF-κB, c-Fos/AP-1, or NFATc1 leads to osteoclast dysfunction, thereby demonstrating the essential role of these proteins in osteoclastogenesis.

Osteoblasts and osteocytes also secrete OPG which like RANK, is a member of the TNF receptor family. However, unlike RANK or other members of this family of receptors, OPG is a secreted protein with no transmembrane domain and therefore has no direct signaling capability. Instead, OPG binds to its natural ligand, RANKL, thereby impeding RANKL interaction with the RANK receptor on osteoclasts. As a result, osteoclast differentiation is inhibited. Thus, the ratio of

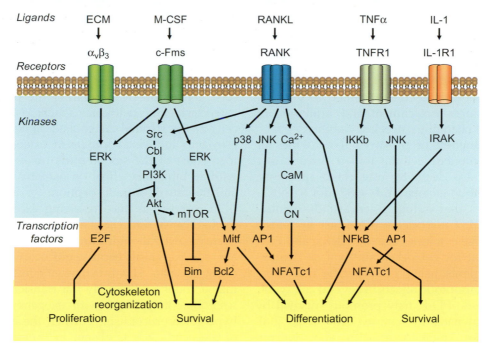

FIGURE 2.6   Signaling pathways regulating proliferation, differentiation, and survival of osteoclasts. Receptors on osteoclast precursors or on the basolateral membrane of mature osteoclasts regulate osteoclast formation or function, respectively. Macrophage colony-stimulating factor 1 (M-CSF) binds to the c-Fms receptor leads to the phosphorylation of several signaling proteins, including Src, ERK, JAK/STAT, and Akt. c-Fms is ubiqitinated by CBL and internalized, and interacts with PI3-K, thereby activating Akt and promoting osteoclast survival. Src activation also promotes cytoskeletal reorganization, migration, and cell spreading. M-CSF/c-Fms also activates mTOR, which inhibits the proapoptotic actions of Bim. Integrin signaling via $\alpha_5\beta_3$ engaged by extracellular matrix (ECM) proteins, as well as signaling by activated receptors for TNF-α and IL-1, promote proliferation and differentiation of osteoclastic cells. RANK ligand [tumor necrosis factor ligand superfamily member 11/receptor activator of the NF-κB ligand (RANKL)] binds its receptor RANK on osteoclast precursors and promotes differentiation through activation of NFATc1. In addition, RANKL induces osteoclast survival through activation of p38, Mitf and Bc1-2.

RANKL to OPG is critical for determining osteoclast numbers in vivo. The secretion of OPG in osteoblasts is also regulated by many factors. Studies using transgenic mice have demonstrated that increases in OPG secretion by signaling through the lymphoid enhancer-binding factor 1 (LEF-1/TCF1) transcription factor leads to a decrease in osteoclast differentiation and, consequently, to an increase in bone mass. A direct role for β-catenin in osteoclasts also has been reported, revealing that β-catenin directly controls osteoclastogenesis as well as bone resorption. Deletion of the gene encoding β-catenin impairs osteoclast precursor proliferation, while constitutive activation of β-catenin in the osteoclast lineage sustains proliferation but blocks osteoclast differentiation, leading to osteopetrosis in mice.

Thus, osteoclast differentiation which involves activating the RANKL-RANK, M-CSF-M-CSF 1 receptor (CSF-1 R-c-Fms), and β-catenin-Wnt pathways, regulates the expression of genes required for osteoclast function. Mutation or deletion of genes encoding key signaling proteins in these pathways results in the failure of osteoclasts to properly form or function, which

leads to the development of osteopetrosis, or high bone mass.

Osteoclast Apoptosis

After completing bone resorption, osteoclasts undergo programmed cell death or apoptosis. The signals that trigger osteoclast apoptosis in vivo are not completely understood. High concentrations of extracellular calcium, similar to levels present in bone resorption lacunae, have been found to induce osteoclast apoptosis in vitro and may be a triggering event. Fas ligand, which is secreted by osteoblasts, stimulates osteoclast apoptosis in vitro and Fas-deficient mice have increased numbers of osteoclasts and decreased bone mass, suggesting that the activation of this pathway also may be involved in the control of osteoclast life span in vivo. It has long been known that cell detachment induces apoptosis (termed anoikis, the Greek word for *homeless*) due to loss of integrin-ligand associations. Indeed, detachment of osteoclasts induces apoptosis in vitro. However, osteoclasts lacking β3 integrins have increased survival compared to wild-type cells, suggesting that the unoccupied

$\alpha_v\beta_3$ receptor in wild-type cells transmits a death signal. Therefore, it is possible that upon completion of resorption, inside-out signaling leads to integrin disengagement that may render osteoclasts more susceptible to apoptosis. However, osteoclast detachment is insufficient to induce apoptosis, as demonstrated by the fact that calcitonin induces detachment and inhibition of resorption, without changing osteoclast viability.

Decreased production of prosurvival cytokines or growth factors in the extracellular milieu surrounding osteoclasts may also lead to osteoclast apoptosis. Potential antiapoptotic factors are M-CSF and RANKL, the same cytokines that induce osteoclast differentiation. TNF-α and IL-1 also delay osteoclast apoptosis (Fig. 2.6). All of these cytokines activate ERKs, and ERK activation is crucial for osteoclast survival as demonstrated by experiments using ERK-specific inhibitors. PI3-K also has antiapoptotic effects in several cell types, including osteoclasts. PI3-K may promote survival by activating the downstream kinase, Akt (Fig. 2.6). Unexpectedly, Akt knockdown with short-hairpin RNAs (or shRNAs) revealed that Akt is required for osteoclast differentiation, but not for survival. However, this finding might be partly explained by recent reports that Akt is a negative regulator of ERK signaling. Mammalian target of rapamycin (mTOR) is another PI3-K-Akt target required for the antiapoptotic actions of M-CSF, RANKL, and TNF-α and acts by inhibiting the actions of the proapoptotic protein Bim. However, ERKs also regulate the prosurvival proteins, microphthalmia-associated transcription factor (Mitf) and Bcl$-$2. Since mTOR is also activated by ERKs, it appears these proteins may be a point of convergence in the action of prosurvival kinases in osteoclasts.

RANKL, TNF-α, and IL-1 also activate NF-κB in osteoclasts, and this transcription factor has been shown to inhibit apoptosis in various cell types (Fig. 2.6). In osteoclasts, downregulation of NF-κB mRNA inhibits IL-1-dependent survival, and inhibition of NF-κB binding to DNA with specific oligonucleotides induces apoptosis. However, osteoclast precursors lacking NF-κB subunits exhibit normal survival rates, and inhibition of NF-κB activation via a dominant-negative inhibitor of NF-κB kinase subunit beta (IKK2) does not affect the ability of IL-1 to promote osteoclast survival. Thus, the relevance of NF-κB signaling in osteoclast survival is controversial.

Regulation of Osteoclast Generation and Survival

The number and life span of osteoclasts influences the extent of bone resorption in BMUs. An increase in the life span of osteoclasts is likely to lead to an increase

in the depth of resorption pits on bone and may be important in the control of bone remodeling rates.

Several steroid hormones are known to regulate osteoclast survival, as well as osteoclast formation and activity. Both estrogens and androgens inhibit osteoclast generation by regulating the production of pro-osteoclastogenic cytokines (such as IL-1 and IL-6) by cells of the stromal/osteoblastic lineage. Furthermore, estrogens induce apoptosis in osteoclasts. The inhibitory effect of the sex steroids on osteoclasts, together with an inhibitory effect on osteoblastogenesis, attenuates the rate of bone resorption.

In contrast, in mice receiving excess glucocorticoids, the number of osteoclasts on trabecular surfaces does not decrease, even though numbers of osteoclast progenitors are reduced. This is because glucocorticoids prolong osteoclast life span. This prosurvival effect of glucocorticoids may account for the early but transient increase in bone resorption in patients with exogenous or endogenous hyperglucocorticoidism. Further, the antiapoptotic effect of glucocorticoids is absent in transgenic mice overexpressing corticosteroid 11-beta-dehydrogenase isozyme 2 (11β-HSD2), the enzyme responsible for inactivating glucocorticoids, specifically in osteoclasts, thereby confirming that these steroids act directly on osteoclasts to prolong their life span. In contrast to the rapid prosurvival effect of glucocorticoids on mature osteoclasts, glucocorticoids decrease osteoclastogenesis. This is probably a consequence of a reduction in the pool of osteoblastic cells that support osteoclast formation, and leads to the typical low remodeling rate observed during prolonged exposure to excess glucocorticoid.

OSTEOBLASTS

Osteoblasts are the cells responsible for bone formation. They originate from mesenchymal progenitors, which also give rise to chondrocytes, muscle cells, and adipocytes. Commitment of mesenchymal cells to the osteoblastic lineage depends on the specific activation of transcription factors induced by morphogenetic and developmental proteins. Osteoblasts carry out the functions of bone matrix protein secretion and bone mineralization (Fig. 2.7). Upon completion of bone matrix formation, some mature osteoblasts remain entrapped in bone as osteocytes, some flatten to cover quiescent bone surfaces as bone lining cells, and the remainder die by apoptosis.

Osteoblast Morphology and Function

The main function of osteoblasts is to synthesize bone matrix. Mature osteoblasts actively engaged in this process are recognized by their location on the

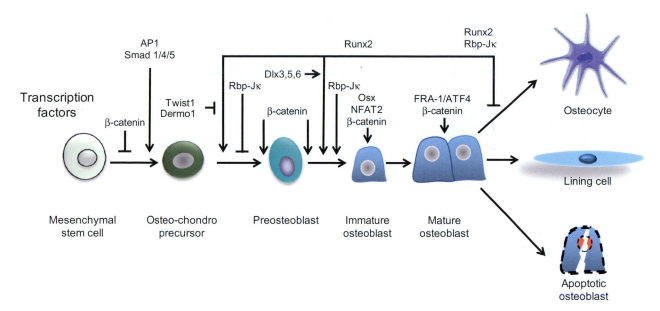

FIGURE 2.7 **Osteoblast generation and fate.** The scheme illustrates the stages of osteoblastogenesis and main transcription factors that affect the proliferation and differentiation of osteoblast precursors. Mature osteoblasts can surround themselves by bone matrix and become osteocytes, flatten to cover the quiescent bone surface as lining cells, or die by apoptosis.

bone surface and their morphological features, which are typical of cells secreting high levels of proteins: cuboidal cells with large nuclei located close to the basal membrane of the cell, enlarged Golgi apparatus on the apical surface of the nuclei, and extensive endoplasmic reticulum (Figs 2.8 and 2.9). Osteoblasts express high levels of alkaline phosphatase (ALP) and osteocalcin, and the circulating concentrations of these proteins reflect the rate of bone formation. Osteoblasts secrete abundant type I collagen and other specialized matrix proteins, which form osteoid. This organic phase of bone serves as a template for the subsequent deposit of mineral in the form of hydroxyapatite.

Interaction of osteoblasts among themselves, with lining cells, and with bone marrow cells are established by adherent junctions, tight junctions, and gap junctions. Adherent junctions, mainly mediated by cadherins, together with tight junctions serve to join cells and facilitate their anchorage to the extracellular matrix through the cytoskeleton. Alterations in the major cadherins expressed in osteoblasts, cadherin-2 (N-cadherin) and cadherin-11, influence osteoblast differentiation and survival. Intercellular communication among osteoblasts and neighboring cells is maintained via gap junctions. The opening of gap junction channels contributes to coupling and the coordination of responses within a cell population. The major gap junction protein expressed in bone cells, and in particular in cells of the osteoblastic lineage, is gap junction alpha-1 protein/connexin 43 (Cx43). Its absence or dysfunction leads to impaired osteoblast differentiation, premature apoptosis, and a deficient response to hormones and

pharmacotherapeutic agents. Furthermore, gap junction communication is fundamental to the maintenance of a continuum from the mineralized matrix, where osteocytes reside, to the bone surface cells, where osteoblasts and osteoclasts are located, and then continuing to cells in the bone marrow and to endothelial cells of the blood vessels. This functional syncytium might be responsible for the coordinated response of the bone tissue to changes in physical and chemical stimuli (see Osteocytes section below).

Interactions between osteoblasts and the bone matrix via integrins also modulate osteoblast differentiation, function, and survival. In particular, the loss of antiapoptotic signals provided by the extracellular matrix causes detachment-induced cell death or anoikis. Moreover, antibodies that neutralize the matrix protein, fibronectin, induce osteoblast apoptosis and transgenic mice expressing collagenase-resistant type 1 collagen exhibit increased osteoblast and osteocyte apoptosis compared to age-matched wild-type controls. Collectively, this evidence supports the notion that interactions between cellular integrins with domains of matrix proteins that become exposed by proteinases are required for *outside-in* signaling that preserves viability.

Osteoblast Formation and Differentiation

Osteoblast growth and differentiation is governed by an array of transcription factors, resulting in the temporal expression of proteins involved in bone

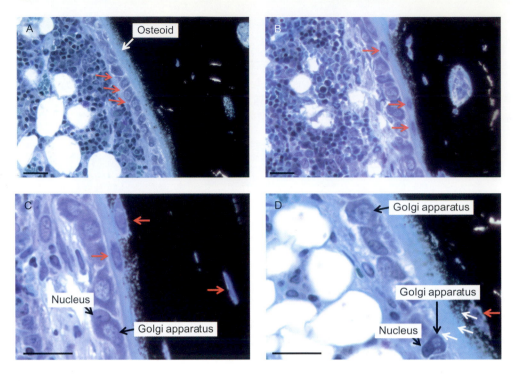

FIGURE 2.8 **Osteoblast and osteocyte morphology by optical microscopy.** Osteoblasts present on bone surfaces covered by osteoid in rat bone sections stained with von Kossa and McNeal. (A) Row of osteoblasts (red arrows) on an osteoid seam. (B) An osteocyte recently embedded (top arrow) and two osteocytes completely embedded in osteoid (bottom arrows). (C and D) Higher magnification images showing the morphology of osteoblasts, with prominent Golgi apparatus on the apical membrane facing the bone surface and a nucleus located away from the bone surface. Images also show an osteoid osteocyte (C, left red arrow), an osteocyte partially embedded in mineralized bone (C, top red arrow), an osteocyte partially surrounded by mineral (D, red arrow), and an osteocyte completely embedded in mineralized bone (C, right red arrow). Scale bars, 40 μm.

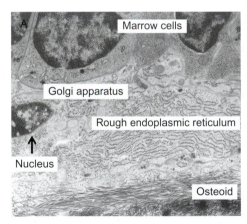

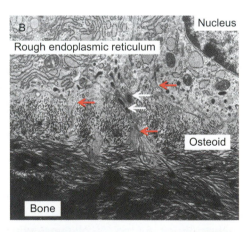

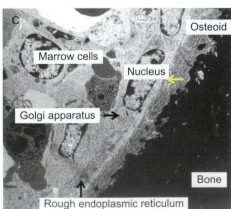

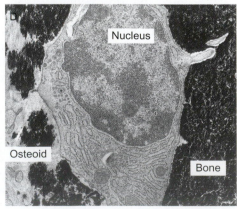

FIGURE 2.9 **Osteoblast and osteocyte morphology by transmission electron microscopy.** (A) Active osteoblast on the bone surface in which the matrix is still unmineralized (osteoid). Golgi vesicles are located adjacent to the nucleus and abundant rough endoplasmic reticulum is observed throughout the cytoplasm. Magnification, 8000×. (B) Osteoblast with cytoplasmic projections into underlying bone matrix (red arrows) and containing procollagen-filled dense secretory granules (white arrows) being released into areas of new collagen deposition. Magnification, 8000×. (C) Osteoblast on the bone surface (yellow arrow) that appear to have finished the synthesizing face and have morphological features of a lining cell, including a small amount of rough endoplasmic reticulum and a small, poorly organized Golgi apparatus. Magnification, 3000×. (D) Early osteocyte being embedded in rat bone matrix that is partially (left) and completely (right) mineralized. Magnification, 10,000×.

matrix production and matrix mineralization (Fig. 2.7). The entire process can be divided into steps comprising proliferation, extracellular matrix development and maturation, mineralization, and apoptosis. Each stage is characterized by activation of specific transcription factors and genes leading to a succession of osteoblast phenotypic markers. Thus, transcription factors of the helix-loop-helix family [DNA-binding protein inhibitor ID (Id), twist-related protein 1 (Twist), and twist-related protein 2 (Dermo-1)] are expressed in proliferating osteoblast progenitors and are responsible for maintaining the osteoprogenitor population by inhibiting the expression of genes that characterize the mature osteoblast phenotype. Transcription factors of the AP-1 family such as c-Fos, c-jun, jun-D, are expressed during proliferation as well as later in the differentiation pathway and may activate or repress transcription. RUNX2 and transcription factor Sp7/osterix are essential for establishing the osteoblast phenotype. In mice, an absence of the *Runx2* or *Sp7* genes (or double knockouts) results in a lack of skeletal mineralization and perinatal lethality. RUNX2 is expressed at an earlier stage than is osterix, but both of these proteins regulate the expression of genes that control bone formation and remodeling, including *BGLAP* (which encodes osteocalcin), *SSP1* (which encodes osteopontin), *MMP13* (which encodes matrix metalloprotease 13/collagenase 3), *OPG*, and *RANKL*. RUNX2 regulates osteoblast differentiation and function by several signaling pathways, including those activated by Wnts and bone morphogenetic proteins (BMPs), as well as the differentiation and survival of osteoblasts induced by integrins and the PTH receptor.

Osteoblast Apoptosis

It is estimated that upon completing the process of bone formation, 60−80% of osteoblasts die by apoptosis (Fig. 2.7).

Apoptosis has been extensively studied in cultured osteoblasts using several methods, including increased activity of initiator or effector caspases, presence of cleaved genomic DNA by TUNEL or in situ end-labeling, as well as nuclear fragmentation and chromatin condensation using fluorescent dyes that bind to DNA. Examination of the nuclear morphology of cells transfected with fluorescent proteins containing a nuclear localization sequence has proven a particularly useful tool for studying apoptosis in cells cotransfected with genes of interest. Cell detachment from the substrate, changes in the plasma membrane composition, and cell shrinkage are also features that have been used to detect and quantify apoptotic cells.

Apoptosis appears to begin at the early stages of osteoblast differentiation and continues throughout all stages of osteoblast life. Thus, apoptotic mesenchymal progenitors have been found in primary spongiosa of developing long bones of chicks and rabbits and in murine calvarial bone, at sites of fracture healing, as well as bone formation during distraction osteogenesis.

Regulation of Osteoblast Generation and Survival

Most major regulators of skeletal homeostasis influence the generation as well as the survival of osteoblasts.

The two major signaling pathways that promote osteoblast differentiation are those activated by BMPs and Wnts (see Chapter 3 for details of these signaling pathways). BMPs play an important role in osteoblast differentiation, and also induce apoptosis in mesenchymal osteoblast progenitors in interdigital tissues during the development of hands and feet, as well as in mature osteoblasts. Wnt signaling has a profound effect on bone, as shown by the high bone mass phenotype of mice and humans with activating mutations of low-density lipoprotein receptor-related protein 5 (LRP5), which together with frizzled proteins are receptors for Wnt ligands. Wnts stimulate differentiation of undifferentiated mesenchymal cells toward the osteoblastic lineage and also stimulate differentiation of preosteoblasts. Remarkably, Wnt signaling in osteoblasts (and osteocytes) also affects osteoclasts. Thus, Wnt signaling in osteoblastic cells increases expression of the RANKL decoy receptor OPG, leading to inhibition of osteoclast development. In addition, Wnt signaling inhibits apoptosis in mature osteoblasts and osteocytes. The increased bone formation exhibited by mice lacking the Wnt antagonist secreted frizzled-related protein 1 (sFRP-1) is associated with decreased osteoblast numbers and osteocyte apoptosis. Likewise, the prevalence of osteoblast and osteocyte apoptosis is decreased in mice expressing the high bone mass activating mutation of *Lrp5* (G171V). The resulting mutant LRP5 protein exhibits a reduced ability to bind sclerostin (encoded by the *Sost* gene), which is a Wnt antagonist specifically secreted by osteocytes. Consistent with this, sclerostin induces osteoblast apoptosis in vitro. Moreover, the ability of both PTH and mechanical loading to reduce sclerostin synthesis is responsible for their stimulation of osteoblast differentiation and extension of osteoblast life span. Activation of Wnt signaling in vitro also prevents apoptosis in uncommitted C2C12 osteoblast progenitors and the more differentiated MC3T3-E1 and OB-6 osteoblastic cell models. Wnt ligands known to activate the so-called canonical (mediated by β-catenin-induced gene transcription) as well as noncanonical (independent of β-catenin-induced gene transcription) pathways prevent apoptosis in osteoblastic cells by a mechanism that involves the Src/ERK and PI3-K/AKT prosurvival kinase signaling pathways.

Glucocorticoids induce rapid bone loss, resulting from a transient increase in resorption due to delayed osteoclast apoptosis. This initial phase is followed by a sustained and profound reduction in bone formation and turnover caused by decreased osteoblast and osteoclast generation and increased osteoblast apoptosis (as well as osteocyte death; see next paragraph).

Both chronic excess of PTH, as in hyperparathyroidism, and intermittent elevations in PTH (by daily injections) increase the number of osteoblasts. However, whereas the former condition can lead to bone catabolism, intermittent administration of PTH causes bone anabolism. In both situations, the rate of bone turnover is increased due to the simultaneous increase in osteoclast and osteoblast numbers and delayed osteoblast apoptosis. However, intermittently administered PTH also causes de novo bone formation not coupled to previous resorption. This occurs early, creating the so-called *anabolic window*. Additionally, PTH causes an overfilling of erosion pits, thereby creating net bone formation.

The increased number of osteoblasts can be achieved by either an increase in their rate of production from progenitors, a decrease in the rate of their death by apoptosis, or a combination of the two. Studies in mice indicate that chronic and intermittent PTH elevations increase osteoblast number by different mechanisms. The increase in osteoblast number and the anabolic effect of intermittent PTH can be accounted for by attenuation of osteoblast apoptosis and increased recruitment of osteoblast progenitors into the differentiation pathway. Chronic elevation of the endogenous hormone resulting from dietary calcium deficiency or continuous infusion of exogenous PTH has no effect on osteoblast survival, but accelerates osteoblast generation driven by a reduction in the expression of sclerostin, the inhibitor of bone formation expressed by osteocytes (as described in Chapter 15).

Bone Lining Cells

After completion of their bone matrix-synthetizing function, some osteoblasts become flattened and cover the inactive bone surface as bone lining cells. Lining cells may recover the ability to produce matrix in response to PTH, a mechanism that may contribute to the rapid increase in bone formation observed upon daily administration of the hormone. Lining cells have been also proposed to participate in the regulation of calcium exchange between mineralized bone and the bone marrow extracellular compartment. Although the mechanisms underlying this function are unknown, lining cells are in close contact with osteocytes embedded in the bone matrix through gap junctions, suggesting their potential involvement in the interchange of minerals and other metabolites between bone and cells on the bone surface or the bone marrow. Accumulating evidence supports the notion that lining cells might play an important function in bone remodeling by retracting from the bone surface and creating a canopy over osteoclasts and osteoblasts in the BMU. This canopy of lining cells presumably encases bone marrow osteoblast precursors and is penetrated by blood vessels that provide hematopoietic osteoclast progenitors. Thus, the lining cell canopy, associated capillaries, osteocytes, osteoclasts, and osteoblasts form a compartment separated from the rest of the marrow named the *bone remodeling compartment* (BRC; Fig. 2.10). The signals that trigger lining cell detachment to form the BRC in a particular area of the bone surface are unknown. Premature apoptosis in osteocytes has been shown to precede osteoclast accumulation and resorption. Thus, apoptotic osteocytes might release molecules that induce lining cell retraction, facilitating the access of osteoclast precursors to bone surfaces. Osteocytes also express M-CSF, which stimulates proliferation of preosteoclasts, as well as RANKL, the master cytokine inducer of osteoclast differentiation, which may reach the BRC. It is still uncertain whether membrane-bound RANKL in the osteocyte dendritic processes or soluble RANKL released from osteocytes through the canalicular circulation is responsible for osteoclast differentiation within the BRC. In turn, factors released from the bone matrix upon resorption stimulate proliferation of preosteoblasts and their differentiation to mature osteoblasts. Thus, the BRC provides a supportive environment for the differentiation of osteoclast and osteoblast progenitors, and is under the influence of hormonal and mechanical stimuli, thus permitting tight regulation of bone remodeling.

OSTEOCYTES

Osteocytes are former osteoblasts that become entombed during the process of bone deposition and are regularly distributed throughout the mineralized bone matrix (Fig. 2.7). Osteocytes are the most abundant cells in bone, comprising more than 90% of cells within the matrix or on the bone surfaces. Increasing evidence supports the notion that osteocytes coordinate the function of osteoblasts and osteoclasts in response to both mechanical and hormonal cues.

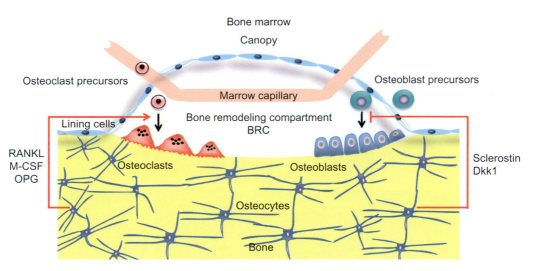

FIGURE 2.10 **The bone remodeling compartment.** When the need for bone resorption is sensed by osteocytes, they send signals to lining cells, which retract from the bone surface to form a structure named the bone remodeling compartment (BRC). Osteoclast precursors are transported to the BRC by marrow capillaries, where they differentiate to mature osteoclasts under the influence pro and anti-osteoclastogenic cytokines [tumor necrosis factor ligand superfamily member 11/receptor activator of the NF-κB ligand (RANKL)], macrophage colony-stimulating factor 1 (M-CSF), and osteoprotegerin (OPG) derived from osteocytes, and initiate bone remodeling. Osteoblasts precursors from the bone marrow or the circulation differentiate into mature, bone synthesizing cells in response to factors released from the bone matrix by resorption. The differentiation and function of osteoblasts is controlled by molecules derived from osteocytes, including sclerostin and Dkk-1.

Osteocyte Morphology and Functions

Osteocyte bodies are individually encased in lacunae and exhibit cytoplasmic dendritic processes that run along narrow canaliculi within the mineralized matrix (Fig. 2.11). Osteocyte morphology is dictated by the expression of genes involved in dendrite formation and branching, such as *E11/PDPN/GP38* (which encodes podoplanin), *CD44*, and *PLS3* (which encodes plastin/fimbrin), which are also expressed in neurons and give osteocytes their characteristic morphology in vivo as well as in culture. Quantitative analysis using microscopy determined that each osteocyte exhibits an average of 50 cytoplasmic projections emerging from its cell body. Projections from neighboring osteocytes touch each other and establish communication through gap junctions within canaliculi. Canaliculi also reach both periosteal and endocortical bone surfaces in cortical bone, as well as surfaces adjacent to the bone marrow in cancellous bone. The lacunar-canalicular system also allows the transport of proteins that are produced and secreted by osteocytes and exert their action on cells on the bone surface or the bone marrow. Osteocytes, but not other cells in bone, produce sclerostin, the product of the *SOST* gene. As expected for an osteocyte-derived secreted protein, high levels of sclerostin are detected in canaliculi (Fig. 2.11). Sclerostin potently antagonizes several members of the BMP family of proteins and also binds to LRP5/6, preventing activation of Wnt signaling. Both BMPs and Wnts are critical for osteoblastogenesis as they provide the initial, essential stimulus for the commitment of multipotential mesenchymal progenitors to the osteoblast lineage; they also regulate osteoblast activity.

Today, it is accepted that osteocytes are the mechanosensory cells. Osteoblasts and osteoclasts are present on bone only transiently, in low numbers, and in variable locations. In contrast, osteocytes are present in the entire bone volume and are long-lived. Osteocytes are the core of a functional syncytium that extends from the mineralized bone matrix to the bone surface and the bone marrow, and which also reaches the blood vessels. Their strategic location permits osteocytes to detect variations in mechanical signals (either through strain or fluid flow), as well as levels of circulating factors (ions or hormones), and allows signal amplification leading to adaptive responses of the skeleton to environmental changes.

Increasing evidence demonstrates that osteocytes regulate the function of osteoblasts and osteoclasts. In response to mechanical and hormonal cues, osteocytes produce and secrete factors (such as OPG, RANKL, and sclerostin) that affect other bone cells by paracrine or autocrine mechanisms, and hormones (such as FGF-23) that affect other tissues by endocrine mechanisms.

Osteocytes detect fatigue-induced microdamage and signal to osteoclasts to induce replacement of damaged bone through remodeling. They also respond to changes in mechanical load by inducing local changes in bone mass and geometry through modeling.

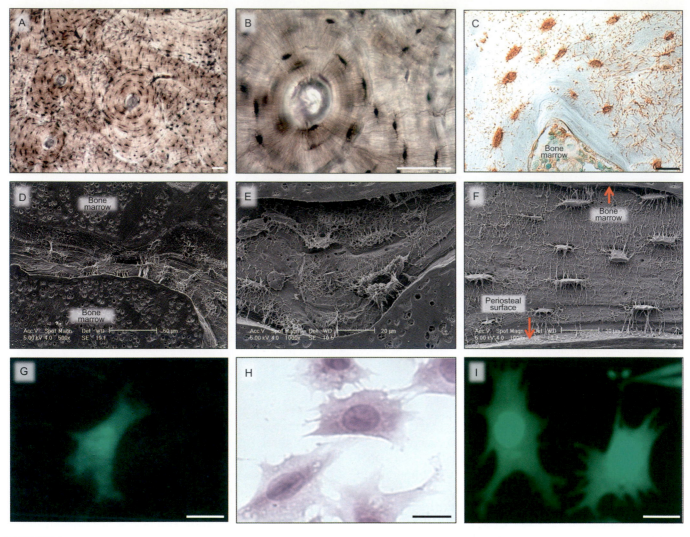

FIGURE 2.11   **Osteocyte morphology.** (A and B) Spatial location of osteocytes in cortical bone showing organized lamellae surrounding the haversian canals, and osteocyte lacunae and canaliculi; dry human bone. (C) Sclerostin expression in osteocyte bodies and its presence in canaliculi detected with an anti-sclerostin antibody in murine vertebral bone. (D–F) Scanning electron microscopy of acid etched murine vertebral bone showing the extensive osteocyte network connecting osteocytes with each other and the bone marrow in cancellous bone (D and E) and the bone marrow and the periosteal surface of cortical bone (F). (G) Osteocyte within neonatal calvarial bone from DMP-1-8 kb-GFP mice in which osteocytes express green fluorescent protein. (H and I) MLO-Y4 osteocytic cell line stained with hematoxylin-eosin (H) or stably transfected with green fluorescent protein targeted to the nucleus but also seen in the cytoplasm (I). Scale bar, 50 μm for A–D and 20 μm for E–I.

Furthermore, osteocytes detect alterations in the levels of circulating hormones and respond by changing the rate of bone formation and resorption.

Osteocytogenesis and Osteocyte Maturation

Between 5% and 20% of mature osteoblasts become entombed in the matrix that they generate and that subsequently mineralizes (Fig. 2.11). The process of osteocyte formation was long thought to be stochastic. However, it is now recognized that some osteoblasts may be prompted to extend cytoplasmic projections and make contact with already embedded cells,

resulting in their differentiation into osteocytes. In particular, expression of the membrane-associated proteins *E11/PDPN/GP38* and MMP-14 is required for the formation of osteocyte dendritic processes and canaliculi. This evidence supports the notion that osteocytogenesis is an active process driven by changes in gene expression. However, the mechanisms that determine which osteoblasts will become osteocytes remain obscure.

Osteocyte formation is one of the three possible fates of mature osteoblasts, the other two being becoming lining cells or undergoing apoptosis. It is therefore expected that stimuli that alter an osteoblast's fate would

TABLE 2.1 Osteocytic Gene Expression

Group	Protein	Expressed in	Function
1	*E11/podoplanin/gp38*	Early, embedding osteocytes	Dendrite formation
	CD44	Highly expressed in osteocytes compared to osteoblasts	Hyaluronic acid receptor associated with E11 and linked to cytoskeleton
	Plastin/fimbrin	All osteocytes	Dendrite branching
	MMP-14	Matrix degradation	Canaliculi formation
2	PEX	Early and late osteocytes	Phosphate metabolism
	MEPE/OF45	Late osteoblasts through to osteocytes	Inhibitor of bone formation and involved in phosphate metabolism
	DMP-1	Early and mature osteocytes	Phosphate metabolism and mineralization
	FGF-23	Early and mature osteocytes	Phosphate metabolism
3	Dkk-1	Osteoblasts and early and mature osteocytes	Inhibitor of bone formation
	Sclerostin	Late osteocytes	Inhibitor of bone formation
4	RANKL	Osteocytes	Osteoclast differentiation and survival
	M-CSF	Osteocytes	Preosteoclast and osteoclast proliferation and survival factor
	OPG	Osteocytes	Inhibitor of osteoclast differentiation

The osteocyte phenotype is characterized by the expression of proteins related to their morphology and function, which may change at different stages of osteocyte development and maturation. There are four main categories: (1) Proteins related to dendritic morphology and canaliculi formation that facilitate cell embedding into the bone matrix; (2) proteins related to phosphate metabolism and matrix mineralization; (3) proteins that regulate bone formation; and (4) proteins that regulate bone resorption.

DMP-1, dentin matrix acidic phosphoprotein 1; FGF-23, fibroblast growth factor 23; M-CSF, macrophage colony-stimulating factor 1; MEPE, matrix extracellular phosphoglycoprotein; OPG, osteoprotegerin; PEX, phosphate-regulating neutral endopeptidase; RANKL, tumor necrosis factor ligand superfamily member 11/receptor activator of the NF-κB ligand.

affect osteocyte formation. Consistent with this notion, inhibition of osteoblast apoptosis by intermittent administration of PTH leads to increased osteocyte density. However, it is still unknown whether this effect of the hormone is accompanied by changes in the expression of genes required for the osteoblast–osteocyte transition.

Osteocytes express most of the genes expressed by osteoblasts, including osteoblast-specific transcription factors and proteins, although their levels of expression may differ slightly. Thus, ALP and type I collagen expression is lower, whereas osteocalcin expression is higher in osteocytes. Keratocan (KTN), an extracellular matrix protein that belongs to the small leucine-rich proteoglycan family, has emerged as an osteoblast marker because its expression is greatly reduced in osteocytes.

Osteocytes are richer than osteoblasts in proteins related to mineralization and phosphate metabolism (Table 2.1), including phosphate-regulating neutral endopeptidase (PEX), dentin matrix acidic phosphoprotein 1 (DMP-1), matrix extracellular phosphoglycoprotein (MEPE), and fibroblast growth factor 23 (FGF-23). Osteocytes also express high levels of the inhibitor of bone formation Dickkopf-related protein

1 (Dkk-1), and the *SOST* gene (encoding the Wnt antagonist and bone formation inhibitor, sclerostin) is expressed in osteocytes but not in osteoblasts (Fig. 2.11; Table 2.1).

Osteocyte Apoptosis: Consequences and Regulation

Osteocytes are long-lived cells. However, like osteoblasts and osteoclasts, osteocytes die by apoptosis; and decreased osteocyte viability accompanies the bone fragility syndromes that characterize glucocorticoid excess, estrogen withdrawal, and mechanical disuse. Conversely, preservation of osteocyte viability might explain at least part of the antifracture effects of bisphosphonates, which cannot be completely accounted for by increases in bone mineral density.

Preservation of Osteocyte Viability by Mechanical Stimuli

Osteocytes interact with the extracellular matrix (ECM) in the pericellular space through discrete sites in their membranes, which are enriched in integrins and vinculin, as well as through transverse elements that tether osteocytes to the canalicular wall. Fluid movement in the canaliculi resulting from mechanical loading may induce ECM deformation, shear stress, and/or tension in the tethering elements. The resulting change in circumferential strain in osteocyte membranes is hypothesized to be converted into intracellular signals by integrin clustering and integrin interaction with cytoskeletal and catalytic proteins at focal adhesions. Physiologic levels of mechanical strain imparted by stretching or pulsatile fluid flow prevent apoptosis in cultured osteocytes. Mechanistic studies indicate that the transduction of mechanical forces into intracellular signals is accomplished by molecular complexes assembled at caveolin-rich domains of the plasma membrane and composed of integrins, cytoskeletal proteins, and kinases including focal adhesion kinase (FAK) and Src, resulting in activation of the ERK pathway and osteocyte survival. Intriguingly, a ligand-independent function of the estrogen receptor (ER) is indispensable for mechanically induced ERK activation in both osteoblasts and osteocytes. Accordingly, mice lacking the ERα and ERβ exhibit a poor osteogenic response to loading.

In vivo mechanical forces also regulate osteocyte life span. Apoptotic osteocytes are found in unloaded bones or in bones exposed to high levels of mechanical strain. In both cases, increased apoptosis in osteocytes is observed before any evidence of increased osteoclast resorption, and apoptotic osteocytes accumulate in areas that are subsequently removed by osteoclasts. These findings suggest that dying osteocytes in turn become the signals for osteoclast recruitment and the resulting increase in bone resorption. In support of this notion, targeted ablation of osteocytes in transgenic mice is sufficient to induce osteoclast recruitment and resorption, leading to bone loss. It remains to be determined whether living osteocytes continually produce molecules that restrain osteoclast recruitment, or whether in the process of undergoing apoptosis osteocytes produce pro-osteoclastogenic signals. Taken together with evidence that osteocyte apoptosis is inhibited by estrogens and bisphosphonates, these findings raise the possibility that preservation of osteocyte viability contributes to the antiremodeling properties of these agents.

Aging and Osteocyte Apoptosis

One of the functions of the osteocyte network is to detect microdamage and trigger its repair. During aging, there is an accumulation of microdamage and a decline in osteocyte density accompanied by decreased prevalence of osteocyte-occupied lacunae, an index of premature osteocyte death. Reduced osteocyte density may be a direct consequence of increased osteoblast apoptosis, whereas increased osteocyte apoptosis may result from the decline in physical activity with old age, leading to reduced skeletal loading, accumulation of reactive oxygen species in bone and/or increased levels of endogenous glucocorticoids with age (see below). In view of the evidence already discussed on the role of osteocytes in microdamage repair, age-related loss of osteocytes may be partially responsible for the disparity between bone quantity and quality that occurs with aging.

Hormonal Regulation of Osteocyte Life Span

Estrogen, as well as androgen, deficiency leads to an increased prevalence of osteocyte apoptosis. Conversely, estrogens and androgens inhibit apoptosis in osteocytes, as well as in osteoblasts. This antiapoptotic effect is due to the rapid activation of the Src/Shc/ERK signaling pathway through nongenotropic actions of the classical receptors for sex steroids. This effect requires only the ligand-binding domain of the receptor and, unlike the classical genotropic action of the receptor protein, it is eliminated by nuclear targeting of the receptors.

Increased glucocorticoid action in bone may also contribute to induction of osteocyte apoptosis. This might result from treatment with the steroids, which have immunosuppressive effects, from endogenous elevation of the hormones with age, or from increased expression in bone of estradiol 17-beta-dehydrogenase 1/11β-hydroxysteroid dehydrogenase type 1 (11β-HSD1), the enzyme that amplifies glucocorticoid action by converting inactive into active steroids. The apoptotic effect of glucocorticoids is reproduced in cultured osteocytes and osteoblasts in a manner strictly dependent on the glucocorticoid receptor (GR). The induction of osteocyte and osteoblast apoptosis by glucocorticoids results from direct actions of the steroids on these cells, as overexpression of the 11β-HSD2 enzyme that specifically inactivates glucocorticoids in osteoblastic cells abolishes the increase in apoptosis. The proapoptotic effect of glucocorticoids in cultured osteocytic cells is preceded by cell detachment due to interference with FAK-mediated survival signaling generated by integrins. In this mechanism, Pyk2 becomes phosphorylated and subsequently activates proapoptotic mitogen-activated protein kinase (MAPK)/c-Jun N-terminal kinase (JNK) signaling. In addition, the proapoptotic actions of glucocorticoids may involve suppression of the synthesis of locally produced antiapoptotic factors, including insulin-like growth factor I (IGF-I) and IL-6-type cytokines, as well as MMPs, and stimulation of the proapoptotic Wnt antagonist sFRP-1.

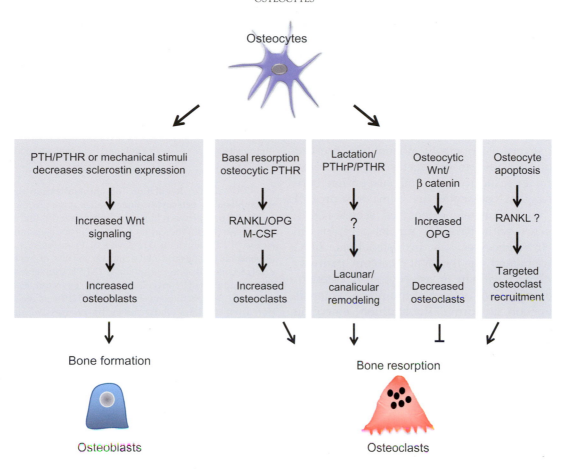

FIGURE 2.12 **Regulation of osteoblast and osteoclast production and function by osteocytes.** Osteocytes regulate bone formation through sclerostin. Thus, bone formation induced by systemic elevation of parathyroid hormone (PTH) or local mechanical loading is associated with decreased expression of sclerostin. Osteocytes regulate bone resorption through pro- and anti-osteoclastogenic cytokines. Resorption under basal conditions, induced by PTH elevation or by parathyroid hormone-related peptide (PTHrP) increased during lactation, is regulated by RANK ligand [tumor necrosis factor ligand superfamily member 11/receptor activator of the NF-κB ligand (RANKL)] through the PTH receptor (PTHR) expressed in osteocytes. Activation of Wnt signaling in osteocytes increase OPG expression leading to inhibition of resorption. Osteocyte apoptosis induced by immobilization, fatigue loading, sex steroid deficiency, or genetically induced by activating diphtheria toxin receptor signaling is sufficient to recruit osteoclasts to specific bone areas and increase resorption, probably through a mechanism that increases RANKL expression in osteocytes.

Regulation of Bone Formation by Osteocytes: The Sclerostin Paradigm

Osteocytes express sclerostin, which antagonizes several members of the BMP family of proteins and also binds to LRP5/6, thereby preventing canonical Wnt signaling. Both BMPs and Wnts are critical for osteoblastogenesis as they provide the initial and essential stimulus for commitment of multipotential mesenchymal progenitors to the osteoblast lineage. Loss of sclerostin expression in humans causes the high bone mass disorders van Buchem syndrome and sclerosteosis. In addition, administration of an anti-sclerostin antibody increases bone formation and restores the bone lost upon ovariectomy in rodents. Conversely, transgenic mice overexpressing sclerostin

exhibit low bone mass. Taken together, these lines of evidence indicate that sclerostin derived from osteocytes—the most differentiated cells of the osteoblastic pathway—exerts a negative feedback control at the earliest step of mesenchymal stem cell differentiation toward the osteoblast lineage (Fig. 2.12).

Regulation of Bone Resorption by Osteocytes: RANKL and OPG

The cues that signal bone resorption are not completely understood. One important factor in the regulation of remodeling appears to be osteocyte apoptosis following local bone damage or microdamage, which signals to osteoblast lining cells to form the BRC (Fig. 2.10). Apoptotic osteocytes may regulate the recruitment of osteoclast precursors and their differentiation in two ways. First, osteocyte apoptosis may indirectly stimulate osteoclastogenesis by inducing stromal or osteoblastic cells to secrete RANKL. Second, osteocytes can directly secrete RANKL. Indeed, in vitro, purified osteocytes express higher levels of RANKL than do osteoblasts and BMSCs. The severe osteopetrotic phenotype observed in mice lacking RANKL in osteocytes and their resistance to bone loss induced by tail suspension supports the idea that osteocytes are a major source of RANKL in vivo. Further, osteocytes secrete OPG, which competes with RANKL for its receptor on osteoclasts. In osteocytes, as in osteoblasts, OPG secretion is regulated by the Wnt-β-catenin pathway and mice lacking β-catenin in osteocytes are osteoporotic due to increased osteoclast numbers, whereas osteoblast function is normal. In addition, emerging experimental evidence also points to osteocytes as an additional source of secreted M-CSF in bone. Together, these new findings suggest that osteocytes control the bone remodeling process through direct and indirect regulation of osteoclast and osteoblast differentiation and function (Fig. 2.12).

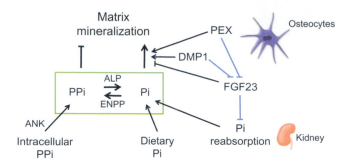

FIGURE 2.13 Regulation of mineralization and phosphate metabolism by osteocytes. Bone matrix mineralization is controlled by the balance between inorganic phosphate (Pi) and inorganic pyrophosphate (PPi). Whereas an excess in Pi induces mineralization, an excess of PPi inhibits it. The levels of Pi in the extracellular matrix depend on the dietary intake and on its rate of synthesis from PPi catalyzed by alkaline phosphatase (ALP). Conversely, the levels of PPi depend on its conversion from Pi by ectonucleotide pyrophosphatase/phosphodiesterase enzymes (ENPP), as well as by its transport from the intracellular milieu by progressive ankylosis protein (ANK). Proteins expressed in osteocytes regulate mineralization. Specifically, PEX and DMP-1 induce mineralization, whereas FGF-23 inhibits mineralization. Inhibition of mineralization by FGF-23 is believed to be caused by inhibition of Pi reabsorption in the kidney, which would reduce Pi levels, whereas induction of mineralization by PEX and DMP-1 might be secondary to inhibition of FGF-23 and thus increase circulating Pi.

Regulation of Bone Mineralization by Osteocytes

Approximately 50–70% of the bone matrix is mineral. As mature osteoblasts are surrounded by the collagenous matrix and differentiate into osteocytes, mineral is deposited to transform osteoid into mineralized bone. Studies using genetically modified mice have demonstrated that osteocytes actively participate in the regulation of bone mineralization. In particular, DMP-1 and MEPE, proteins of the small integrin-binding ligand N-linked glycoprotein (SIBLING) family, are produced by late-stage osteoblasts and osteocytes and can be detected in the canalicular and lacunar walls. DMP-1 appears to be dispensable for bone mineralization during development, but adult DMP-1-deficient mice show defective osteocyte morphology and altered bone mineralization (Fig. 2.13). In contrast, MEPE appears to be an inhibitor of mineralization, as MEPE-deficient mice exhibit increased bone density and ASARM, a cleavage product of MEPE, can block mineralization in vitro and in vivo. Both DMP-1 and MEPE are encoded by mechanoresponsive genes and changes in their expression may be responsible for the reduced mineralization of the matrix surrounding osteocytes induced by mechanical stimulation.

Osteocytes also express and secrete FGF-23, a hormone that regulates phosphate reabsorption in the kidney and affects bone mineralization by changing circulating levels of phosphate. FGF-23 directly activates intracellular signaling in osteocytes and osteoblasts mediated through binding to the fibroblast growth factor receptor 1 (FGFR-1)/Klotho receptor complex and has been shown to suppress osteoblast differentiation and matrix mineralization in vitro, suggesting a role for FGF-23 not only in the regulation of systemic phosphate levels but also in the local control of bone mineralization.

STUDY QUESTIONS

1. Describe two mechanisms that control osteoclast apoptosis and how these may or may not be affected in old age.
2. Osteoblast numbers can be regulated by Wnt signaling. How does the Wnt signaling pathway contribute to the increased osteocyte density observed after bone is mechanically loaded?
3. Describe the features of osteocyte morphology, lifecycle, and gene expression that contribute to the mechanosensory function of bone.
4. β-Catenin signaling in bone cells is necessary for bone homeostasis. How does sclerostin production by osteocytes affect osteoblasts and osteoclasts?

5. Osteoclastogenesis is regulated in large part by the RANKL-OPG system. Briefly describe the regulation of osteoclastogenesis by the RANKL-OPG system and discuss the sources of RANKL and OPG.

6. Describe how osteoclasts generate an acidic environment for resorption and disease that can result from disruption of this process.

Suggested Readings

Aubin, J.E., 2008. Mesenchymal stem cells and osteoblast differentiation. In: Bilezikian, J.P., Raisz, L.G., Martin, T.J. (Eds.), Principles of Bone Biology. Academic Press, San Diego, San Francisco, New York, London, Sydney, Tokyo, pp. 85–107.

Balemans, W., Van Hul, W., 2006. Human genetics of SOST. J. Musculoskelet. Neuronal Interact. 6, 355–356.

Bellido, T., 2006. Downregulation of SOST/sclerostin by PTH: a novel mechanism of hormonal control of bone formation mediated by osteocytes. J. Musculoskelet. Neuronal Interact. 6, 358–359.

Bellido, T., 2007. Osteocyte apoptosis induce bone resorption and impairs the skeletal response to weightlessness. BoneKEy-osteovision. 4, 252–256.

Bonewald, L.F., 2011. The Amazing Osteocyte. J Bone Miner. Res. 26, 229–238.

Bruzzaniti, A., Baron, R., 2006. Molecular regulation of osteoclast activity. Rev. Endocr. Metab. Disord. 7, 123–139.

Eriksen, E.F., 2010. Cellular mechanisms of bone remodeling. Rev. Endocr. Metab. Disord. 11, 219–227.

Feng, J.Q., Ye, L., Schiavi, S., 2009. Do osteocytes contribute to phosphate homeostasis? Curr. Opin. Nephrol. Hypertens. 18, 285–291.

Jilka, R.L., Bellido, T., Almeida, M., Plotkin, L.I., O'Brien, C.A., Weinstein, R.S., et al., 2008. Apoptosis in bone cells. In: Bilezikian, J.P., Raisz, L.G., Martin, T.J. (Eds.), Principles of Bone Biology. Academic Press, San Diego, San Francisco, New York, London, Sydney, Tokyo, pp. 237–261.

Kramer, I., Halleux, C., Keller, H., Pegurri, M., Gooi, J.H., Weber, P. B., et al., 2010. Osteocyte Wntβ-catenin signaling is required for normal bone homeostasis. Mol. Cell Biol. 30, 3071–3085.

Lian, J.B., Stein, G.S., Javed, A., Van Wijnen, A.J., Stein, J.L., Montecino, M., et al., 2006. Networks and hubs for the transcriptional control of osteoblastogenesis. Rev. Endocr. Metab. Disord. 7, 1–16.

Marotti, G., Ferretti, M., Muglia, M.A., Palumbo, C., Palazzini, S., 1992. A quantitative evaluation of osteoblast-osteocyte relationships on growing endosteal surface of rabbit tibiae. Bone. 13, 363–368.

Paszty, C., Turner, C.H., Robinson, M.K., 2010. Sclerostin: a gem from the genome leads to bone-building antibodies. J. Bone Miner. Res. 25, 1897–1904.

Quarles, L.D., 2012. Skeletal secretion of FGF-23 regulates phosphate and vitamin D metabolism. Nat. Rev. Endocrinol. 8, 276–286.

Rowe, P.S., 2012. Regulation of bone-renal mineral and energy metabolism: The PHEX, FGF23, DMP-1, MEPE ASARM pathway. Crit. Rev. Eukaryot. Gene Expr. 22, 61–86.

Teitelbaum, S.L., Ross, F.P., 2003. Genetic regulation of osteoclast development and function. Nat. Rev. Genet. 4, 638–649.

Xiong, J., O'Brien, C.A., 2012. Osteocyte RANKL: new insights into the control of bone remodeling. J. Bone Miner. Res. 27, 499–505.

Local Regulation of Bone Cell Function

Lilian I. Plotkin and Nicoletta Bivi

Department of Anatomy and Cell Biology, Indiana University School of Medicine, Indianapolis, Indiana, USA

OVERVIEW OF CYTOKINES AND GROWTH FACTORS AND THEIR RECEPTORS

Bone mass is maintained throughout life by the coordinated activities of osteoblasts and osteoclasts. To achieve a balance, bone cells and cells of the bone marrow produce factors that trigger intercellular signaling and modulate bone resorption and formation. In addition to factors produced locally in bone, hormones also participate in the regulation of bone homeostasis. The main difference between hormones and cytokines/growth factors is that hormones are produced by a gland, released into the circulation, and exert a widespread sphere of influence, while cytokines/growth factors are made in many different sites in the body and have a local effect. Another important difference is that hormones have low biological redundancy and loss of one hormone has profound consequences in the organism. On the other hand, cytokines and growth factors often have overlapping actions and deficiency in one can be at least partially compensated by others.

Definition: Cytokines versus Growth Factors

Although locally produced molecules have been named cytokines or growth factors, the difference between the two terms is vague. Originally, these terms were coined to distinguish factors that affect immune cells (cytokines) from those that affect nonimmune cells (growth factors). However, this distinction can no longer be made. By definition, a cytokine is a low molecular weight regulatory protein that is produced by a cell and involved in receptor-mediated cell-cell communication. Cytokines typically affect cell development and function by autocrine (acting on the same cell that produces the factor) or paracrine (acting on a cell near to that which produces the factor) mechanisms and act at very low concentrations (nano- to picomolar). Growth factors were first described as molecules that act locally to induce cell proliferation. Thus, a cytokine that induces cell proliferation could be called a growth factor. For the purpose of this chapter, both cytokines and growth factors will be considered together and referred to as locally produced factors.

Soluble Ligands

Molecules produced by bone cells and cells in the bone marrow are released and can act in an autocrine or paracrine fashion (Fig. 3.1A). Some cytokines are produced and released to the extracellular milieu, where they bind to the matrix. Cells expressing the corresponding receptor are able to bind to these matrix-bound cytokines. Other cytokines bind to soluble receptors, and the cytokine/soluble receptor complex formed can trigger signaling cascades in cells that lack the receptor. Soluble receptors can be produced by enzymatic cleavage from membrane-anchored receptors by actions of proteases [such as tumor necrosis factor receptor (TNFR) and growth hormone (GH) receptor], phospholipases [such as ciliary neurotrophic factor receptor (CNTFR)], or can be generated by alternative splicing of the mRNA for the receptor [such as interleukin-6 Rα (IL-6 Rα) and glycoprotein 130 (gp130)]. These soluble receptors may be (1) carriers for the ligands, which bring them into close proximity to the cell-membrane receptor; (2) agonists of the cytokines, which associate with a coreceptor in the cell membrane to allow signal transduction; or (3) antagonists, which block the interaction of the cytokine with its membrane-bound receptor (Fig. 3.1B).

Basic and Applied Bone Biology.
DOI: http://dx.doi.org/10.1016/B978-0-12-416015-6.00003-4

A

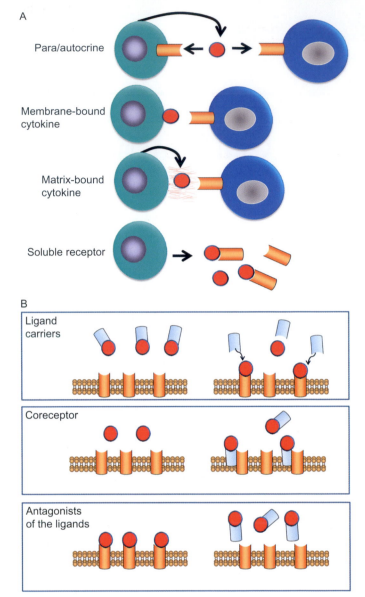

Para/autocrine

Membrane-bound
cytokine

Matrix-bound
cytokine

Soluble receptor

B

Ligand
carriers

Coreceptor

Antagonists
of the ligands

FIGURE 3.1 Local regulation of cell signaling. (A) Scheme illustrating the diverse modes by which molecules produced by a cell can induce signaling in the same cells or in others. In addition to associating with membrane bound receptors, these molecules can associate with soluble receptors. (B) Scheme showing how soluble receptors can modulate cell signaling by carrying the ligands close to the cells expressing the corresponding receptor, as coreceptors activating signals in cells devoid of the receptor (agonists), or by blocking the interaction of the ligand with the receptor in the target cell (antagonists).

Membrane-Bound Ligands

Local regulation of bone cells can also be achieved by membrane-bound ligands. In this case, cell-cell contact is required, and therefore, the actions of the ligands are limited to neighboring cells (paracrine signaling). The activation of these receptor-ligand systems can be unidirectional, in which case a cell expressing

the ligands induces signaling in the cells carrying the receptor. Examples of this are the TNF receptor superfamily member 11A [receptor activator of nuclear factor κB (RANK)]-RANK ligand [tumor necrosis factor ligand superfamily member 11/receptor activator of the NF-κB ligand (RANKL)] system and the Notch system. Membrane-bound signaling can also be bidirectional, with intracellular signaling activated in both the cell expressing the ligand and the one expressing the receptor. An example of this is the ephrin-ephrin receptor (Eph) system.

Receptor Classification

Local factors bind to one of four general types of cellular receptors that are classified based on their structure and signaling pathways.

1. Receptors with intrinsic protein kinase activity. These receptors exhibit tyrosine activity [such as insulin, epidermal growth factor (EGF), platelet-derived growth factor (PDGF), and macrophage colony-stimulating factor 1 (M-CSF) receptors] or serine kinase activity [such as transforming growth factor beta (TGF-β), activin, inhibin, and bone morphogenetic protein (BMP) receptors].
2. TNF/FasL receptors. This family of receptors share extracellular and intracellular domains. The intracellular domain contains the death domain, which is required for induction of apoptosis. Examples include the TNF and RANK receptors.
3. G protein-coupled receptors. These are seven-transmembrane domain-containing receptors. Ligand binding results in activation of downstream kinases and regulation of ionic channels. Examples include the interleukin-8 (IL-8) and endothelin receptors.
4. Hematopoietin/interferon receptors. A large, heterogeneous group called the cytokine receptor superfamily that lacks intrinsic kinase activity. These receptors associate with cytoplasmic kinases. Examples include the interferon and IL-6-type cytokine receptors.

Signal Transduction Cascades

Binding of a soluble factor or a membrane-bound molecule to its cell surface receptor triggers biochemical transformations in the recipient cell that are cell type specific and determine the response of the target cell. Ligand-receptor association can result in increased enzymatic activity, conformational changes that allow interaction between molecules, or changes in their subcellular localization. In addition to activation steps, these signaling pathways are regulated by intracellular inhibitors that inactivate enzymes, or bind to and sequester active components of the cascade. The series of molecular changes that follow ligand-receptor interaction constitute a signal transduction cascade. The ultimate consequence of these signaling cascades is the activation or inhibition of transcription factors and regulation of gene transcription, resulting in cell proliferation, differentiation, and/or cell death. Activation of signaling cascades can also result in transcription-independent events, including the generation and mobilization of second messengers such as Ca^{2+}, reactive oxygen species, cyclic AMP (cAMP), and inositol-3-phosphate.

LOCAL FACTORS THAT REGULATE OSTEOCLAST AND OSTEOBLAST DIFFERENTIATION AND FUNCTION

The spatial and temporal regulation of the differentiation and activity of osteoblasts and osteoclasts is achieved by the activation of membrane receptors induced by factors produced locally. Although some of these factors affect one specific cell type, most can modulate the activity of both osteoblasts and osteoclasts (Table 3.1).

FACTORS THAT REGULATE OSTEOCLAST DIFFERENTIATION AND FUNCTION

The RANKL-RANK-OPG System

The RANKL-RANK-OPG system is crucial for the regulation of osteoclast formation and bone resorption. It is composed of the ligand, RANKL, and its two receptors, RANK and osteoprotegerin/tumor necrosis factor receptor superfamily member 11B (OPG). RANKL and OPG are expressed by stromal cells, osteoblasts, and osteocytes. RANKL is present both as a transmembrane protein and as a secreted protein, following proteolytic cleavage by metalloproteases. RANK is expressed as a transmembrane protein by osteoclast precursors, while OPG is a soluble, secreted molecule. The N-terminal region of RANK and OPG share a similar structure, which allows OPG to function as a decoy receptor for RANKL. Because RANKL and OPG can be found as soluble molecules, their levels can be quantified in plasma to gauge bone resorption.

The binding between RANKL and RANK leads to the activation of the transcription factors NF-κB, c-Fos [a subunit of activator protein 1 (AP-1)], and nuclear factor of activated T-cells (NFATc1) that in turn drive the differentiation of the mononuclear precursors into multinucleated active osteoclasts. The crucial role exerted by these transcription factors for osteoclast formation is supported by the observation that mice lacking p50 and p52 (subunits of NF-κB), c-Fos, or NFATc1 completely lack mature osteoclasts. Besides promoting osteoclast differentiation, activation of RANK signaling by RANKL is involved in osteoclast activation and survival through a mechanism that requires proto-oncogene tyrosine-protein kinase Src. The prodifferentiation activity of RANKL is controlled by OPG, which protects bone from excessive resorption by neutralizing RANKL. The RANKL:OPG ratio is thought to be the critical determinant of bone resorption. Several studies have proposed that the RANKL:OPG ratio is increased to favor osteoclastogenesis through both increasing RANKL and decreasing OPG.

Global RANK- and RANKL-deficient mice, as well as transgenic mice expressing high levels of OPG, share a very similar phenotype, characterized by a severe osteopetrosis due to lack of osteoclasts. On the other hand, mice deficient in OPG develop osteopenia due to increased osteoclast number and activity. Exogenous administration of OPG suppresses bone resorption in mice, although very high doses are required. A more potent OPG derivative was generated by fusing OPG to the Fc fragment of human immunoglobulin G1 (IgG1), which increased the activity of OPG by about 200-fold. However, when administered to patients it was found to generate an immune response, and its production was stopped due to fears of triggering an autoimmune response against OPG. Similar efforts were carried out to develop a RANK-Fc fusion protein to treat conditions of low bone mass. However, administration of this compound to humans also generated an immune response with high titers of activating anti-endogenous RANK antibody and hypercalcemia. For this reason, the production of RANK-Fc was discontinued. More recently, a fully humanized anti-RANKL antibody was developed, and it is currently approved for the treatment of osteoporosis and cancer-associated bone loss.

Since the early 1990s, even before the discovery and cloning of RANKL and OPG, the widely accepted view was that stromal cells and osteoblasts were the main producers of an osteoclast-activating factor and, therefore, the main players in the orchestration of bone resorption. This view was recently challenged by two simultaneous reports that RANKL is more than 10 times higher in osteocytes, compared to osteoblasts, and that osteocyte-specific deletion of *RANKL* is sufficient to cause an osteopetrotic phenotype. These discoveries demonstrate the in vivo importance of osteocyte-derived RANKL for basal bone remodeling. It remains to be clarified whether osteocytic RANKL functions mainly as a transmembrane protein, thus

TABLE 3.1 Actions of Selected Locally Produced Factors on Bone Cells

Factor	Osteoblast/Osteocyte			Osteoclast		
	Differentiation	Function (bone formation)	Proliferation or Survival	Differentiation Generation	Function (resorption)	Survival
BMPs (2, 4, and 6)	↑	↑	↓	↑ (BMP-2)	↑	Unknown
Ephrin-Eph	↑	Unknown	Unknown	↓	Unknown	Unknown
FGF-2	↓	↑	↑	↑, ↓	↑, ↓	Unknown
IGFs	↑	↑	↑	↑	↑	
IL-1	↑	↓	↓	↑	↑	↑
IL-6	↑(through IL-6R and gp130)	Unknown	↓	↑ (via ↑RANKL and ↓OPG)	Unknown	Unknown
IL-8	Unknown	Unknown	Unknown	↑	↑	Unknown
IL-11	↑ (through IL-11R and gp130)	↓	↓	↑ (via ↑RANKL and ↓OPG)	Unknown	Unknown
IL-15	Unknown	Unknown	Unknown	↑	Unknown	Unknown
IL-17	Unknown	Unknown	Unknown	↑ (via ↑RANKL)	↑ (via ↑RANKL)	↑ (via ↑RANKL)
IL-18	Unknown	Unknown	↑	↓ (via ↓OPG and ↑GM-CSF)	Unknown	Unknown
IL-33	Unknown	↑ no effect	Unknown	↓	Unknown	Unknown
M-CSF	Unknown	Unknown	Unknown	↑	↑	↑
Notch	↓	↓	Unknown	↓	Unknown	Unknown
PDGF	↓		↑	Unknown	Unknown	Unknown
Prostaglandins	↑	↑,↑ (intermittent),↓ (continuous), ↑	↑	↑ (via ↑RANKL)	↑ (via ↑RANKL)	↑ (via ↑RANKL)
PTHrP	↑	↑	↑	↑ (via ↑RANKL and ↓OPG)	↑ (via ↑RANKL and ↓OPG)	↑ (via ↑RANKL and ↓OPG)
RANK-RANKL	Unknown	Unknown	Unknown	↑	↑	↑
TGF-β	↑	↑	↓ ↑	↑ (via ↑RANKL) ↓ (via ↓RANKL, ↑OPG)	↑ (via ↑RANKL) ↓ (via ↓RANKL, ↑OPG)	↑ (via ↑RANKL) ↓ (via ↓RANKL, ↑OPG)
TNF-α	↓	↓	↓	↑	↑	↑
VEGF	↑	Unknown	Unknown	↑	↑	Unknown
Wnts	↑	↑	↑	↓ (via ↑OPG, ↓RANKL)	↓ (via ↑OPG, ↓RANKL)	↓ (via ↑OPG, ↓RANKL)

↓, decreased; ↑, increased.

BMP, bone morphogenic protein; Eph, ephrin receptor; FGF, fibroblast growth factor; IGF, insulin-like growth factor; IL, interleukin; M-CSF, macrophage colony-stimulating factor 1; OPG, osteoprotegerin; PDGF, platelet-derived growth factor; PTHrP, parathyroid hormone-related protein; RANK, receptor activator of nuclear factor-κB/tumor necrosis factor receptor superfamily member 11A; RANKL, RANK ligand/tumor necrosis factor ligand superfamily member 11; VEGF, vascular endothelial growth factor.

requiring cell-cell contact with RANK-expressing cells to promote osteoclastogenesis, or as a soluble molecule.

Similar to RANKL, OPG was long thought to be produced mainly by osteoblasts. However, recent studies have demonstrated that osteocytes express high levels of OPG, comparable to those detected in osteoblasts. Moreover, OPG expression can be regulated by the activation of Wnt signaling or the specific deletion of *GJA1* [encoding gap junction alpha-1 protein/connexin 43 (Cx43)] in osteocytes. Importantly, the small size of OPG (60 kDa) could allow its diffusion through the lacunocanalicular system to influence osteoclast precursors that are present in the bone marrow. Besides functioning as a decoy receptor for RANKL, OPG has been proposed to suppress the shedding (cleavage from the cell membrane) of RANKL, based on the finding that low levels of OPG are associated with high circulating RANKL.

Macrophage Colony-Stimulating Factor

M-CSF (encoded by the *CSF1* gene), is a secreted molecule that promotes the proliferation, survival, and differentiation of mononuclear phagocytic cells and the cytoskeletal reorganization, cell spreading, and migration of mature osteoclasts. M-CSF is produced by osteoblasts and osteoblast precursors, but large amounts of M-CSF are also produced by osteocytes. M-CSF is present in two distinct biologically active forms, a membrane-bound and a secreted form, that result from alternative splicing of the *CSF1* gene. Osteocytes, osteoblasts, and stromal cells express both forms. A soluble form of M-CSF can also be generated by cleavage of the ectodomain of membrane-bound cytokine through a mechanism thought to require the metalloprotease disintegrin and metalloproteinase domain-containing protein 17/TNF-α converting enzyme (ADAM 17/TACE). Mice with a mutation in the coding region of the *CSF1* gene, known as op/op mice, develop severe osteopetrosis due to a markedly reduced number of osteoclasts. The phenotype, however, disappears as the animal ages, suggesting that other cytokines, such as granulocyte-macrophage colony-stimulating factor (GM-CSF), can compensate for the absence of M-CSF.

The receptor for M-CSF is CSF-1R/c-Fms. c-Fms is expressed by multipotent hematopoietic cells, mononuclear phagocyte progenitor cells, monocytes, tissue macrophages, and osteoclasts. M-CSF-dependent osteoclastogenesis can be inhibited by reducing the level of the receptor c-Fms at the cell surface by shedding via the action of metalloproteinases. Binding of M-CSF to its receptor induces the association of Src and phosphoinositide 3-kinase (PI3-K) with c-Fms, which in turn triggers the activation of the prosurvival pathways mitogen-activated protein kinase (MAPK) and Akt and reorganization of the actin cytoskeleton through activation of Ras-related C3 botulinum toxin substrate (Rac)-Rho. In addition, the transcription of RANK is promoted, thus fully committing the precursor to the osteoclast lineage. In mature osteoclasts, M-CSF also cooperates with integrin αvβ3 to remodel the actin cytoskeleton and enable formation of the actin ring, a crucial structure for osteoclast activity.

Recently, a mouse model in which the *CSF1* gene was globally deleted early during embryogenesis not only reproduced the osteopetrotic phenotype but also revealed that lack of M-CSF impairs osteoblast and osteocyte function. In particular, in M-CSF-deficient mice, osteoblasts lacked the polarity required to deposit matrix and collagen fibrils appeared disorganized; bone mineralization was patchy and the lacunocanalicular network was poorly developed; osteocyte apoptosis was increased and osteocytic expression of dentin matrix acidic phosphoprotein 1 (DMP-1), which promotes mineralization, was markedly reduced. These intriguing findings suggest that M-CSF can influence osteocyte function in an autocrine or paracrine fashion, probably by regulating their production of DMP-1.

FACTORS THAT AFFECT OSTEOBLAST DIFFERENTIATION AND FUNCTION

Wnts

Wnts are a family of secreted proteins that participate in the regulation of cell differentiation, proliferation, and apoptosis, and, through these mechanisms, play a key role in the development and homeostasis of the whole organism. Wnts were first identified independently in flies and mice, in *Drosophila*, as the Wingless (*Wg*) gene that is involved in wing development and in mice, as the *Int-1* gene that is involved in breast cancer development. The name Wnt was created by combining the name of the two homologous genes, *Wg* and *Int*. Wnts are lipid-modified glycoproteins. There are 19 members of the human Wnt ligand family, which share approximately 35% homology and have some overlapping and some distinct functions. All Wnts contain 23–24 conserved cysteine residues and a palmitate residue bound to the first conserved cysteine, close to the N-terminus of the molecule. The role of Wnts in bone has been elucidated through genetic studies in which the different components of the signaling pathways have been modified either in cells of the osteoblastic lineage or in the germline.

Wnts induce intracellular signaling by binding to one or more of the 10 members of the frizzled (FZD) family of receptors (Fig. 3.2A). FZDs are seven-transmembrane proteins, with a structure similar to G protein-coupled receptors. In addition, all members of this family contain a cysteine-rich domain in the N-terminus that constitutes the ligand-binding site and a cytoplasmic tail that interacts with intracellular molecules.

Wnts also interact with the low-density lipoprotein receptor protein 5 (LRP5) and LRP6 single-pass transmembrane proteins that act as coreceptors for Wnts. They contain a large extracellular domain that binds to

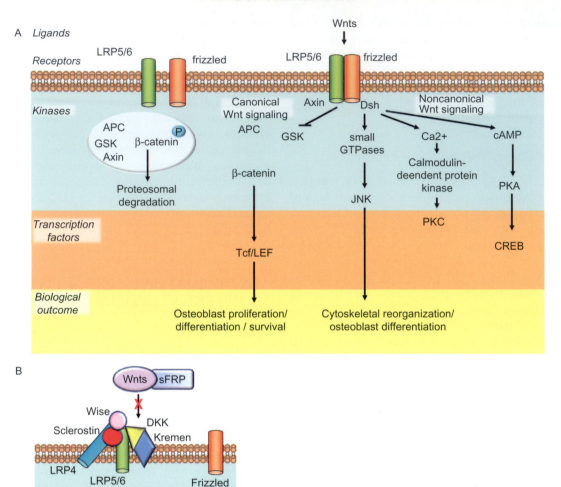

FIGURE 3.2   Canonical and noncanonical signaling pathways activated by Wnts. (A) In the absence of Wnt ligands, β-catenin associates with inhibitory molecules, is phosphorylated, and targeted for proteasomal degradation. When Wnts bind to LRP5/6 and frizzled receptors, GSK is inactivated, and the canonical Wnt signaling pathway is activated. β-Catenin then translocates to the nucleus where Tcf/LEF-induced transcription leads to osteoblast proliferation, differentiation, and survival. Formation of certain Wnt-frizzled complexes leads to the activation of the β-catenin-independent noncanonical signaling pathway that results in cytoskeletal organization and may induce osteoblast differentiation. (B) Wnt signaling is regulated by inhibitors that bind to Wnts, inhibiting their interaction with LRP5/6-frizzled (sFRPs). Inhibitors may also abolish Wnt signaling by binding to LRP5/6, an interaction that is facilitated by LRP4 (sclerostin and Wise) or by Kremen (Dkks).

Wnts and an intracellular domain required for activation of signaling. The preeminent role of LRPs on the control of bone mass has been uncovered by gain-of-function and loss-of-function mutations in humans and animals. Osteoporosis-pseudoglioma syndrome is a consequence of mutations in the *LRP5* gene that render proteins inactive and result in osteoporosis and blindness. Moreover, polymorphisms in the *LRP5* gene have been associated with reduced bone mass and increased fracture risk in numerous studies. Conversely, individuals expressing a mutated form of LRP5 characterized by impaired binding to the Wnt inhibitors Dickkopf (Dkk-1) and sclerostin (Fig. 3.2B) and reduced internalization of the receptor exhibit high bone mass and resistance to fractures. The

skeletal consequences of gain and loss of function of the *LRP5* gene observed in humans have been recapitulated in mouse models with germline deletion of *Lrp5* and LRP5 overexpression, respectively, both ubiquitously and in cells of the osteoblastic lineage.

LRP6 is approximately 70% homologous to LRP5. Both a mutation and polymorphisms in human *LRP6* are associated with reduced bone mass. Mouse models have shown that complete *Lrp6* deletion is lethal but that LRP6$^{+/-}$ and compound LRP5$^{-/-}$/LRP6$^{+/-}$ are viable, but show low bone mineral density. Similarly, mice that carry a spontaneous hypomorphic *Lrp6* mutation exhibit reduced bone mineral density due to increased bone resorption, but have normal osteoblast number and mineralization. The difference in the phenotypes between LRP5-deficient and LRP6-deficient mice suggests that LRP5 mediates the effect of canonical Wnt signaling on bone formation (see below), whereas LRP6 is required for normal bone resorption.

Contradictory data argues for and against LRP5 expression in osteoblastic cells having a direct role in controlling bone mass. Some investigators have found that deletion of *Lrp5* in osteocytes results in decreased bone mass and that overexpression in osteocytes results in elevated bone mass. On the other hand, another group showed that LRP5 affects bone mass through an extraskeletal function, mediated by increased serotonin production in the gut. The inconsistencies between the two studies could be due to differences in the genetic background of the mice, the nature of the pharmacological inhibitors, or the methods used to evaluate the bone phenotype.

LRP4 is another member of the LRP family of proteins with a role in bone homeostasis. LRP4 binds to the Wnt antagonists sclerostin and Wise (see below), facilitating their function. Therefore, unlike LRP5 and LRP6, LRP4 functions as an inhibitor of Wnt signaling. *LRP4* gene polymorphisms are associated with altered bone mineral density and fracture incidence. *Lrp4*-null mice and mice expressing hypomorphic mutant LRP4 show defective limb development and polysyndactyly, and mice expressing a truncated form of LRP4 lacking the transmembrane domain exhibit reduced bone mass and increased bone turnover. Mutations affecting the LRP4 extracellular region that impair its association with sclerostin were found in patients with high bone mass.

Wnts are extracellular proteins, but have poor transmembrane mobility due to the presence of lipid modifications. The extracellular transport of all Wnts is achieved by Wntless, a seven-transmembrane protein indispensable for Wnt secretion. In the absence of Wntless, Wnts are retained inside cells within components of the secretory pathway, resulting in Wnt loss-of-function phenotypes. In particular for bone, deletion of the mouse ortholog of Wntless using Wnt-1-Cre causes craniofacial defects and defective anterior-posterior axis formation. However, the role of Wntless in the adult skeleton has not been investigated.

Binding of Wnts to FZD can result in the activation of four different signaling pathways: the Wnt-β-catenin pathway (known as canonical Wnt signaling); the noncanonical planar cell polarity (PCP) pathway; the Wnt-Ca^{2+} pathway; and the protein kinase A (PKA) pathway (Fig. 3.2A). Wnts bind to FZD-LRP5/6 coreceptor complex and, through a mechanism that is not completely understood, FZD induces the phosphorylation of Disheveled (Dsh), which becomes activated. In addition, Dsh can be phosphorylated as a result of

Wnt binding to FZD independently of LPR5-LRP6, resulting in the activation of noncanonical signaling. Thus, Wnt canonical and noncanonical signaling pathways branch downstream Dsh activation.

Canonical Wnt Signaling

The canonical Wnt signaling pathway is normally inactive. In the absence of Wnts, β-catenin levels are kept low through its association with casein kinase I (CKI), the adenomatous polyposis coli (APC) tumor suppressor, and the scaffolding protein axin, in the presence of active glycogen synthase kinase-3β (GSK-3β). Upon phosphorylation by GSK-3β and CKI, β-catenin is degraded by the 26S proteasome complex. Nuclear levels of β-catenin are then low and the lymphoid enhancer-binding factor (LEF) transcription factor forms a complex with GROUCHO, repressing the transcription of Wnt target genes. When canonical Wnts (such as Wnt-1, Wnt-3, Wnt-3a, Wnt-7a, and Wnt-7b) bind to FZD-LRP5/6, Dsh is activated, leading to the phosphorylation and inactivation of GSK-3β. At the same time, the cytoplasmic domain of LRP5/6 becomes phosphorylated and binds axin. This leads to disassembly of the APC-axin-β-catenin complex and the release of β-catenin, which then translocates into the nucleus. In the nucleus, β-catenin binds to the transcription factor LEF and increases the transcription of Wnt target genes encoding axin, Cx43, cyclin D1, and Smad6.

Interestingly, whereas the decreased bone mass induced by targeted deletion of the gene encoding β-catenin in osteoblasts and osteocytes results from enhanced bone resorption due to low OPG, low bone mass in mice lacking the LRP5 coreceptor results from decreased bone formation. This has led some investigators to conclude that LRP5 does not mediate Wnt signaling in osteoblasts. However, it is possible that by removing β-catenin not only Wnt signaling is lost; β-catenin function in tight junctions and cytoskeletal organization may also be lost, thus explaining differences in the phenotype between the two genetic manipulations. Whether β-catenin indeed mediates the effect of LRP5 in bone remains controversial.

Noncanonical Wnt Signaling

Stimulation of the PCP pathway results from the binding of certain Wnts to particular FZD receptors followed by activation of Dsh, the small GTPases Rho and Rac, and MAPK/c-Jun N-terminal kinase (JNK). This results in cytoskeletal reorganization, cell migration, the establishment of cellular asymmetries within the tissue plane, and, ultimately, tissue morphogenesis. This pathway is involved in embryogenesis, and in bone it has been associated with the control of limb shape and dimensions. Thus, using a mouse model of loss-of-function mutations of the PCP gene *Vangl2*, it was demonstrated that PCP activation downstream of Wnt-5a is required for limb bud distal extension and to restrict the limb expansion in width and thickness. This is consistent with the phenotype of individuals with brachydactyly type B and Robinow syndrome, skeletal disorders caused by mutations in the *ROR2* gene (encoding the Wnt-5a coreceptor, ROR2). The Wnt-Ca^{2+} pathway is activated by specific Wnt ligand-FZD receptor pairs and results in intracellular Ca^{2+} release and activation of Ca^{2+}/calmodulin-dependent PKA and protein kinase C (PKC). Wnts can also activate the cAMP-PKA pathway in embryonic muscle, leading to phosphorylation of the transcription factor cAMP-response element-binding protein (CREB) and transcription Paired of box protein 3 (Pax3) and myogenic factor 5 (Myf-5).

Activation of noncanonical Wnt signaling has been associated with osteoblast differentiation. Thus, noncanonical Wnt-4, Wnt-5a, and Wnt-11 increase the differentiation of mesenchymal stem cells (MSCs) toward the osteoblastogenic pathway. In particular, Wnt signaling induced by Wnt-5a results in stimulation of osteogenic differentiation and inhibition of adipogenesis in human adipose tissue-derived mesenchymal stromal cells. This pro-osteoblastogenic effect is mediated by the Rho-associated protein kinase ROCK. Similarly, mechanical stimulation increases the expression of Wnt-5a and its coreceptor ROR2, resulting in RhoA activation, which is required for upregulation of RUNX2 and osteogenic lineage commitment induced by mechanical signals.

Inhibitors of Wnt Signaling

The Wnt signaling pathway is highly regulated by several proteins, including the inhibitors Dkk, Kremen, sclerostin (the product of the gene Sost), secreted frizzled-related protein 4 (sFRP-4), and Wise.

Dickkopf

Four members of the Dkk family of proteins (Dkk-1-4) have been described, all containing two cysteine-rich domains and an N-terminal signaling peptide, but with different patterns of proteolytic processing and N-linked glycosylation. Dkk-1 and Dkk-2 inhibit Wnt signaling by simultaneously binding to LRP5/6 and Kremen, a single transmembrane protein. Dkk-1 is one of the best studied Wnt pathway inhibitors, and Dkk-1 knockout mice lack head formation and have abnormal limb morphogenesis. Deletion of a single *Dkk1* allele in Dkk-1$^{+/-}$ mice is not lethal and the mice exhibit increased bone mass due to increased osteoblast number and bone formation, without changes in bone

resorption. Conversely, mice overexpressing Dkk-1 in osteoblasts develop osteopenia due to reduced osteoblast number and bone formation. Humanized anti-Dkk-1 antibodies have been generated and have shown to prevent the inhibitory action of Dkk-1 on osteoblast differentiation in vitro and to increase bone mass in growing female mice

Even though Dkk-2 has been shown to inhibit Wnt signaling in the presence of Kremen, *Dkk2*-null mice, unexpectedly, are osteopenic and show defective mineralization, increased osteoid, and reduced expression of the osteoblastic genes, *Bglap* (encoding osteocalcin) and *Opg*. Dkk-2 deficiency also results in upregulation of RANKL and increased osteoclastogenesis. Although Dkk-2 can act as a very weak Wnt-β-catenin activator in the absence of Kremen, this probably does not explain the reduction in bone mass due to Dkk-2 removal. Whether the phenotype of Dkk-2$^{-/-}$ mice is due to a lack of inhibition of Wnt signaling or to absence of other Dkk-2 functions in bone remains to be determined. Dkk-4 has also been shown to bind to LRP5/6 and to inhibit canonical Wnt signaling in the presence of Kremen, although its function in bone has not been investigated.

Kremen

Kremen proteins 1 and 2 function as coreceptors for Dkk-1 and enhance Dkk-1-induced inhibition of Wnt signaling. It was originally reported that single Kremen 1 or Kremen 2 knockout mice do not exhibit an altered bone phenotype, suggesting that the two forms of Kremen are redundant. A later report showed that Kremen 2-deficient mice develop a high bone mass phenotype with age, with increased bone formation and bone mass at 24 weeks of age. This study also showed that Kremen 1 is widely expressed in several tissues, while Kremen 2 is predominantly expressed in bone. Transgenic mice expressing Kremen 2 in cells of the osteoblastic lineage exhibit reduced Wnt signaling activity in osteoblasts and low bone mass due to reduced osteoblast maturation and bone formation, as well as increased bone resorption, probably resulting from reduced OPG expression. Compound Kremen 1 and 2 knockout (Kremen 1/2$^{-/-}$) mice exhibit increased Wnt signaling, bone mass, and bone formation, a phenotype similar to that of Dkk-1$^{+/-}$ mice. Moreover, deletion of one *Dkk1* allele in Kremen 1/2$^{-/-}$ mice does not further increase bone mass, suggesting that Dkk-1 acts through Kremen to inhibit bone formation. However, complete deletion of *Dkk1* is lethal, whereas mice lacking both Kremen proteins are viable, indicating that some of the functions of Dkk-1 are independent of Kremen. In addition, it has been shown that, in the absence of Dkk-1, Kremen 1 and 2 may act as Wnt agonists by binding LRP6 and increasing its levels in the plasma membrane.

Sclerostin

Sclerostin is the secreted protein product of the *SOST* gene. It was identified in 2001 as the gene mutated in individuals with sclerosteosis, a condition characterized by syndactyly and overgrowth and sclerosis of the skeleton, mainly involving the skull. Another genetic disease with similar characteristics to sclerosteosis, Van Buchem disease, is also associated with *SOST* mutations, although in this case an enhancer element, and not the coding sequence, is mutated, resulting in reduced sclerostin expression. A mutation in the secretion signal of sclerostin leading reduced levels of secreted sclerostin has also been found in patients with autosomal dominant craniodiaphyseal dysplasia. These *SOST* mutations lead almost exclusively to skeletal phenotypes, suggesting that sclerostin is expressed only in bone. Although reverse transcription PCR studies show *SOST* mRNA expression in different tissues, sclerostin protein is expressed only in bone, in particular in osteocytes (see Chapter 2).

Sclerostin was originally described as a BMP inhibitor, due to its homology with BMP antagonists. However, it was later discovered that sclerostin binds with very low affinity to BMPs. Instead, it is now believed that sclerostin inhibits Wnt signaling by binding to LRP5/6, although the precise mechanism for inhibiting Wnt signaling by sclerostin is unknown. In addition, it has been shown that sclerostin stimulates the osteoclastogenic signaling of osteocytes by increasing RANKL expression in vitro. Whether this effect is mediated by LRP5/6 and whether it is dependent on canonical or noncanonical Wnt signaling remains to be determined.

Secreted Frizzled-Related Proteins

Secreted frizzled-related proteins (sFRPs) are structurally related to the FZD coreceptors. sFRPs bind to Wnts and prevent their association with FZD on the cell surface, thereby blocking Wnt signaling. Five members of this family of proteins have been described (sFRP-1-5). sFRP-1 has been shown to inhibit osteoblast differentiation from MSCs and to induce apoptosis in osteoblasts and osteocytes. Consistent with this, targeted deletion of *Sfrp1* results in increased bone mass, and sFRP-1 overexpression decreases bone mass in transgenic mice. In addition to binding to Wnts, sFRP-1 binds to RANKL and blocks osteoclastogenesis in vitro. Similar to Dkk-1, inhibition of sFRP-1 has been a target for therapies to treat bone loss. A small molecule that binds to and inhibits sFRP-1 in vitro has been described, as have neutralizing sFRP-1 antibodies that block bone resorption.

sFRP-4 also negatively regulates bone formation by inhibiting osteoblast proliferation. Overexpression of SFRP-4 in bone or liver result in high circulating levels and reduced bone mass. In humans, *SFRP4* polymorphisms have been associated with changes in bone mass in the hips and spine.

sFRPs have also been associated with the bone manifestations of certain cancers. For example, sFRP-2 and sFRP-3, as well as Dkk-1, have been implicated in the suppression of bone formation induced by myeloma cells. sFRP-2 is also involved in reduced bone formation induced by ameloblastoma.

Wise

Wise binds to LRP5/6 and prevents Wnt binding and, therefore, Wnt signaling. Wise also interacts with LRP4, and mutations the genes encoding each of these proteins lead to similar tooth phenotypes in mice. Wise is required for normal tooth development and removal of Wise in mice results in activation of Wnt signaling and changes in the components of the fibroblast growth factor (FGF) and sonic hedgehog pathways.

Insulin-Like Growth Factor

Insulin-like growth factors (IGFs) were originally described as soluble factors induced by somatotropin/growth hormone (GH) that were able to increase the incorporation of sulfate into cartilage explants. Over the years, the growth factors were named sulfation factor, thymidine factor, nonsuppressible insulin-like activity (NSILA), somatomedin-C, somatomedin-A, basic somatomedin, and multiplication stimulating activity (MSA). Due to their structural similarity with insulin and their insulin-like properties, the name *insulin-like growth factor* was finally adopted to identify these growth factors.

The IGF system is composed of two ligands, IGF-I and IGF-II, and two main receptors (IGF-IR and IGF-IIR). IGFs are single-chain polypeptides produced mainly in the liver (the liver produces approximately 75% of the total circulating IGFs), although they are also synthesized in the skeleton and in adipose tissue. Together, these three tissues produce 95% of the circulating IGF pool. IGF-I and IGF-II share approximately 70% of their sequence and have 47% identity with insulin. IGF-I is expressed embryonically, and also has a major role in postnatal growth. IGF-II is also important during embryonic growth, but its expression is downregulated after birth.

Two main receptors that bind IGFs are IGF-IR and IGF-IIR, although IGFs also bind weakly to the insulin receptor. IGF-IR is a tetramer formed by two identical extracellular α subunits, which bind to the ligand, and two identical transmembrane β subunits, which have tyrosine kinase activity. Activation of IGF-IR results in phosphorylation of adaptor proteins belonging to the insulin receptor substrate (IRS) family or phosphorylation of Shc. Activation of IRS and Shc leads to activation of extracellular signal-regulated kinases (ERKs) 1 and 2 of the MAPK cascade via the growth factor receptor-bound protein 2 (GRB2)-Sos-Ras-Raf-MAPK-extracellular signal-regulated kinase kinase (MEK) pathway (Fig. 3.3). IRS proteins also bind to the p110 subunit of PI3-K, leading to the generation of phosphatidylinositol 3,4,5-triphosphate (PIP3) and phosphorylation of Akt by 3-phosphoinositide-dependent kinase 1 (PDK1). Akt phosphorylation leads to subsequent activation of mammalian target of rapamycin (mTOR), eukaryotic translation initiation factor 4E (eIF4E), and p70S6 kinase (S6K). Activation of these downstream signaling pathways enhances proliferation, survival, and metastatic potential in cancer cells.

The IGF-IIR, also known as cation-independent mannose-6-phosphate receptor, has no intrinsic kinase activity and binds to IGF-II with high affinity, inducing IGF-II lysosomal degradation. In contrast, IGF-IIR binds IGF-I with very low affinity. The biological functions of IGF-IIR were demonstrated in IGF-IIR-deficient mice, which exhibit high levels of circulating IGF-II, fetal overgrowth, and perinatal lethality due to cardiac abnormalities. The lethality in IGF-IIR mice is rescued by deleting IGF-II, indicating that in the absence of IGF-IIR, there is an excess of IGF-II that activates IGF-IR. Consistent with this, removal of IGF-IR also rescues the phenotype of IGF-IIR-deficient mice.

Most of the IGFs in the circulation are bound to proteins called IGF-binding proteins (IGFBPs), which regulate IGF bioavailability. Six members of this family of proteins have been identified and they have both common and distinct features. IGFBPs can bind to IGFs and block their interaction with the receptor because they have higher affinity for IGF than do IGF receptors, thereby suppressing IGF actions. However, they can also prevent the proteolytic degradation of the growth factors and increase the delivery of IGF to the cell surface receptor, thereby improving their local bioavailability. In addition, some IGFBPs have IGF-independent effects and can directly regulate cell functions. The IGF-IGF-R system can also be controlled by degradation of IGFBPs through the actions of different proteases, including metalloproteinases and cathepsins. IGFs regulate the activity of proteases that degrade IGFBPs, thereby adding a further level of regulation. For example, IGF-II increases the activity of the protease that targets IGFBP-4, a binding protein that inhibits IGF activity. IGFBP synthesis and the activity of the IGFBP proteases are also affected by hormones and cytokines, such as GH, glucocorticoids, and IL-6.

Another component of the IGF-IGF-R system is the acid-labile subunit (ALS), which forms a ternary complex with IGF-I and IGFBPs in the blood and is required for the circulation of IGF-IGFBP and for IGF function. This is demonstrated by the low levels of IGF and IGFBP-3 in serum and the consequent growth retardation observed in patients with ALS deficiency. Similarly, mice deficient in ALS show decreased IGF-I levels in serum, reduced cortical bone volume, and growth retardation.

In addition to the supply of IGF-I and IGF-II from the circulation, these growth factors are produced locally in bone by osteoblasts and stored in the bone matrix in their inactive form, bound to several IGFBPs. During the resorption process, IGFs are released through acidification of the bone surface. The presence of IGFs on the resorbed bone is thought to be a signal

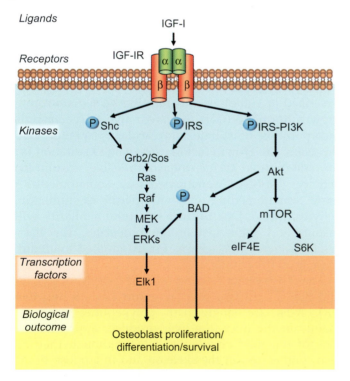

FIGURE 3.3 **Signaling pathways activated by IGF-I.** Association of insulin-like growth factor I (IGF-I) with a tetrameric receptor formed by two IGF receptor Iα (IGF-IRα) and two IGF-IRβ subunits results in activation of the extracellular signal-regulated kinase (ERK)/mitogen-activated protein kinase (MAPK) and phosphoinositide 3-kinase (PI3-K)-Akt pathways, ultimately leading to increased osteoblast proliferation, differentiation, and survival.

for recruiting osteoblast precursors to the bone surface. IGFBPs are also produced by osteoblasts and the pattern of IGFBP expression depends on the stage of osteoblast differentiation. Thus, IGFBP-2 and -5 are highly expressed during the proliferation phase, whereas expression of IGFBP-3, -4, and -6 is higher during the maturation phase of cultured osteoblastic cells. IGFs promote proliferation and differentiation and inhibit apoptosis in osteoblasts.

IGF-I levels in the circulation are positively associated with bone mass in humans, suggesting an important role for the growth factor in the skeleton. This was confirmed in several animal models in which the levels of IGF-I and IGFBPs were genetically manipulated. Mice with global deletion of *Igfi* show growth retardation, and IGF-I heterozygous deficient mice exhibit decreased cortical bone mass. Interestingly, *IgfI* deletion results in increased trabecular bone density and connectivity, probably due to defective osteoclast function resulting from decreased expression of M-CSF, RANKL, and RANK. Global deletion of *Igfir* results in delayed skeletal calcification, and growth retardation, with perinatal lethality. The phenotype of compound IGF-I/IGF-IR knockout mice is similar to that of the *Igf1r*-null mice, suggesting that IGF-I acts exclusively through IGF-IR.

Igf1r-null mice die at birth of respiratory failure, probably due to muscle hypoplasia, but growth of IGF-IR-deficient embryos is severely impaired and ossification is delayed. The severe growth deficiency and delayed ossification were reproduced in mice lacking Akt1 and Akt2, thus demonstrating the importance of this signal transduction pathway for bone growth. To circumvent neonatal lethality, mice with osteoblast-specific deletion of *Igf1r* were generated; these mice exhibit reduced cancellous bone mass, with decreased trabecular bone volume and trabecular number and thickness, and reduced numbers of osteoblasts on the bone surface. These IGF-IR-deficient mice also show increased osteoid, suggesting that in the absence of IGF-IR osteoblasts can deposit bone matrix normally but mineralization is impaired.

In an attempt to better understand the role of local versus systemic IGF-I, tissue-specific genetically modified mice were generated. Mice lacking IGF-I in the liver exhibit a 75% reduction in total serum IGF-I, without changes in free IGF-I levels, and a 26% decrease in cortical bone volume. On the other hand, the trabecular compartment is unaffected. This contrasts with the cancellous bone phenotype of mice lacking the IGF-IR in osteoblastic cells. Nevertheless, when mice lacking IGF-I in the liver are crossed with mice deficient in ALS, the levels of IGF-I in the circulation are reduced by 90% and bone mass is reduced in both cancellous and cortical compartments,

highlighting the importance of IGF-I transport through the circulation for bone mass acquisition. However, the bone phenotype of global IGF-I knockout mice is much more dramatic, suggesting that locally produced IGF-I contributes significantly to osteoblast differentiation and function. Consistent with this, mice in which IGF-I is deleted in osteoblasts have a 70% decrease in IGF-I expression in bone and only a 20% decrease in the liver, and exhibit decreased bone formation in both cancellous and cortical compartments. Similarly, overexpression of inhibitory IGFBP-4 in osteoblast leads to decreased bone volume and cortical BMD.

Overexpression of IGF-I in the liver of mice lacking IGF-I in all tissues does not rescue the bone phenotype of IGF-I knockout mice in young (1-month-old) mice, but it does so in adults (2-4-month-old mice), indicating that local IGF-I production is essential for neonatal and early postnatal body size and bone mass acquisition, even in the presence of high circulating IGF-I levels. On the other hand, mice overexpressing IGF-I under the control of the *Mt* (metallothionein) promoter, which leads to a 1.5-fold increase in circulating IGF-I levels, do not exhibit skeletal abnormalities. Using an osteoblast-specific promoter (from the human *BGLAP* gene), it was shown that overexpression of IGF-I results in increased bone formation and decreased mineralization lag time without changes in osteoblast numbers; and high cancellous bone volume without changes in cortical bone in young (6-week-old) mice, but not in older (24-week-old) mice. Consistent with a local effect of IGF-I, circulating levels of the growth factor are unchanged in these transgenic mice. Overexpression of IGF-I under the control of a 2.3 kb fragment of the *Col1a1* [type I collagen α-1(I)] promoter, on the other hand, results in increased osteoblast number and cortical thickness, but decreased cancellous bone volume, probably due to high turnover.

Parathyroid Hormone-Related Peptide

Parathyroid hormone-related peptide (PTHrP) was originally described as the factor responsible for hypercalcemia of malignancy. It was later found that, in addition to its role in cancer complications, PTHrP is involved in the physiologic regulation of several tissues, including mammary glands, pancreas, skin, vascular smooth muscle, and the skeleton. In bone, PTHrP is expressed in osteoblasts, lining cells, and osteocytes and it is expressed in preosteoblasts in culture. As indicated by its name, PTHrP is related to parathyroid hormone (PTH): 8 of the first 13 amino acids are identical in the two proteins and the first 34 amino acids are responsible for their biological activity. However, even though both PTHrP and PTH activate the parathyroid hormone/parathyroid hormone-related peptide receptor (PTH1-R), the two proteins are products of different genes and differ in their site of production (parathyroid gland versus ubiquitous production by normal and transformed cells), sequence (the two molecules only share the short sequence similarity in the N-terminal region), and function.

In bone, PTHrP activates PTH1-R expressed in osteoblasts and osteocytes. PTHrP is required for normal skeletal development, as evidenced by the multiple defects observed in PTHrP ubiquitous knockout mice, which die at birth due to defective rib cage formation resulting in respiratory failure. On the other hand, removal of PTH results in a milder phenotype, adding support to the existence of different functions for the two proteins. PTHrP$^{+/-}$ mice are viable and have reduced bone mass, altered recruitment of bone marrow progenitors, and increased osteoblast apoptosis. A similar phenotype has been described for mice with an osteoblast-specific deletion of *Pthlh* (encoding PTHrP). These in vivo studies, as well as in vitro experiments, have demonstrated that PTHrP is required for the commitment of precursors to the osteoblastic lineage and their subsequent maturation. In addition, PTHrP stimulates bone formation by increasing osteoblast differentiation and inhibiting apoptosis in these cells. Osteoclasts and their progenitors lack PTH1-R. However, PTHrP controls osteoclast differentiation indirectly via increasing RANKL expression in osteoblastic cells. Consistent with this, osteoblast-specific *Pthlh* deletion results in decreased osteoclast formation.

PTHrP binding to PTH1-R results in phosphorylation of the receptor leading to conformational changes that enable the activation of trimeric GTP-binding proteins (G proteins). Most of the effects of PTHrP on osteoblasts result from activation of the cAMP-PKA pathway, downstream of activation of the Gs-α subunit of the G protein. This is followed by activation of transcription factors such as CREB, members of the AP-1 family of transcription factors (c-Fos and c-jun), and RUNX2, as well as regulation of gene transcription.

Activation of PTH1-R by PTHrP can also result in activation of the G_q subunit, leading to activation of 1-phosphatidylinositol 4,5-bisphosphate phosphodiesterase [phospholipase C (PLC)] and the PI3-K-diacylglycerol pathway, with elevation of intracellular Ca2$^+$ and activation of PKC. PTHrP can also activate or inhibit the MAPK ERK1/2 signaling pathway, depending on whether cells are undergoing proliferation or differentiation, respectively.

Signaling triggered by activation of PTH1-R is short-lived. Upon ligand binding, the receptor becomes phosphorylated, and associates with the scaffolding proteins β-arrestin-1 and-2. β-arrestins reduce the affinity of PTH1-R to Gs-α, thus suppressing cAMP responses. PTH1-R-β-arrestin complexes are internalized, resulting in desensitization of the cell to PTH and PTHrP actions. Subsequently, the receptor is either recycled to the cell surface or degraded. In addition to their function in receptor desensitization, β-arrestins mediate the actions of the other PTH1-R ligand, PTH, independent of trimeric G proteins, leading to activation of MAPK and, in many cases, sequestration of the activated enzymes in the cytoplasm. Whether this pathway is also activated by PTHrP remains to be determined.

In addition to activating PTH1-R, PTHrP is able to exert nuclear actions. PTHrP has a nuclear localization signal and by translocating to the nucleus is able to regulate apoptosis and cell proliferation by a so-called *intracrine* mechanism. Moreover, the C-terminal domain of PTHrP has been found to increase bone formation and inhibit bone resorption in vivo, and it has been shown to interact with β-arrestin in the cytoplasm. However, a receptor for the C-terminal domain of PTHrP has yet to be found.

Fibroblast Growth Factors

The FGF family of proteins comprises 22 members grouped into subfamilies based on sequence homology. FGFs control a variety of cellular processes, including cell proliferation, migration, and differentiation. Most FGFs act in a paracrine manner; only three of them, FGF-19, FGF-21, and FGF-23, function as endocrine factors. FGFs bind to the extracellular matrix until they are activated by proteases and carried to their respective receptors by extracellular chaperone FGF-binding proteins.

FGFs exert their functions upon their binding to the FGFR family of tyrosine kinase receptors (FGFR-1-4) that are present in several alternative splice isoforms. Among the FGFR isoforms, FGFR-1-3 are expressed in osteoblastic cells. Following receptor activation, the main signal transduction pathways are mediated by PLC-γ1, MAPK, and PI3-K-Akt (Fig. 3.4). FGFs also bind to cell surface heparin sulfate proteoglycans, such as syndecans, expressed in bone cells. In addition to direct actions of FGFs through their receptors, these growth factors affect bone cells indirectly through upregulating other local factors, such as TGF-β, IGF-I and vascular endothelial growth factor (VEGF).

FGFR-2 is normally expressed by osteoprogenitor cells, as well as by mature osteoblasts. Deletion of *Fgfr2* in osteochondroprogenitors results in skeletal abnormalities, including a shorter axial and appendicular skeleton and a markedly reduced mineral apposition rate. However, osteoblast differentiation is

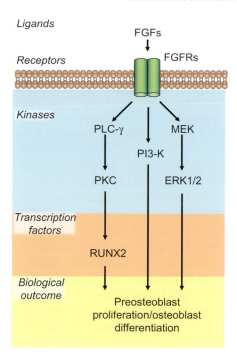

FIGURE 3.4 **Activation of FGF receptor signaling in osteoblasts.** Fibroblast growth factors (FGFs) bind to the FGF receptor (FGFR) family of tyrosine kinase receptors (FGFR-1-4), resulting in activation of the intracellular kinases protein kinase C (PKC), phosphoinositide 3-kinase (PI3-K), and extracellular signal-regulated kinases (ERKs). These kinases stimulate preosteoblast proliferation and osteoblast differentiation.

unaffected, while proliferation is diminished. On the other hand, mice lacking FGFR-1 in osteoprogenitors or in differentiated osteoblasts exhibit increased bone mass, suggesting that FGFR-1-mediated signaling has an inhibitory effect on bone formation.

FGF-2-mediated signaling has been shown to be important for bone formation. Thus, *Fgf2*-null mice exhibit reduced trabecular bone volume, mineral apposition and bone formation rates due to increased differentiation of bone marrow stromal cells (BMSCs) toward the adipocyte lineage. Of note, FGF-2 exists in two main forms: a high molecular weight (HMW) nuclear isoform and an extracellular low molecular weight protein (LMW). Targeted overexpression of each of these isoforms in osteoblasts results in distinct bone phenotypes: overexpression of the LMW isoform confers increased bone mass, whereas overexpression of HMW isoforms results in decreased bone mass.

Promotion of osteoblast differentiation by FGF-2 is potentiated by Cx43. In the presence of FGF-2, Cx43 interacts with PKC-δ via its C-terminal tail and this, together with activation of ERKs, increases RUNX2 activity and, ultimately, *BGLAP* gene transcription and osteoblast differentiation. It has been shown that FGF-2 promotes survival of osteoblasts by activating PI3-K-mediated signaling.

FGF-18, expressed in mesenchymal cells and differentiating osteoblasts, also induces osteogenic differentiation of BMSCs; this is achieved through activation of FGFR-1 and FGFR-2, followed by activation of MAPK and PI3-K signaling.

FGF-23 is produced by mature osteoblasts and osteocytes and released in the circulation to regulate vitamin D and phosphate homeostasis. The overall effect of FGF-23 is to reduce phosphate levels, and this is achieved through multiple mechanisms. FGF-23 downregulates sodium-dependent phosphate transport proteins 2 A and 2C (NaPi-2a and NaPi-2c) on the epithelial cells of the renal proximal tubules, thus suppressing renal phosphate reabsorption and promoting urinary phosphate excretion. FGF-23 also reduces the levels of active 1,25-dihydroxyvitamin D [1,25-$(OH)_2$D; also known as calcitriol] by repressing 25(OH)D 1α-hydroxylation and increasing the expression of the 24-hydroxylase that inactivates calcitriol. FGF-23 production is in turn regulated by 1,25$(OH)_2$D and phosphate levels in a tightly regulated feedback loop. To date, the mechanisms by which osteocytes sense phosphate levels resulting in regulation of FGF-23 production are unknown. FGF-23 levels also increase in response to elevated PTH. *Fgf23*-null mice exhibit hyperphosphatemia and increased 1,25$(OH)_2$D levels, whereas transgenic mice overexpressing FGF-23 are hypophosphatemic.

The affinity of FGF-23 for FGF receptors (in particular FGFR-1) and for heparin sulfate is increased by the simultaneous binding of the coreceptor Klotho. The importance of Klotho in FGF-23 activity is demonstrated by the phenotype of combined FGF-23$^{-/-}$;Klotho$^{-/-}$ mice, which resembles that of mice with a single deletion in either *Fgf23* or *Klotho*. Klotho is expressed in the kidney (in the distal convoluted tubule), the parathyroid gland, the brain, and skeletal muscle. Klotho also has been detected by immunohistochemistry in osteocytes, suggesting that FGF-23 may exert autocrine and/or paracrine signaling in bone. The active FGF-23 is the full length molecule, and its inactivation by proteolytic cleavage is induced by an unknown enzyme. Antibodies directed against FGF-23 have proven effective in ameliorating hypophosphatemic rickets in a mouse model characterized by high FGF-23 levels.

LOCAL FACTORS THAT AFFECT BOTH OSTEOCLASTS AND OSTEOBLASTS

Interleukins

Interleukins are a family of proteins that were originally isolated from leukocytes, from which their name originates. It is now known that interleukins are produced by practically all cell types including osteo-clasts, osteoblasts, and osteocytes. Interleukins are soluble proteins or peptides that exert overlapping and antagonistic actions. As a family, interleukins are involved in the regulation of cell activation, differentiation, proliferation, and survival, as well as cell-cell communication. Interleukins shown to have a role in bone are listed in Table 3.1.

IL-1 is a proinflammatory cytokine produced in bone by cells of both the hematopoietic and mesenchymal lineages, although the exact nature of the cells that produce them is unknown. Two isoforms of IL-1, α and β, have been described, encoded by two separate genes with an identical function. IL-1 acts on bone by increasing bone resorption, at least partially by stimulating transcription of the *RANKL* gene. IL-1α and β also act on early progenitors to induce osteoclast differentiation, and they enhance the resorptive activity of mature osteoclasts via activation of NF-κB. Interestingly, activation of RANK by RANKL is required for osteoclast precursors to respond to the prodifferentiating effect of IL-1. On the other hand, even though IL-1 increases osteoblast proliferation, it has an inhibitory effect on collagen synthesis in vitro and bone formation in vivo.

Two IL-1 receptors have been described: type I and type II. Activation of the type I receptor mediates the actions of IL-1 and results in the hydrolysis of sphingo-myelin and the production of the second messenger ceramide, as well as NF-κB activation. On the other hand, the type II receptor acts as a decoy receptor, blocking IL-1 effects. Moreover, the type II receptor can be cleaved and acts in a soluble form as an inhibitor of IL-1. IL-1 is also antagonized by the IL-1 receptor antagonist protein, IL-1ra, a naturally occurring inhibitor.

IL-6-type cytokines are members of a family of cytokines that share the common signal transducer molecule, gp130. This family of proteins includes IL-6, IL-11, leukemia inhibitory factor (LIF), oncostatin-M (OSM), CNTF, and cardiotrophin-1 (CT-1). The signal specificity for CNTF, IL-6, and IL-11 is achieved by the interaction of the ligand with an α subunit of the receptor (Fig. 3.5A). CT-1, LIF, and OSM lack α subunits and bind to β1 subunits. This leads to the formation of homodimers of the signal transducer gp130 (also known as the β2 subunit), or gp130 heterodimers with an additional β1 subunit. The α subunits of CNTF and IL-6 receptors (CNTFR and IL-6R) also exist in soluble forms. However, unlike the soluble extracellular domains of most receptors, which function as antagonists, soluble IL-6 and CNTFR-α function as agonists by binding their cognate cytokines and then interacting with the signal transducing components of the receptor [gp130 and LIF receptor beta (LIF-Rβ)] on the cell surface.

Ligand-induced dimerization of the β subunits initiates intracellular signaling by activating members of a family of receptor-associated tyrosine kinases, known as the Janus kinases (Jaks), which in turn phosphorylate several proteins, including the β components of the receptor complex, the kinases, and a series of cytoplasmic proteins termed signal transducers and activators of transcription (STATs; Fig. 3.5B). STAT phosphorylation causes the formation of protein complexes that migrate to the nucleus and initiate gene transcription. In addition, stimulation of IL-6-type cytokine receptors results in activation of the ERK pathway and gene transcription.

As a group, IL-6-type cytokines stimulate RANKL synthesis, leading to increased osteoclast differentiation and activity. These cytokines have a direct effect on osteoclasts by increasing the expression of M-CSF receptor, c-Fms, in osteoclast precursors. However, some members of this family of interleukins have opposite effects on osteoclasts. Thus, mice lacking CT-1, LIF, and LIF-R have increased osteoclast formation, whereas IL-11R and OSM-specific receptor (OSMR)-deficient mice exhibit fewer osteoclasts. On the other hand, mice lacking IL-6 have normal osteoclast number, although IL-6 contributes to the osteoclastogenic response to sustained PTH elevation, vitamin D_3, thyroid hormone, and ovariectomy.

The effect of IL-6-type cytokines on osteoblasts is unclear. In vitro, IL-6 cytokines enhance osteoblast differentiation; however, overexpression of IL-6 in two different transgenic mouse models has shown decreased osteoblast number. Further studies show mice lacking CT-1, IL-11R, and OSMR exhibit reduced bone formation, whereas mice lacking IL-6 or LIF have no defects in bone formation. Some of the effects of the cytokines on bone formation may be indirect, since CT-1, LIF and OSM (but not CNTF, IL-6, and IL-11) reduce expression of the Wnt inhibitor sclerostin. This might explain the opposite effects of the two groups of cytokines on bone formation. IL-6-type cytokines have also been shown to prevent osteoblast apoptosis via activation of the STAT signaling pathway and transcriptional activation of the *CDKN1A* gene (which encodes p21^{WAF1,CIP1,SDI1}).

The relevance of IL-6-type cytokines for the skeleton has been demonstrated in animal models in which the coreceptor gp130 is deleted. Loss of signaling by gp130 results in multiple abnormalities in osteoblast function and an increase in osteoclast number; however, these cells have poorly developed ruffled borders, suggesting a decrease in individual osteoclast activity.

Tumor Necrosis Factor

TNF, originally described as an antitumor agent, is a proinflammatory cytokine that mediates immunologic

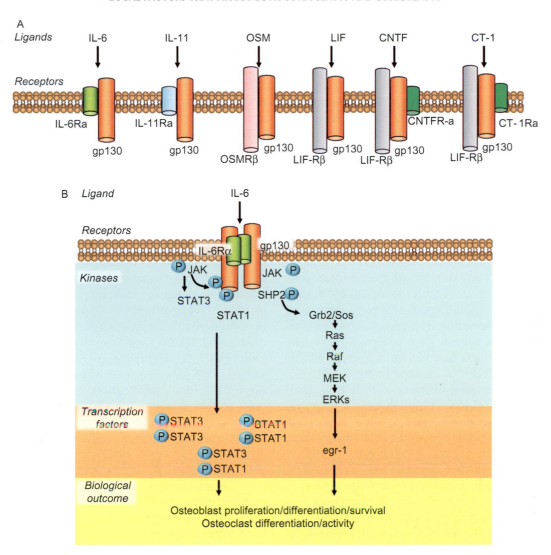

FIGURE 3.5   **IL-6-type cytokine-induced signaling.** (A) The receptors for the IL-6-type cytokines share a common signal transduction unit, gp130. The specificity for a particular cytokine is given by either the α or β subunit that associates with gp130. IL-6 associates with a tetramer composed of two IL-6 Rα and two gp130 subunits. (B) Binding of gp130 cytokines, exemplified here by IL-6, to the receptor results in the activation of the JAK-STAT signaling pathway and the JAK-SHP2-ERK pathway, affecting both osteoblast and osteoclast function and activity.

reactions by activating lymphocytes, inducing the expression of endothelial cell adhesion molecules, and promoting angiogenesis. TNF is involved in the pathogenesis of inflammatory and autoimmune disorders and, in addition, is a pro-osteoclastogenic cytokine. The two variants of TNF, TNF-α and TNF-β [now known as lymphotoxin-alpha (LT-α)], are products of separate genes but have similar biological action, although TNF-β/LT-α is often less potent. TNF-α and TNF-β/LT-α are transmembrane proteins that can be cleaved by proteases to form soluble cytokines. Both membrane-bound and soluble TNFs are biologically active. TNFs are produced by a wide variety of cell types of the immune system, including monocytes and macrophages, dendritic cells, B cells, and T cells, but also by adipocytes, keratinocytes, mammary cells, and active osteoblasts.

There are also two different TNF receptors, TNFR-1 (also called p55) and TNFR-2 (also called p75), which belong to the same cell membrane-bound receptor family as RANK. Similar to TNFs, TNFRs can be cleaved; however, in the case of the receptors, the fragments inhibit TNF actions by binding to TNF without transducing signals. TNFR-1 and TNFR-2 have 28% homology in their extracellular domain, but no homology is found in the cytoplasmic domains. Interestingly, murine TNFR-1 binds both murine and human TNF-α, but murine TNFR-2 only binds the murine ligand—not the human. TNFR-1 is constitutively expressed in all nucleated cells, whereas the expression of TNFR-2 is induced and is higher in endothelial and hematopoietic cells. Another difference between the two receptors is that TNFR-1 can be activated by intact or soluble TNF, while TNFR-2 is only activated by the membrane-bound cytokine. In addition, TNFR-1, but not TNFR-2, is a member of the death receptor family, because it contains a death domain in the cytoplasmic region and activates apoptosis by recruiting protein FADD (or FAS-associated death domain protein). Moreover, whereas activation of TNFR-1 results in increased inflammation, it has been proposed that activation of TNFR-2 leads to anti-inflammatory signals.

TNFRs do not possess intrinsic kinase activity. Instead, a signaling cascade is activated when a TNF trimer binds to the extracellular domain of the receptor, leading to activation of intracellular kinases. For TNFR-1, the first step is the release of BAG family molecular chaperone regulator 4 [BAG-4; also known as silencer of death domains (SODD)] from its association with the intracellular domain of the receptor, thus allowing the binding of TNFR1-associated DEATH domain protein (TRADD). TRADD, in turn, acts as an adaptor molecule, recruiting receptor-interacting serine/threonine-protein kinase 1 [or receptor-interacting protein-1 (RIP-1)], which has kinase activity, and the E3 ubiquitin ligase TNFR-associated factor 2 (TRAF2). This leads to the activation of canonical NF-κB signaling and of members of the MAPK family of proteins. The proapoptotic pathway is mediated by the formation of two different complexes: one at the plasma membrane, comprising RIP-1, TNFR-1, TRADD, TRAF2, and baculoviral IAP repeat-containing protein 3 [inhibitor of apoptosis 1 (C-IAP1)]; and the second in the cytoplasm, composed of FADD and caspases 8 and 10. The TNFR-2 signaling cascade is much less well understood. Binding of membrane-bound TNF results in the recruitment of TRAF1 and TRAF2, C-IAP1 and C-IAP2, leading to activation of NF-κB. TNFR-2 also may induce MAPK activation and noncanonical NF-κB signaling.

Exogenous administration of TNF-α or TNF-β/LT-α to mice results in increased serum calcium, osteoclast number, and percentage of the bone surface undergoing resorption. The mechanism for the increase in osteoclast activity appears to be secondary to elevation of RANKL. Thus, although TNF induces osteoclast formation in vitro even when osteoclast precursors are obtained from RANK-deficient mice, administration of TNF to RANK-deficient mice only induces the formation of a few, if any, osteoclasts. In addition, TNF, through activation of TNFR-1, stimulates the synthesis of both RANKL and RANK, which can be responsible for the increased osteoclastogenesis in individuals in which TNF levels are increased.

TNF actions are not required for normal bone development, since mice deficient in TNFR-1 or TNFR-2 do not show any developmental deficiencies, and only exhibit absence of a normal immune response and apoptotic mechanisms. On the other hand, TNF appears to be involved in pathologies with increased bone resorption. Thus, TNF levels are elevated in patients with different inflammatory conditions, such as ankylosing spondylitis, juvenile idiopathic arthritis, and psoriasis. Further, therapies with anti-TNF agents such as neutralizing antibodies and soluble receptors ameliorate the clinical manifestations of these diseases in bone. Moreover, transgenic mice overexpressing human TNF develop spontaneous lesions in the joints reminiscent of features of rheumatoid arthritis, including bone destruction. In addition, elevated TNF has been found in postmenopausal women; administration of soluble TNFR-1 lessens bone loss induced by sex steroid removal in rats and mice and reduces resorption markers in postmenopausal women who were withdrawn from estrogen therapy. Consistent with a potential pathogenic role for TNF in postmenopausal women, TNF synthesis by lymphocytes is suppressed by estrogens and it has been proposed that TNF production by T cells is required for ovariectomy-induced bone loss to occur. However, this evidence is controversial. TNF has also been implicated in the pathogenesis of calcium mobilization that results from certain tumors, based on evidence showing that antibodies against TNF reduced circulating calcium in animals with hypercalcemia of malignancy.

In addition to its actions on osteoclasts, TNF blocks osteoblast function by blocking osteoblast differentiation, suppressing the activity of mature osteoblasts, and inducing osteoblast apoptosis. TNF is not produced by osteoblastic cells in culture under unstimulated conditions, but its expression is increased in the presence of proinflammatory factors such as IL-1, GM-CSF, and bacterial lipopolysaccharide. Local TNF production by osteoblasts might contribute to the local dysregulation of bone formation and resorption due to chronic inflammation. Moreover, estradiol inhibits IL-1-induced TNF production by osteoblasts, suggesting that lack of the sex steroid might exacerbate the effects of inflammation on bone resorption in postmenopausal women.

Bone Morphogenetic Proteins

BMPs are growth factors that belong to the TGF-β superfamily of proteins. There are at least 20 BMPs, but only five of them are known to induce bone formation: BMPs 2, 4, 5, 6, and 7. BMPs were discovered in 1965 through the pioneering work of Urist as agents present in bone extracts that can induce the formation of ectopic cartilage and bone. Structurally, most of the BMPs exist as dimers consisting of two identical monomers linked by a disulfide bond. BMPs exert a potent osteogenic effect by promoting osteoblast differentiation and function.

BMP action is mediated by the binding of one molecule of BMP to type I (ALK-1, −2, −3, and −6) and type II (BMPR-II, ACTR-IIA, and ACTR-IIB) serine/threonine kinase receptors. The subsequent formation of the oligomeric complex results in activation of the type I receptor through its phosphorylation by the type II receptor. In canonical BMP signaling, the activated receptor in turn phosphorylates and activates the receptor-regulated Smad transcription factors 1, 5, and 8 that, after association with Smad4, translocate to the nucleus. Within the nucleus, the complex interacts with other key molecules such as RUNX2, and promotes the transcription of target genes, such as the inhibitor of differentiation (Id) gene family (Fig. 3.6). The gene expression program induced by BMP-Smad signaling includes genes encoding proteins important for osteoblast differentiation and function, such as osteocalcin, collagen alpha-1(I) chain, and alkaline phosphatase.

Noncanonical BMP signaling comprises Smad-independent intracellular events that follows activation of type I receptors. In this signaling cascade, BMPs promote the activation of the MAPKKK 7/TGF-β-activated kinase 1 (TAK1)-p38 pathway through recruitment and ubiquitylation of TRAF6. Noncanonical BMP signaling has been shown to be required for tooth and palate development and for induction of PGE2.

BMP activity is tightly regulated by specific extracellular antagonists including chordin, follistatin, gremlin, and noggin. These antagonists function by binding and sequestering BMP in the extracellular space, inhibiting intracellular signaling, or blocking receptor binding and activation by ligands. The role of BMP inhibitors is particularly important during development, when they allow the formation of patterning gradients. Mice overexpressing the BMP inhibitor gremlin or noggin in osteoblasts exhibit spontaneous bone fractures, osteopenia, and decreased osteoblast number and function.

The importance of BMPs for skeletogenesis and bone homeostasis in the adult skeleton has been demonstrated in vivo in several mouse models of BMP gain or loss of function. Deletion of both Bmp2 and Bmp4 in the limb mesenchyme results in a severe impairment of osteogenesis, with reduced differentiation of both chondrocytes and osteoblasts. However, removal of BMP-4 alone in the same mouse strain does not impair skeletogenesis or fracture repair, demonstrating the dispensability of BMP-4. Lack of BMP-2 alone, on the other hand, blocks the initiation of fracture repair and this effect cannot be compensated by any other BMPs, although bone formation still occurs. Similarly, deletion of Bmp7 has no effect on bone development or mass. BMP-3 is the only member of the BMP family that acts as an antagonist of BMP signaling and a negative regulator of bone mass. Thus, mice lacking BMP-3 exhibit increased bone mass, while its overexpression is associated with spontaneous rib fractures.

BMPs have been shown to play a crucial role in bone anabolism induced by a variety of stimuli, including intermittent PTH administration and mechanical loading. BMPs potentiate the effect of PTH. Furthermore, it has recently been demonstrated that PTH enhances BMP-Smad signaling in an LRP6-dependent manner, resulting in increased differentiation of MSCs. Further, pretreatment with PTH reverses the increase in BMP antagonists induced by dexamethasone. Mechanical stimulation upregulates BMP-2 and BMP-4 in osteoblastic cells and BMP-7 in osteocytic cells in vitro and induces the phosphorylation of Smad1/5.

Recent studies have focused on the effects of BMP signals from and to osteoclasts. Messenger RNA and protein for BMP-2, -4, -6, and -7 can be detected in osteoclast-derived cocultures of bone marrow cells and calvaria cells. It has been proposed that osteoclasts promote the recruitment of osteoprogenitors through BMP-6. BMP-6 was found to be secreted by mature

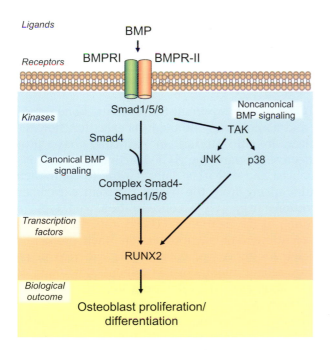

FIGURE 3.6 **BMP signaling in osteoblasts.** Bone morphogenic protein (BMP) action is mediated by the binding of one molecule of BMP to a dimer composed of one molecule of type I and one of type II serine/threonine receptor kinases. This leads to the activation of the canonical signaling pathway that results in translocation of a Smad4-Smad1/5/8 complex to the nucleus, the activation of RUNX2 and osteoblast proliferation and differentiation. Noncanonical BMP signaling is mediated by Smad-independent intracellular events that follow activation of type I receptors.

osteoclasts, and neutralization of this osteoclast-derived BMP-6 attenuated the stimulation of MSC nodule formation and chemokinesis by osteoclast conditioned medium. In vitro studies have shown that BMPs promote osteoclastogenesis when administered to marrow-derived macrophages or hematopoietic cells. In addition, deletion of *Bmpr1a* in mature osteoclasts increases bone volume and bone formation rate with increased osteoblast number and reduced osteoclast number. Deletion of *Bmpr1a* in osteoblasts on the other hand, increases the number of osteoblasts and osteoclasts, suggesting that BMPs are involved in bidirectional communication between osteoblasts and osteoclasts and may regulate coupling during bone remodeling.

Transforming Growth Factor Beta

TGF-βs are members of the transforming growth factor beta superfamily that includes several other growth factors, such as BMPs. There are three different TGF-β isoforms, called TGF-β1 (also known as simply TGF-β), TGF-β2, and TG-F-β3, and they control key cellular processes such as differentiation, proliferation, and survival. TGF-β1 is the most abundant isoform in bone and therefore the focus of most skeletal research.

Similar to the BMPs, signaling mediated by TGF-β1 can be canonical (Smad-dependent) or noncanonical (Smad-independent) (Fig. 3.7). In canonical signaling, TGF-β1 binds to type I (TGFR-1; or ALK-5) and type II (TGFR-2; or Tgfbr2) receptors, followed by activation of Smad2 and 3. In noncanonical signaling, binding of TGF-β1 to its receptors triggers activation of TAK1, which in turn promotes the activation of the MAPK/JNK cascade, the p38 MAPK cascade, and others. The cascade downstream of p38, in particular, results in increased expression of type I collagen. Recently, a novel function of TGFR-2 has been demonstrated that involves the formation of an endocytic complex between this receptor and the PTH1-R that regulates endocytosis of the latter in response to PTH.

Tissue-specific genetic manipulations of TGF-β signaling in mice have shown that TGF-β is involved in cartilage formation. Deletion of *Tgfr2* from chondrocytes in mice does not affect the size of long bones, but results in alterations in the size, spacing, and appearance of vertebrae. On the other hand, deletion of *Tgfbr2* from limb mesenchyme leads to shorter limbs and fusion of the joints in the phalanges. Deletion of *Tgfbr1* from osteochondroprogenitors results in bone growth retardation associated with reduced proliferation and differentiation of osteoblasts and defects in the perichondrium.

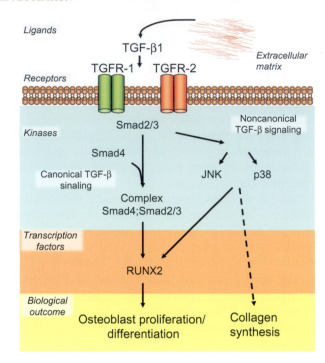

FIGURE 3.7   TGF-β-activated signaling pathway in osteoblasts. Binding of TGF-β1 to its receptors leads to the activation of canonical (Smad-dependent) or noncanonical (Smad-independent) signaling pathways. In canonical signaling, TGF-β1 binds to type I and type II receptors, followed by activation of Smad2/3. Activated Smad2/3 complexes with Smad4 and translocates to the nucleus, where it promotes the transcription of osteoblast-specific genes through RUNX2. TGF-β1 noncanonical signaling leads to activation of TAK1, which activates MAPK-JNK and p38 MAPK. The cascade downstream of p38, in particular, increases expression of type I collagen.

TGF-β is also important for postnatal bone remodeling and homeostasis, although its exact function is still unclear. In vitro studies, in particular, have proven ineffective at clarifying TGF-β action on cells of the osteoblast and osteoclast lineage, due to contradictory results arising from the use of different experimental conditions (e.g. TGF-β concentration, cell type, and duration of the treatment). TGF-β appears to promote the recruitment of osteoprogenitors, their proliferation, and the production of bone matrix, whereas in more mature cells it inhibits matrix mineralization and osteoblast apoptosis, thus favoring their differentiation into osteocytes.

In vivo studies have demonstrated that TGF-β is required for differentiation and function of osteoblasts. Tibiae from young mice lacking TGF-β1 are shorter and exhibit decreased growth plate width, no osteoblasts, and diminished bone formation on the periosteum. Consistent with an inhibitory function for TGF-β on osteoblasts at a later stage of differentiation, inhibition of TGF-β receptor signaling in mature osteoblasts by overexpression of truncated TGFR-2 leads to an age-dependent increase in trabecular bone mass.

However, the mechanisms underlying this phenotype are not completely understood, as bone formation is unaffected, and osteoclast number and eroded surface are diminished only in the skulls of the transgenic mice; not in long bones. Interestingly, these mice also exhibit reduced osteocyte density, indicating that TGF-β is required for terminal differentiation of osteoblasts.

Disruption of signal transduction pathways downstream of activation of the TGF-β receptors results in low bone mineral density and osteopenia. Removal of Smad1 or Smad4 in osteoblastic cells results in decreased bone mass, with reduced bone formation rate, osteoblast proliferation, and differentiation. Since Smad1 and Smad4 are activated by both the BMP and the TGF-β pathways, these studies demonstrate the importance of BMP and TGF-β signals for the maintenance of postnatal bone mass. The noncanonical pathway is important for cartilage development. Thus, removal of TAK1 in chondrocytes results in decreased chondrocyte proliferation and survival of proliferating cells. In addition, removal of TAK1 in limb mesenchyme results in defects in joint fusions.

Regulation of Transforming Growth Factor Beta Bioavailability

All three isoforms of TGF-β are translated as pro-TGF-β homodimers associated noncovalently with pro-regions known as latency associated peptides (LAPs). This complex is secreted as the small latent complex (SLC), in which LAPs confer latency by preventing the binding of TGF-β to its receptors. Large latent complexes (LLCs) also form inside the cell, in which SLCs are bound to a latent TGF-β-binding protein (LTBP) through LAPs. Bone cells release TGF-β in approximately equal proportions of SLCs and LLCs. LTBPs (LTBP-1, -2, and -3) are embedded into the extracellular matrix and, as a consequence, latent TGF-β also is targeted to the matrix. The matrix, in turn, functions as a deposit of TGF-β and modulates the bioavailability of TGF-β through its release from the LLCs in a process known as *TGF-β activation*. The release and activation of latent TGF-β is carried out by multiple mechanisms, including proteolytic digestion of LAP by metalloproteinases (e.g. MMP-2 and MMP-9) or interaction with integrins (e.g. integrins $\alpha_v\beta_6$ and $\alpha_v\beta_8$). The exact mechanisms responsible for TGF-β activation in bone, however, remain to be elucidated. Osteoblasts have been shown to promote TGF-β activation mediated by MMPs, such as MMP-14. Osteoclasts, in addition to protease expression, also possess the unique ability to activate TGF-β by lowering the pH in the resorption lacuna during matrix resorption. Latent TGF-β released during osteoclastic resorption has been demonstrated to be necessary for the recruitment of BMSCs to the sites of resorption, and couple bone formation to bone resorption. The importance of latent TGF-β for bone is demonstrated by the osteopetrosis-like phenotype of mice lacking LTBP-3 in every tissue. The underlying mechanism is still not fully understood and it may involve a decrease in osteoclast function, as indicated by the persistence of cartilage remnants. Interventions that alter the rate of bone resorption or bone formation can potentially affect the amount of TGF-β that is bioavailable. An example is intermittent PTH administration, which concomitantly increases bone mass and the levels of TGF-β.

Prostaglandins

Prostaglandins, molecules that contain 20 carbon atoms and a 5-carbon ring, are synthesized from the fatty acid arachidonic acid through the action of constitutive (COX-1) or inducible (COX-2) prostaglandin G/H synthases/cyclooxygenases. There are five main prostaglandin species PGD, PGE1 and 2, PGF2α, PGI2 (also known as prostacyclin), and thromboxane A2 (TXA2). Osteoblasts and osteocytes produce both PGE2 and PGI2. In bone cells, prostaglandins can be released through Cx43 hemichannels or through activation of $P2 \times 7$ purinergic receptors in response to mechanical stimulation. Prostaglandins act in an autocrine and paracrine fashion through binding to specific receptors. Currently, there are 10 known receptors for prostaglandins. Osteoblasts, osteocytes, and osteoclasts have been shown to express four PGE2 receptor subtypes: EP1 to EP4 (Fig. 3.8). IP, the receptor for PGI2 is expressed in osteoblasts and this expression diminishes as they become osteocytes.

Prostaglandins and their receptors are expressed by most tissues; for this reason the interpretation of global knockouts lacking COX-2 or a specific receptor in every tissue has been challenging. It is clear, however, that PGE2 stimulates bone formation and bone resorption, and that the overall effect of PGE2 is due to the relative activation of the different receptor subtypes.

EP receptors for PGE2 are G protein-coupled receptors that activate different downstream signaling events depending on the G protein. The EP1 receptor is coupled to PLC signaling that triggers the mobilization of intracellular calcium. EP1$^{-/-}$ mice exhibit accelerated fracture healing with increased osteoblast differentiation, although they do not show differences in mechanical properties. EP2 and EP4 receptor activation leads to production of cAMP and stimulates both bone formation and bone resorption. The bone-forming activity of PGE2 has been shown to depend on the EP4 receptor. Thus, the decline of bone volume following ovariectomy or immobilization in rats is prevented by administration of a selective agonist of EP4, and EP4$^{-/-}$ mice

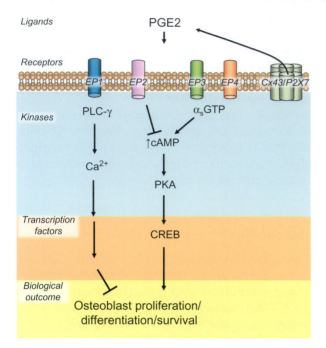

FIGURE 3.8 **Signaling pathway activated by PGE2 in osteoblasts.** Binding of PGE2 to G protein-coupled EP receptors leads to the activation of different downstream signaling events that depend on the particular G protein that is activated. EP1 is coupled to PLC signaling that triggers the mobilization of intracellular calcium. EP2 and EP4 receptor activation leads to production of cAMP, whereas activation of EP3 results in a reduction in cAMP levels. Thus, depending on the receptor that is activated, PGE2 may increase or decrease osteoblast proliferation, differentiation or survival.

develop osteopenia and impaired fracture healing as they age. However, mice in which the EP4 is specifically removed from osteoblasts cannot fully reproduce the osteopenic phenotype, even though osteoblast differentiation is impaired. Femurs from $EP2^{-/-}$ mice exhibit a larger diaphyseal total cross-sectional area and marrow area, suggesting that PGE2 controls both periosteal apposition and endosteal resorption via EP2 receptors. PGE2 can also stimulate bone resorption by inhibiting OPG secretion and stimulating RANKL production by osteoblasts. Moreover, PGE2 enhances the differentiation of osteoclasts from hematopoietic precursors and increases RANK expression by osteoclasts. Activation of the EP3 receptor results in a reduction in cAMP levels. Stimulation of EP1 and EP3 receptors inhibits bone formation in vitro, but has no effect on resorption.

Although information about the role of PGI2 on bone remodeling and development remains limited, it has been demonstrated that PGI2 plays an important role in the maintenance of bone mass in adult animals. Thus, global deletion of the gene encoding PGI2 synthase, the enzyme responsible for PGI2 sythesis, results in a biphasic effect, with reduced bone mass in young mice (5 weeks of age) and higher bone mass in

old mice (34 weeks old) due to a higher remodeling rate with a positive balance at the BMU-level in favor of bone formation.

Mechanical stimulation of animals, cells, and organ cultures has been demonstrated to be a potent stimulator of PGE2 and PGI2 production. Despite all of this evidence, the role of PGE2 in the response to loading is still controversial. In vitro evidence has demonstrated that osteocytes release PGE2 upon mechanical stimulation through Cx43 hemichannels. However, it has also been proposed that activation of $P2 \times 7$ purinergic receptors without the opening of Cx43 hemichannels is required for prostaglandin release induced by mechanical forces. In addition, PGE2 decreases sclerostin expression in vitro. These studies are supported by earlier evidence that pharmacological inhibition of COX-2 impairs the osteogenic response to loading. In contrast, however, $COX-2^{-/-}$ mice do not show a reduced bone apposition after ulnar loading, suggesting that the role of COX-2-PGE2 in vivo may not be essential for bone adaptation to mechanical stimulation.

REGULATION OF BONE CELL DIFFERENTIATION AND FUNCTION THROUGH CELL-CELL CONTACT

Notch

Mammalian Notch proteins (Notch 1–4) are heterodimeric receptors that bind two classes of ligands, Delta-like proteins (Delta1 and 3) and Jagged (1 and 2). Since both receptors and ligands are transmembrane proteins, the Notch system mediates communication between cells that are in contact with each other. In bone, Notch proteins are expressed in osteoblasts, osteocytes, and osteoclasts.

Canonical Notch signaling starts with the binding of the ligand to its receptor, followed by two proteolytic cleavages. In the first one, operated by the metalloproteinase ADAM 17/TACE, the extracellular domain of the Notch receptor is removed; the second, carried out by the γ-secretase complex (containing presenilin-1/2, nicastrin, PEN-2, and Aph-1), cleaves the Notch intracellular domain (ICD; Fig. 3.9). The Notch ICD can then translocate into the nucleus, where it forms a ternary complex with recombining binding protein suppressor of hairless (RBP-JK) and mastermind-like (MAML) transcription factors; the complex recruits other coactivators and promotes the transcription of the HES (HES-1/Hairy-like protein and HES2-5) and HRT [hairy/enhancer-of-split related with YRPW motif-like protein 1(HeyL)/HRT/HESR] families of transcriptional repressors. Most of the work on Notch has focused on this

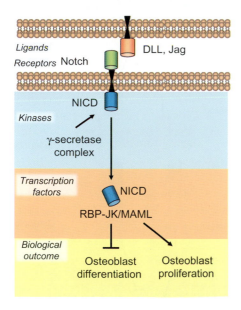

FIGURE 3.9 **Notch signaling.** Notch signaling is activated by direct contact between a cell expressing Notch receptor and a cell expressing the ligands [Delta-like 1, 3 (DLL) or Jagged 1 and 2 (Jag)]. Upon ligand-receptor binding, the γ-secretase complex (containing presenilin-1/2, nicastrin, PEN-2, and Aph-1) cleaves the Notch intracellular domain (NICD); which then translocates into the nucleus, where it forms a ternary complex with the transcription factors RBP-JK and MAML. The complex recruits other coactivators and promotes the transcription of members of the HES and HRT families of transcriptional repressors, leading to inhibition of differentiation and increased differentiation of osteoblasts.

so-called canonical Notch signaling that depends on RBP-JK; however, preliminary work on *Drosophila* suggests that a noncanonical Notch also exists.

Notch signaling plays a crucial role during skeletal development as well as during adult life. During embryonic skeletogenesis, disruption of Notch signaling by inactivating mutations of the ligands or by inhibiting cleavage causes severe skeletal malformations. Several in vivo and in vitro studies have reported an effect of Notch on osteoblastogenesis. However, the conclusions that can be drawn from in vitro studies are conflicting. Whereas some studies report enhancement of osteoblastogenesis by expression of Notch ICD, others show inhibition of osteoblast differentiation. In vivo studies have demonstrated that Notch signaling action on osteoblast differentiation and function is dependent on the stage of cell maturation. Thus, forced expression of Notch ICD in preosteoblasts, mature osteoblasts, and osteocytes using the 2.3 kb fragment of the *C011A1* promoter confers a high bone mass phenotype, with increased osteoblast number and proliferation. The massive osteoblast proliferation, however, is not followed by proper maturation, leading to accumulation of immature cells, disorganized woven bone, and decreased expression of osteoblastic

markers. Consistent with this, overexpression of Notch under the control of the 3.6 kb fragment of the *C011A1* promoter, which is active earlier during osteoblast formation, causes low bone mass because it arrested osteoblast differentiation starting from an early stage.

Deletion of *Psn1* and *Psn2* (encoding presenilin-1 and -2), or *Notch1* and *Notch2*, from the limb mesenchyme results in a high bone mass phenotype. Interestingly, the same mice undergo a rapid and progressive bone loss as they age due to decreased osteoblastogenesis and concomitant increased osteoclastogenesis. Deletion of *Psn1* and *Psn2* in osteoblasts results in low bone mass that develops as the mice age and is caused by decreased OPG levels and increased osteoclastogenesis. Thus, Notch maintains mesenchymal cells in an undifferentiated stage during early osteoblastogenesis but, once the cells are committed to the osteoblastic lineage, Notch promotes osteoblast proliferation and inhibits osteoblast maturation.

Notch can regulate osteoclastogenesis directly by suppressing the differentiation of osteoclast precursors or indirectly by decreasing the levels of OPG and thus increasing the ratio of RANKL:OPG produced by osteoblasts. Thus, suppression of Notch signaling by deleting *NOTCH1* and/or *NOTCH3* in cells of the osteoclastic lineage enhance osteoclast differentiation; and potentiation of Notch signaling by overexpressing Jagged 1 in osteoclast-supporting cells inhibits osteoclast formation in vitro. However, other studies have reported that silencing of Notch2, but not Notch1, inhibits osteoclastogenesis. A possible explanation for the discrepancy between these studies is that, as demonstrated in osteoblasts, Notch signaling also has a dual effect on osteoclasts that depends on the stage of differentiation of the cell. According to this hypothesis, Notch signaling promotes the late stage of osteoclast differentiation, while inhibiting the initial phase.

Notch signaling has also been shown to mediate cross-talk between the hematopoietic stem cell niche and osteoblasts. Thus, constitutive activation of the PTH receptor in osteoblasts results in the expansion of the population of hematopoietic stem cells that is blocked by inhibition of Notch signaling. The increased production of Jagged 1 by osteoblasts in these mice is thought to activate Notch on hematopoietic cells and promote their expansion.

Ephrin-Ephrin Receptor

Ephrin-Eph signaling is based on the binding of ephrin molecules to their cognate ephrin receptors on the surface of adjacent cells. There are two families of ephrins: ephrin-A proteins (A1-A6), which bind to ephrin type-A receptors (EphA1-A10), and ephrin-B proteins (B1-B3), which bind to ephrin type-B receptors (EphB1-EphB6). EphA and EphB receptors share a similar structure with an intracellular tyrosine kinase domain; phosphorylation of tyrosine residues upon ligand binding results in a conformational change and clustering of the receptors. Ephrins A and B, however, can be distinguished by their means of attachment to the plasma membrane. Whereas ephrin-A proteins are anchored through a glycosylphosphatidylinositol tail, ephrin-B ligands have a transmembrane domain.

The role played by the ephrin-B-EphB family in bone has been extensively investigated, while less is known about the contribution of the ephrin-A-EphA system. Osteoblasts express both ephrin-A and B ligands and EphA and EphB receptors, whereas osteoclasts express only ephrins A2, B1, and B2, with no detectable EphB or EphA1, EphA2, and EphA4 receptors. Osteocytes also express EphB4 and ephrins B1 and B2.

The ephrin-B-EphB family plays a critical role in the maintenance of bone homeostasis. Ephrin-Eph signaling is bidirectional; thus, the binding between ligand and receptor triggers a signal transduction cascade downstream of both molecules (Fig. 3.10). The bidirectional nature of ephrin-Eph signaling is the basis of its role in coupling bone resorption to bone formation. Binding of ephrin-B2 expressed by osteoclasts to EphB4 in osteoblasts triggers reverse signaling from ephrin-B2 that suppresses bone resorption by inhibiting expression of the two main osteoclastogenic transcription factors, c-Fos and NFATc1, in hemopoietic precursors. At the same time, forward signaling from EphB4 is triggered and bone formation is stimulated through the upregulation of osteoblast differentiation markers and transcription factors crucial for osteoblast differentiation, such as RUNX2 in calvaria cells. Consistent with these findings obtained in vitro, transgenic mice expressing EphB4 selectively in osteoblasts exhibit high bone mass, due to higher bone formation parameters and reduced osteoclast numbers and function. However, mice lacking ephrin-B2 from macrophages and osteoclasts do not exhibit a bone phenotype, indicating a compensatory effect of ephrin-B1. Ephrin-B2, however, cannot compensate for ephrin-B1; mice in which *Efnb1* is deleted exhibit increased osteoclast differentiation and low bone mass. Ephrin-B2 is also produced by osteoblasts, where it may function in a paracrine or autocrine manner on EphB4 or EphB2 expressed by the same cells and thereby promote osteoblast differentiation. Similarly, targeted disruption of ephrin-B1 in osteoblasts reduces bone size in mice. The role played by the ephrin-Eph system in osteocytes remains to be elucidated.

In contrast to the ephrin-B2-EphB4 system, the ephrin-A2-EphA2 complex has been suggested to

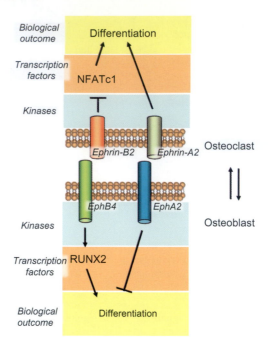

FIGURE 3.10 **Bidirectional signaling activated by Ephrins/Eph in bone cells.** EphA and EphB receptors bind to ephrins A and B on the cell surface, triggering intracellular signaling in both the Eph- and the Ephrin-expressing cells. Osteoclast precursors express ephrin-B2, which binds to EphB4 in osteoblasts. This triggers reverse signaling from ephrin-B2 that suppresses bone resorption, by inhibiting c-Fos and NFATc1. At the same time, forward signaling from EphB4 is triggered and bone formation is stimulated through upregulation of osteoblast differentiation markers and transcription factors. This results in increased osteoblast and decreased osteoclast function. In contrast, activation of ephrin-A2 by association with EphA2 in osteoclast precursors triggers a reverse signaling that stimulates osteoclastogenesis, whereas EphA2 promotes a forward signaling into osteoblasts that suppresses osteoblast differentiation and bone mineralization in vitro.

function as a *coupling inhibitor*. Ephrin-A2 is expressed early during osteoclast differentiation, and activation of ephrin-A2 triggers a reverse signaling that stimulates osteoclastogenesis, whereas EphA2 promotes a forward signaling into osteoblasts that suppresses osteoblast differentiation and bone mineralization.

Connexins

The Cx family of transmembrane proteins comprises 20 members with a high degree of homology. Cx molecules have four transmembrane domains, with both the N- and a C-termini facing the cytoplasm. Six Cx molecules form a connexon or hemichannel, arranged in such a way that a pore is formed in the center. The pores vary in solute charge and size selectivity, depending on the types of connexin molecules that constitute it, although the size limit for a molecule to pass through hemichannels is 1 kDa. Hemichannels can be formed by one type (homomeric) or by different types of Cx molecules (heteromeric). In the latter case, the resulting channel will have properties that are different from those of the channels formed by each particular Cx. Two hemichannels from neighboring cells align to form channels, named *gap junctions* because the plasma membranes from the two cells are not in close contact and are instead separated by a gap of approximately 4 nm. Areas on the plasma membrane containing clusters of gap junction channels are known as gap junction plaques.

Three functions have been ascribed to members of the Cx family of proteins: (1) cell-cell communication via gap junctions; (2) the exchange of small molecules with the extracellular medium via hemichannels; and (3) regulation of intracellular signaling via association with other molecules through their C-terminal tail. While cell-cell communication through gap junctions was originally thought to be the only function of Cxs, mounting evidence supports the role of Cxs in hemichannels and in the regulation of cellular functions through its association with signaling molecules, independently of channel activity.

The expression of three members of the Cx family of proteins has been described in osteoblasts and osteocytes: Cx43, Cx45, and Cx46, although the latter does not form membrane channels and its function in osteoblasts is unknown. Most recently, it has been shown that Cx37 is also expressed in osteoblastic cells and that its expression is higher in osteocytes than in osteoblasts.

Cx43 is the best studied Cx in bone; it is required for full osteoblast differentiation. Moreover, it is necessary for the effect of several bone-acting agents such as bisphosphonates, FGF-2, IGF, and PTH. Mice lacking Cx43 die soon after birth due to cardiac malformation, and these embryos exhibit delayed ossification. Mouse models in which *Gja1* (the gene encoding Cx43) has been specifically deleted in osteochondroprogenitors have a clear bone phenotype, with reduced bone size, mass, and strength. The phenotype becomes progressively less dramatic as Cx43 is removed from more mature preosteoblastic cells, and there is no change in bone mass in mice lacking Cx43 in mature osteoblasts or in osteocytes. Interestingly, however, all these animal models exhibit increased bone marrow cavity area in the long bones, with elevated osteoclast numbers on the endocortical surface and increased periosteal bone formation, suggesting that Cx43 expression in osteocytes may be responsible for controlling cortical bone formation and resorption.

It has been long hypothesized that osteocytes regulate the response of bone to mechanical loading via Cx43. This idea was based on the fact that osteocytes express high levels of the Cx, and, since they are embedded in the bone matrix, they are more likely to communicate among each other and with cells on the bone surface through gap junctions. Moreover, Cx43 expression is increased by mechanical loading in vitro and in vivo, and mechanical stimulation enhances gap junction communication and induces hemichannel opening in vitro. However, recent work from three different laboratories has shown that mice lacking Cx43 from osteochondroprogenitors, from mature osteoblasts and osteocytes, or from osteocytes alone exhibit a greater increase in bone formation in response to loading. The mechanisms that cause the exacerbated bone anabolism in these mice are unknown.

Cx43 is also expressed in osteoclasts and is required for differentiation and fusion of osteoclast precursors. Recent studies showed that deletion of *Gja1* from osteoclast precursors results in decreased osteoclastogenesis and increased cortical thickness.

Less is known regarding the function of other Cxs in bone cells. Cx45 has dominant-negative activity against Cx43 in osteoblasts; however, it is unknown whether it also has actions that are independent of Cx43. Recent evidence demonstrates that mice lacking Cx37 exhibit high bone mass associated with reduced osteoclast number, likely due to defective osteoclast differentiation and fusion.

LOCAL ANGIOGENIC FACTORS THAT REGULATE BONE CELL ACTIVITY

The process of angiogenesis is regulated by a number of factors, such as FGF-2, TGF-β, and hypoxia-inducible transcription factor (HIF). Among these, the members of the VEGF family are thought to be key regulators in skeletal development and repair. The VEGF family comprises several members that result from alternative splicing of exons 6, 7, or 8 of a single *VEGF* gene. The various isoforms are named according to the number of amino acids (e.g. VEGF120, VEGF164, and VEGF188). In bone, VEGFs are highly expressed in hypertrophic chondrocytes. In addition, neovascularization in bone is achieved through the production of VEGF by endothelial cells, osteoclasts, osteoblasts, and osteocytes. Notably, the receptors for VEGF, VEGF receptor 1, 2, and 3 (VEGFR-1, -2, and -3) are found in the same cells that produce the ligand, indicating that an autocrine signaling mechanism exists in addition to its other paracrine signaling mechanisms.

Binding of VEGFs to the respective tyrosine kinase receptors and coreceptors neuropilin-1 and -2 and glypican-1 leads to activation of the Raf-MAPK and PLC-γ-IP3 pathways. The main cellular targets of VEGF are endothelial cells, where the activation of these signaling cascades results in their proliferation, survival, and differentiation. VEGF-VEGF receptor interaction also activates the endothelial nitric oxide synthase (eNOS), resulting in increased nitric oxide production and vasodilation through several pathways, including Akt, Ca^{2+}/calmodulin, and PKC.

It has recently emerged that VEGF, besides promoting neovascularization, also directly induces osteoblast and osteoclast differentiation and therefore bone formation and remodeling. The expression of VEGFs and their receptors increases as preosteoblasts differentiate into mature osteoblasts, and treatment with exogenous VEGF promotes nodule formation in vitro. Moreover, VEGF has been proposed to inhibit osteoblast apoptosis and to act as a chemoattractant for bone marrow-derived MSCs and osteoblastic cells. Similarly, VEGF controls osteoclast differentiation, migration, and activity by upregulating the expression of RANK and RANKL.

VEGF production and vascular development are promoted by oxygen gradients and HIFs. HIFs are important for induction of VEGFs in osteoblasts, where in vivo activation of HIFα signaling pathway leads to higher levels of VEGF and increased vasculature in the bone marrow cavity. Notably, activation of HIFα protects mice from ovariectomy-induced bone loss. Recently, a novel role for osteocytes as a source of VEGF has emerged. In response to in vivo fatigue loading, viable osteocytes located in the vicinity of areas exposed to damage produce increased levels of VEGF, which may be required to promote angiogenesis and osteoclastogenesis, essential steps in the repair of the loading-induced microdamage. Mechanical loading applied to osteocytes in vitro also induces VEGF.

CELL SURFACE ATTACHMENT MOLECULES

The function of bone cells is highly dependent on their interaction with the surrounding matrix, which is achieved by the association of surface receptors with molecules present in the extracellular matrix. Matrix interaction can trigger intracellular signaling that modulates not only cell adhesion and migration, but also differentiation and activation. Molecules that mediate cell attachment to the matrix are known as adhesion receptors and can be classified based on their structure in different families: integrins, immunoglobulin (Ig) superfamily, selectins, cadherins, syndecans, and CD44 antigen (or hyaluronate receptor; Table 3.2). Signaling elicited by the association of these molecules with their counterparts has two components: *inside-out* and *outside-in*. Inside-out signaling is where adhesion molecules become activated by intracellular signals and bind to extracellular ligands. Outside-in signaling is triggered by the association of adhesion molecules with their ligands, resulting in activation of intracellular signaling cascades.

Integrins are transmembrane proteins that participate in bidirectional signaling across the plasma membrane in osteoblasts, osteocytes, and osteoclasts by engaging with molecules in the extracellular matrix (Fig. 3.11). Integrins are heterodimers formed by an α and a β chain, with both contributing to the ligand-binding site. The different α subunits contain seven homologous repeat domains and cation-binding sites. β subunits also have cation-binding sites and a high cysteine content. The three-dimensional structure of the extracellular domains of α/β integrin dimers is highly conserved among different pairs. It is believed that integrins can become

TABLE 3.2 Actions of Surface Attachment Molecules on Bone Cells

Molecule	Osteoblast/Osteocyte				Osteoclast			
	Expression	Differentiation	Function (bone formation)	Proliferation or Survival	Expression	Differentiation Generation	Function (resorption)	Survival
Cadherins	+	+	+	+	+	+	+	Unknown
CD44	+ (more abundant in mature cells)	Unknown	Unknown (may be involved in osteoclast support)	Unknown	+	+/−	+/−	Unknown
Ig family	+	Unknown	Unknown (may be involved in osteoclast support)	Unknown	+	+	Unknown	Unknown
Selectins	Unknown	Unknown	Unknown	Unknown	+ (only in precursors)	+	Unknown	Unknown
Syndecans	+	Unknown	+	Unknown	Unknown	Unknown	Unknown	Unknown

+, reported; +/−, contradictory reports.
Ig, immunogobulin.

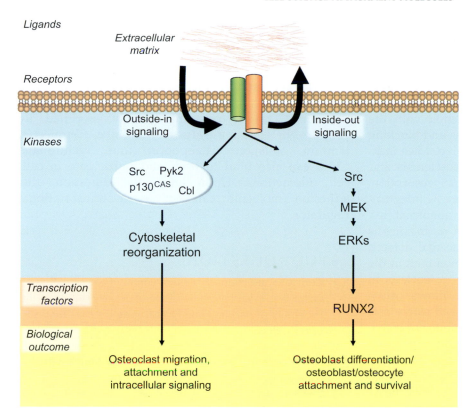

FIGURE 3.11 Signal transduction activated by integrin-mediated cell attachment in osteoblasts and osteoclasts. Cell surface attachment molecules, exemplified in this scheme by integrins, bind to the extracellular matrix and trigger intracellular signaling through the activation of kinases. This is called outside-in signaling and may result in cytoskeletal reorganization, and cell migration, attachment, and differentiation. In addition, activation of intracellular signals can lead to the binding of integrins to the extracellular matrix and cell attachment through inside-out signaling.

activated upon cation binding and change the conformation of their extracellular domains.

Through their large extracellular domain, integrins interact with molecules present in the extracellular matrix such as fibronectin and collagen. Integrins can also interact with soluble molecules or with surface proteins present in neighboring cells. Upon ligand binding, integrins cluster and associate with other intracellular molecules to form complexes called focal adhesions. Integrins lack enzymatic activity and become activated through the assembly of focal adhesion signaling molecules, thereby triggering intracellular signaling cascades and changes in cellular behavior.

Osteoprogenitors, mature osteoblastic cells, and osteocytic cells express several different integrins, and their expression is affected by bone-acting cytokines. In particular, osteoblasts express $\alpha 1\beta 1$ and $\alpha 2\beta 1$, which bind collagen; $\alpha 3\beta 1$, which binds laminin; $\alpha 4\beta 1$ and $\alpha 5\beta 1$, which bind fibronectin; and αv dimers with $\beta 3$, $\beta 5$, $\beta 6$, $\beta 8$, which bind vitronectin. Upon interaction of integrins with the extracellular matrix, molecules involved in focal adhesion are activated, including focal adhesion kinase (FAK) and Pyk2, and the Src and ERK kinases. These signals ultimately lead to cytoskeleton reorganization and RUNX2-mediated transcriptional control of osteoblast differentiation. In osteocytes, engagement of integrins is required for cell survival induced by mechanical stimulation.

Few studies have addressed the role of integrins in osteoblasts in vivo. For example, fracture healing is defective in mice lacking integrin $\alpha 1$, although the mice do not exhibit any gross skeletal abnormalities. This defect in fracture healing is ascribed to decreased proliferation of MSCs and cartilage production. Mice were generated expressing a $\beta 1$ integrin form lacking the extracellular domain that acts as an inhibitor of $\beta 1$ integrin function in mature osteoblasts. Skeletal analysis of these mice revealed a mild decrease in bone mass and strength; defective osteoblast and osteocyte morphology, osteoblast polarity, and matrix secretion in vivo; and reduced adhesion of osteoblasts and osteocytes in vitro. Mice lacking $\beta 1$ integrin in preosteoblasts exhibit an increase in cortical bone volume and strength following skeletal unloading, unlike wild-type mice that show no significant changes in cortical bone. Lastly, mice lacking integrin $\beta 2$ exhibit reduced bone mass due to decreased osteogenic differentiation of MSCs; in contrast, in an in vivo model of ectopic bone formation, bone marrow MSCs overexpressing integrin $\beta 2$ exhibit increased osteoblastogenic potential.

Integrins are essential for osteoclast migration, attachment, and intracellular signaling. Integrin-mediated signaling in osteoclasts raises intracellular calcium, changes phosphoinositide metabolism, and induces the phosphorylation of signaling molecules. Engagement of $\alpha_v\beta_3$ integrin induces the recruitment of the Src kinase and phosphorylation of a group of proteins that includes Pyk2, E3 ubiquitin-protein ligase CBL, paxillin, and breast cancer antiestrogen resistance protein 1 (p130cas), thereby increasing osteoclastic bone resorption. The major integrin expressed in osteoclasts is the vitronectin receptor, formed by integrins $\alpha_v\beta_3$. Blockade of integrin $\alpha_v\beta_3$ interaction with the extracellular matrix using small peptides and blocking antibodies inhibits bone resorption in vivo. Mice lacking β_3 integrin exhibit progressive osteosclerosis and decreased osteoclast adhesion and resorption. However, patients with a *Itgb3* (encoding β_3 integrin)-null mutation have no skeletal abnormalities. Moreover, although deletion of integrin α_v in mice is lethal in 80% of the animals due to placental defects, bone develops normally. It has been shown that, in the absence of $\alpha v\beta3$, $\alpha2\beta1$ is upregulated in osteoclasts in culture, suggesting compensation that allows osteoclast function. Other integrins expressed in osteoclasts are $\beta5$ and $\alpha9$. Integrin $\beta5$ acts as an inhibitor of osteoclast formation, and mice lacking $\beta5$ exhibit an exaggerated response to ovariectomy due to increased osteoclast formation. Double $\beta3/\beta5$-deficient mice do not have a skeletal phenotype. Deletion of integrin $\alpha9$ in mice leads to osteopetrosis.

Members of the Ig superfamily of receptors share a domain consisting of 70–110 amino acids. Ig family proteins bind to other members of the same family and to ligands for integrins. Some members of the family are involved in signal transduction (leading to activation of the MAPK pathway), whereas others only participate in cell adhesion.

Selectins are glycoproteins that contain a calcium-dependent lectin type domain, an epithelial growth factor (EGF)-like domain, complement-binding-like sequences, a transmembrane domain, and a short cytoplasmic tail. Selectins bind to oligosaccharide sequences in sialylated and sulfated glycans. Once bound to the ligands, selectins induce tyrosine phosphorylation and increase intracellular calcium, although the signaling pathways have not been completely elucidated.

Cadherins are calcium-dependent, single-pass transmembrane glycoproteins that contain so-called *cadherin domains*. Cadherins mediate cell-cell adhesion, by association with the same type of cadherin in a neighboring cell. Their cytoplasmic tail binds to β-catenin, thereby participating in Wnt signaling by regulating the amount of free β-catenin. Cadherins also induce the activation of small GTPases and tyrosine kinases.

Syndecans are single transmembrane proteins that contain heparin sulfate and chondroitin sulfate attachment sequences in the extracellular domain. Syndecans act mainly as coreceptors for other receptors such as integrins, FGF, TGF-β, and VEGF. Syndecans bind matrix proteins and members of the FGF family of proteins. Signal transduction molecules induced by syndecans include activation of associated tyrosine kinases and cytoskeletal proteins.

CD44 antigen is a member of a family of transmembrane proteins that share a common N-terminal domain. CD44 is postranslationally modified by chondroitin sulfate residues and binds fibronectin, laminin, collagen, osteopontin, and hyaluronate. CD44 can also bind to itself and participate in cell-cell interactions. Upon enzymatic digestion, this adhesion molecule can act as a transcriptional regulator.

STUDY QUESTIONS

1. Compare and contrast the movement of the signaling factor (e.g. cytokine), time frame of action, and proximity of participating cells among the four modes of cell-cell communication.

2. Describe two examples of communication between bone cells and compare and contrast the functional consequences on the effector (downstream) cell.

3. Compare and contrast the canonical and non-canonical Wnt signaling pathways, including how inhibitors of Wnt signaling function at various points in the pathway. Finally, describe some advantages of therapy with an anti-sclerotin antibody and speculate on the effects of this therapy on bone remodeling.

4. Describe differences in the effect on bone homeostasis of systemic IGF-I (mainly from the liver) with locally produced IGF-I (deposited into the matrix by osteoblasts).

5. Compare and contrast the canonical signaling pathways of BMPs and TGF-β.

Suggested Readings

Baron, R., Kneissel, M., 2013. Wnt signaling in bone homeostasis and disease: from human mutations to treatments. Nat. Med. 19, 179–192.

Bisello, A., Friedman, P., PTH and PTHrP actions in kidney and bone. In: Bilezikian, J.P., Raisz, L.G., Martin, T.J., (Eds.), Principles of Bone Biology. Academic Press, San Diego, San Francisco, New York, London, Sydney, Tokyo, pp. 665–712.

Chen, G., Deng, C., Li, Y.P., 2012. TGF-beta and BMP signaling in osteoblast differentiation and bone formation. Int. J. Biol. Sci. 8, 272–288.

Helfrich, M.H., Stenbeck, G., Nesbitt, S.A., Horton, M.A., Integrins and other cell surface attachment molecules of bone cells. In: Bilezikian, J.P., Raisz, L.G., Martin, T.J., (Eds.), Academic Press, San Diego, San Francisco, New York, London, Sydney, Tokyo, pp. 385–424.

Horiguchi, M., Ota, M., Rifkin, D.B., 2012. Matrix control of transforming growth factor-beta function. J. Biochem. 152, 321−329.

Horowitz, M.C., Lorenzo, J.A., Local regulators of bone: IL-1, RND, lymphotoxin, Interferon-γ, the LIF/IL-6 family, and additional cytokines. In: Bilezikian, J.P., Raisz, L.G., Martin, T. J., (Eds.), Principles of Bone Biology. Academic Press, San Diego, San Francisco, New York, London, Sydney, Tokyo, pp. 1209−1234.

Johnson, M.L., Wnt signaling and bone. In: Bilezikian, J.P., Raisz, L. G., Martin, T.J., (Eds.), Principles of Bone Biology. Academic Press, San Diego, San Francisco, New York, London, Sydney, Tokyo, pp. 121−138.

Long, F., 2012. Building strong bones: molecular regulation of the osteoblast lineage. Nat. Rev. Mol. Cell Biol. 13, 27−38.

Marie, P.J., 2012. Fibroblast growth factor signaling controlling bone formation: an update. Gene. 498, 1−4.

Matsuo, K., Irie, N., 2008. Osteoclast-osteoblast communication. Arch. Biochem. Biophys. 473, 201−209.

Mundy, G.R., Elefteriou, F., 2006. Boning up on ephrin signaling. Cell. 126, 441−443.

Patil, A.S., Sable, R.B., Kothari, R.M., 2012. Occurrence, biochemical profile of vascular endothelial growth factor (VEGF) isoforms and their functions in endochondral ossification. J. Cell Physiol. 227, 1298−1308.

Pilbeam, C.C., Choudhary, S., Blackwell, K., Raisz, L.G., Prostaglandins and Bone Metabolism. In: Bilezikian, J.P., Raisz, L. G., Martin, T.J., (Eds.), Academic Press, San Diego, San Francisco, New York, London, Sydney, Tokyo, pp. 1235−1274.

Plotkin, L.I., Bellido, T., 2012. Beyond gap junctions: connexin43 and bone cell signaling. Bone. 52, 157−166.

Sims, N.A., Gooi, J.H., 2008. Bone remodeling: multiple cellular interactions required for coupling of bone formation and resorption. Semin. Cell Dev. Biol. 19, 444−451.

Sims, N.A., Walsh, N.C., 2010. GP130 cytokines and bone remodeling in health and disease. BMB. Rep. 43, 513−523.

Xiong, J., O'Brien, C.A., 2012. Osteocyte RANKL: new insights into the control of bone remodeling. J. Bone Miner. Res. 27, 499−505.

Yakar, S., Courtland, H.W., Clemmons, D., 2010. IGF-I and bone: new discoveries from mouse models. J. Bone Miner. Res. 25, 2267−2276.

Zanotti, S., Canalis, E., 2012. Notch regulation of bone development and remodeling and related skeletal disorders. Calcif. Tissue Int. 90, 69−75.

Bone Modeling and Remodeling

Matthew R. Allen and David B. Burr

Department of Anatomy and Cell Biology, Indiana University School of Medicine, Indianapolis, Indiana, USA

SKELETAL DEVELOPMENT

Skeletal development begins during the first trimester of gestation and continues into the postnatal years. Development occurs through two distinct processes: intramembranous ossification and endochondral ossification. These modes of development differ in the environment in which ossification is initiated and in the cells that produce the matrix. Intramembranous ossification occurs in a collection or condensation of mesenchymal cells that differentiate directly into osteoblasts, whereas endochondral ossification occurs on a cartilage template produced by chondrocytes. Both processes occur during embryogenesis, as well as postnatally.

Intramembranous Ossification

Most bones of the skull, as well as certain other bones (such as the scapula and clavicle), are formed embryonically through intramembranous ossification. The process of intramembranous ossification is initiated and takes place within a sea of mesenchyme, an embryonic or primitive connective tissue comprised primarily of mesenchymal cells. Although intramembranous ossification is most often associated with embryonic development, it can also occur postnatally (during bone healing, for example). The initial step in intramembranous ossification is the consolidation of mesenchymal cells, commonly referred to as a bone blastema. Cells within the blastema differentiate into osteoblasts and begin to produce matrix (Fig. 4.1). The transcription factor RUNX2 plays an indispensable role in driving cells within the blastema toward the osteoblastic lineage. The initial production of bone matrix by osteoblasts establishes a primary ossification center, i.e. a spatial location where the processes of

bone growth take place. As the osteoblasts within the individual ossification centers produce more and more matrix, some of the osteoblasts become encapsulated, at which point they become osteocytes. The bone matrix produced by these initial osteoblasts is known as *woven bone*, an unorganized collagen structure resulting from rapid production. Once sufficient bone matrix is produced to form a small island of bone, additional osteoblasts are recruited to the surface, where there is continued production of either woven bone or more organized primary lamellar bone. During development, some bones form through the merging of several small bony islands. Some bones formed through intramembranous ossification, such as the jaw, develop marrow cavities. These cavities form once the bone becomes so large that the central osteocytes are too distant from an adequate blood supply, thus stimulating the invasion of blood vessels into the middle of the ossification center to form a marrow cavity. Other bones formed through this process, such as the scapula and clavicle, do not form a marrow cavity.

Endochondral Ossification

The remainder of the bones in the skeleton are formed through endochondral ossification, a process in which a hyaline cartilage template is formed and over time replaced by mineralized bone tissue (Fig. 4.2). Endochondral ossification is not limited to embryonic development; it also has a significant role in fracture healing. Similar to intramembranous ossification, endochondral ossification begins with a condensation of mesenchymal cells. Instead of differentiating into osteoblasts, however, these cells differentiate into chondroblasts. This process is driven by the transcription factor SOX-9. These chondroblasts produce a cartilage matrix that eventually envelops some

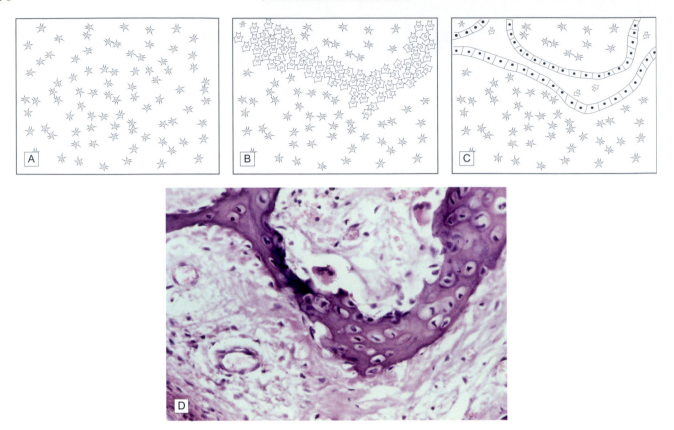

FIGURE 4.1   **The process of intramembranous ossification begins within a sea of mesenchymal cells.** (A). mesenchymal cells begin to consolidate into a blastema and transform into osteoblasts (B), eventually producing bone matrix (C), which over time will be remodeled. (D) Photomicrograph of island of bone forming through intramembranous ossification in the jaw (hematoxylin and eosin stain).

cells, which then become chondrocytes. The hyaline cartilage is surrounded by perichondrium, a cellular/fibrous membrane found on the surface of the cartilage model that serves to provide cells for cartilage growth. This cartilage template strongly resembles the shape of the mineralized bone that will eventually be formed, and so it is sometimes called the *cartilage model* (or anlage).

Early during development, cells of the perichondrium differentiate into osteoblasts and begin forming bone on the surface of the cartilage template. As in intramembranous ossification, this process of osteoblast differentiation is governed by the transcription factor RUNX2. Bone formation is initially localized to the circumference of the midshaft (diaphysis) of the long bone and results in a structure called the *bone collar*. The bone collar is lamellar bone and, once formed, the adjacent fibrous tissue transitions from perichondrium to periosteum, becoming populated with osteogenic precursor cells. Formation of the bone collar limits the ability of nutrients to diffuse into the nearby cartilage resulting in calcification of the local matrix and, eventually, death of the chondrocytes. These processes signal recruitment of a primary blood vessel to

penetrate the bone collar (with the help of osteoclasts) and enter the region of calcified cartilage. This vessel delivers nutrients to the surviving cells and transports osteoclasts that remove the calcified matrix. The result of this vascular invasion is the formation of a marrow space and a primary ossification center—the site for coordinated cell activity for further development. As the bone marrow is slowly formed and populated with cells, additional bone continues to be formed on the periosteal surface of the bone collar. Secondary ossification centers eventually form at the ends of the long bones (epiphyses) through a similar process. Other vessels penetrate into the region to supply the cells and nutrients necessary for additional development.

As the primary ossification center grows, it eventually makes up roughly the middle third of the hyaline cartilage template. This results in the template having two cartilaginous ends with a central diaphyseal region that includes a marrow cavity. At the interface between the marrow and cartilage at each end of the bone is a structure called the *growth plate* (or epiphyseal plate). The growth plate is responsible for longitudinal bone growth. It is comprised of five morphologically distinct zones that are conveniently

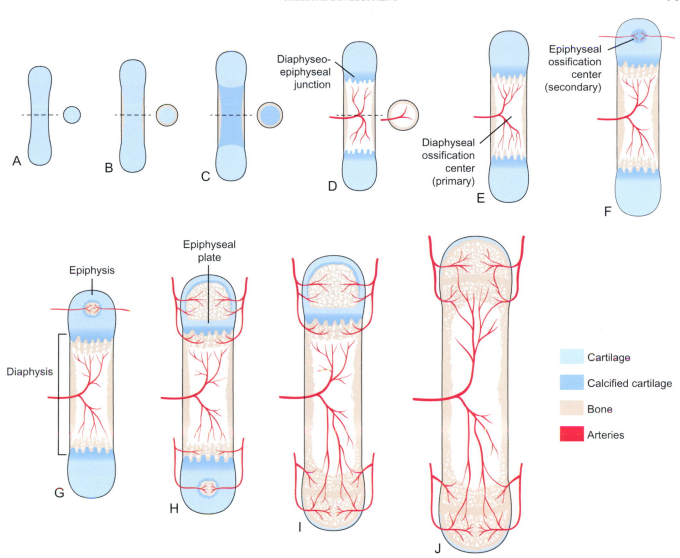

FIGURE 4.2 Endochondral ossification. The process of endochondral ossification begins with a cartilage template that transforms into bone through a series of stages involving the coordinated activity of chondrocytes, osteoblasts, and osteoclasts.

classified according to the main cellular processes that occur at each location (Fig. 4.3). In reality, they exist as a continuum of cells that gradually transition from one zone to another as growth occurs.

The region most distant from the primary ossification center (nearest the ends of the cartilage template) is called the *resting zone* and is comprised of hyaline cartilage matrix with embedded chondrocytes. New resting zone matrix, rich in type II collagen, is continually produced by the chondroblasts near the perichondrium. The chondroblasts that embed themselves in this matrix differentiate into mature chondrocytes. The chondrocytes embedded in matrix also produce new matrix. This region of active production is sometimes referred to as the *reserve zone*, since the term *resting zone* implies that

the cells are inactive. The chondrocytes within the resting zone have similar morphological and physiologic characteristics as those in hyaline cartilage in other regions of the body.

The second region is called the *proliferative zone* and, as the name implies, it is a site of active chondrocyte mitosis (Fig. 4.3B,C). This region is readily identifiable histologically by its stacked coin appearance, which results from cell division in columns along the longitudinal axis. These cells produce modest amounts of matrix rich in type II collagen. The zone of proliferation is regulated by a number of growth factors. Somatotropin/growth hormone (GH), insulin-like growth factors (IGFs), Indian hedgehog protein (IHH), bone morphogenic proteins (BMPs), and the Wnt-β-catenin signaling pathway all play important

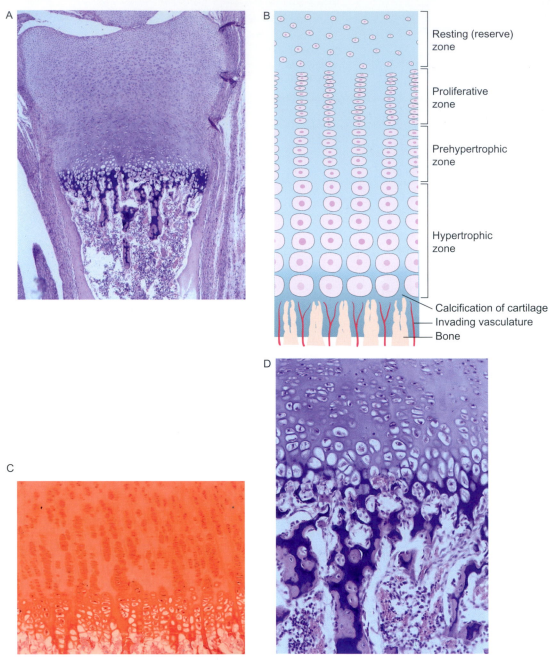

FIGURE 4.3 The epiphyseal growth plate is classified into five zones (A and B) based on the different histologic appearance of the cells and matrix: resting (reserve) zone; proliferative zone; hypertrophic zone; zone of calcified cartilage; and zone of ossification. As long bone growth occurs, cells progress through the different zones. In the low power photomicrograph (A), the individual delineations between zones are not obvious, although the overall structure of the region can be appreciated. (B) The schematic more clearly shows the cell/tissue morphology of the zones within the growth plate. (C) At the top of this photomicrograph is the zone of proliferation, where the chondrocytes rapidly divide and become stacked in a longitudinal orientation. In the zone of hypertrophy the cells grow in size and the matrix produced by the cells begins to change. Toward the bottom of the field is the zone of calcified cartilage, which can be identified by the slightly darker-stained matrix. (D) This photomicrograph shows the hypertrophic, calcified cartilage, and ossification zones. Chondrocytes in the zone of hypertrophy grow in size, while the region in which the matrix begins to calcify (noted by the dark blue-stained matrix) delineates the transition to the zone of calcified cartilage. The zone of ossification begins where bone matrix is produced (pink stain).

roles in stimulating chondrocyte proliferation. Fibroblast growth factor (FGF) is one of the few factors shown to inhibit proliferation of chondrocytes in this region.

The third region is the *hypertrophic zone* (Fig. 4.3). Growth of the long bones is driven by cells in the upper regions of this zone (or prehypertrophic zone). In the lower hypertrophic zone, cells begin to enlarge and die.

Thyroxine appears to be the main promoter of chondrocyte hypertrophy, although other factors (e.g. components of the Wnt-β-catenin pathway) also are important for promoting hypertrophy. IHH and parathyroid hormone-related protein (PTHrP) are two factors known to inhibit chondrocyte hypertrophy in this region. Hypertrophy enhances cell growth and the production of extracellular matrix, which is initially rich in type II collagen. However, this region also features a transition from the production of type II collagen to type X collagen. Type X collagen is exclusive to hypertrophic chondrocytes in the developing bone (although it can also be found in the fracture callus and damaged articular cartilage in adults). It differs from type II collagen in that it contains fibers, which are absent in the type II collagen matrix. Although the fibers of type X collagen provide stiffness to the region, they create a matrix less able to diffuse nutrients to the cells. Type X collagen is intimately connected to vascular invasion, which occurs in the adjacent regions. If no type X collagen is present in the hypertrophic zone, then vascular invasion does not occur and growth in disrupted.

The condensed matrix around the hypertrophied chondrocytes eventually begins to mineralize (or calcify; Fig. 4.3D). Mineralization of the matrix does not occur in the absence of hypertrophy. The region where cartilage calcification can be observed is the fourth region of the growth plate, the calcified cartilage zone. The chondrocytes in this region are either dead, or in the process of dying, due to lack of nutrient diffusion or cellular waste removal. The signal for apoptosis appears to be related to cellular hypoxia, as cells in the lower proliferative zone and upper hypertrophic zone have been shown to be more hypoxic than those in the more superficial regions of the growth plate. Matrix calcification is an active process directed by the chondrocytes, the specifics of which are not well understood. Chondrocytes release vesicles into the extracellular matrix. These vesicles contain alkaline phosphatase (a key contributor to matrix mineralization), ATPase (to provide energy to transport calcium ions into the vesicles), and enzymes that cleave

calcium and phosphate from the surrounding environment. The increase in local mineral concentration leads to formation of calcium-phosphate aggregates and calcification of the matrix. As the region becomes more calcified and loses more cells, local signaling leads to vascular invasion. Chondroclasts, cells that are similar to osteoclasts but specialized for the removal of calcified cartilage, come to the site to begin cartilage resorption.

The final zone of the growth plate is the zone of ossification, where bone is initially formed (Fig. 4.3). This skeletal tissue is formed by osteoblasts that are recruited to the calcified tissue surface to produce new woven bone. There are also osteoclasts in the zone of ossification, working to remove both the calcified cartilage and the newly produced woven bone—the latter being replaced by more mature lamellar bone through bone remodeling.

Longitudinal growth via endochondral ossification occurs until the epiphyseal plate becomes ossified. In humans, for most bones this occurs in the late teens and early twenties. As the skeleton matures, activity within the zone of ossification exceeds chondrocyte repopulation in the reserve zone. This leads to a slow reduction in the size of the growth plate until the process ceases completely, leaving behind a mineralized region of bone separating the epiphysis and metaphysis. This thin plate of bone is called the epiphyseal line. Epiphyseal closure is accelerated by estrogen, which causes a more rapid senescence of chondrocytes. This is the reason for earlier growth plate closure in women than in men.

BONE MODELING

Bone modeling is defined as either the formation of bone by osteoblasts or resorption of bone by osteoclasts on a given surface. This contrasts with bone remodeling (discussed below), in which osteoblast and osteoclast activity occur sequentially in a coupled manner on a given bone surface (Table 4.1).

TABLE 4.1   Important Characteristics of Modeling and Remodeling

	Modeling	Remodeling
Goal	Shape bone, increase bone mass	Renew bone
Cells	Osteoclasts or osteoblasts and precursors	Osteoclasts, osteoblasts, and precursors
Bone envelope	Periosteal, endocortical, trabecular	Periosteal, endocortical, trabecular, intracortical
Mechanism	Activation-formation or activation-resorption	Activation-resorption-formation
Timing	Primarily childhood but continues throughout life	Throughout life
Net effect on bone mass	Increase	Maintain or slight decrease

Modeling by osteoblasts is called *formation modeling*; modeling by osteoclasts is called *resorptive modeling*. The primary function of bone modeling is to increase bone mass and maintain or alter bone shape. Although formation and resorption modeling are locally independent, they are not globally independent since both processes occur simultaneously throughout the skeleton and must be coordinated to shape bone. Modeling always occurs on a preexisting bone surface, which is why the initial stages of intramembranous and endochondral ossification are not considered modeling. Modeling activity can occur on periosteal, endocortical, and trabecular bone surfaces or envelopes.

Events that Signal Modeling

The principal signal for bone modeling is local tissue strain (see Chapters 6 and 9 for more detail). If the local strains exceed a certain threshold, then formation modeling is initiated to add new bone matrix. If strains are low, then resorptive modeling is stimulated and bone is removed.

Cellular Processes

The process of modeling occurs in two stages: activation and either formation or resorption. Activation involves recruitment of precursor cells that differentiate into mature osteoblasts or osteoclasts. Additionally, bone lining cells can be stimulated to differentiate into mature, active osteoblasts that begin producing matrix. Once the appropriate cells are activated, the processes of formation or resorption take place until sufficient bone mass is added or removed to normalize local strains.

Modeling During the Life Cycle

Bone modeling is most prominent during growth and development and primarily serves to reshape the bone or change the position of the cortex relative to its central axis (called *bone drift*). The adult skeleton does undergo modeling but, in the absence of pathology, it is less prominent.

Longitudinal Growth

Both formation and resorption modeling play an essential role in maintaining bone shape during growth associated with endochondral ossification. In order to maintain the proper shape of the long bones, both types of modeling are coordinated in the metaphyseal region (Fig. 4.4). As the bone lengthens,

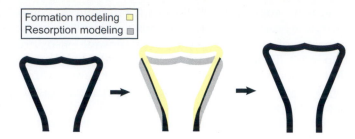

FIGURE 4.4 As the bone grows longitudinally, there is coordinated modeling activity on the metaphyseal bone surfaces that serve to preserve the bone shape.

resorption modeling removes bone on periosteal surfaces, while formation modeling adds new bone to endocortical surfaces. While these processes are highly coordinated, they are distinct from remodeling because formation and resorption occur on different surfaces. Disruption of modeling during growth results in abnormal metaphyseal morphology. This is most evident in cases of osteoclast inhibition (examples include genetic dysfunction or pharmaceutical intervention), which results in the metaphysis developing an *Erlenmeyer flask* or club-shaped morphology.

Radial Growth

Formation modeling is the major mechanism of radial bone growth throughout life, beginning with the initial formation of the bone collar on the cartilaginous diaphysis. The rate of periosteal modeling is highest during growth and then slows during adulthood. The rapid periosteal modeling during growth is countered by resorptive modeling on the endocortical surface, resulting in a relatively consistent cortical thickness over time (Fig. 4.5). The modeling activity on the diaphyseal cortices is sexually dimorphic, both during puberty and with aging. Estrogen acts to inhibit periosteal modeling such that at puberty the amount of formation modeling is decreased in girls relative to boys. Conversely, boys have a spike in growth hormone and IGF-I during puberty and this, along with increasing levels of testosterone, stimulates periosteal growth. These gender-specific hormonal differences result in men having larger bone diameters when peak bone mass is attained. The estrogenic inhibition of formation modeling is released when women go through menopause, resulting in a brief but measurable stimulation of formation modeling and an increase in bone diameter. A number of other factors, such as mechanical loading, parathyroid hormone, and sclerostin, play major roles in dictating radial bone growth through periosteal formation modeling.

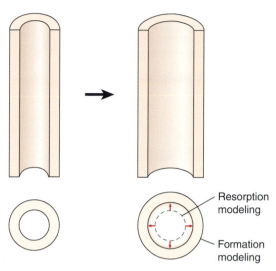

FIGURE 4.5 Radial bone growth involves formation modeling on periosteal surfaces and resorption modeling on the endocortical surface. Over time, these processes preserve cortical thickness, while increasing the width of the bone.

Bone Drift

Bone drift is a process related to radial growth but it has the distinct goal of changing the position of the cortex relative to its central axis. This process is most active during growth, although it can occur in the adult skeleton in the presence of extreme alterations in mechanical loading patterns. Bone drift occurs through the coordinated action of formation and resorption modeling on distinctive bone surfaces. In diaphyseal bone, formation modeling occurs on one periosteal and one endocortical surface, while resorptive modeling occurs on opposing periosteal and endocortical surfaces (Fig. 4.6). Bone drift can be illustrated using the example of bone curvature correction. Children with rickets have long bones that are considerably bowed. Following correction of the underlying condition that causes rickets, the bone will straighten to some extent through the process of cortical drift. Cortical bone drift also represents the main mechanism for orthodontic tooth movement, where application of spatially specific loads can be utilized to move teeth through space. The trabecular network can also undergo modeling, in which individual trabecular struts are moved in order to more effectively accommodate local strain environments (Fig. 4.7).

BONE REMODELING

Bone remodeling involves sequential osteoclast-mediated bone resorption and osteoblast-mediated

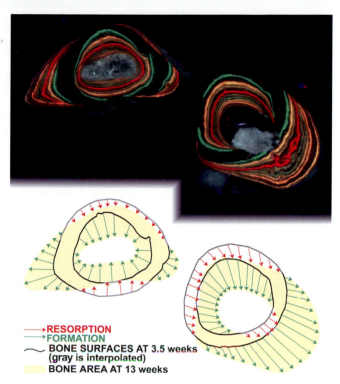

FIGURE 4.6 Using multiple fluorochrome labels, bone drift can be observed in a growing mouse. Several different fluorochromes were administered, roughly 1 week apart, to a growing mouse from age 3.5 weeks to 13 weeks, and then the radius and ulna were sectioned for histologic analysis. Radial growth occurred in the ulna (left bone) as the periosteal surfaces expanded outward, while there was also some formation modeling on the endocortical surface. The radius (right bone) underwent significant drift (down and to the right in the picture) by formation modeling on one periosteal surface and the opposite endocortical surface, along with resorption modeling on the other two surfaces. The schematic below the photomicrograph conceptualizes what the bone probably looked like at 3.5 weeks of age and how it was transformed through modeling to the eventual geometry.

bone formation at the same location. Remodeling is often referred to as a *quantum concept*, whereby discrete locations of the skeleton are replaced by quantum packets of bone through the coupled activity of osteoblasts and osteoclasts. This is the mechanism by which the replacement of bone matrix and the repair of small defects (e.g. microdamage) occur, thereby renewing the skeleton over time. Remodeling can occur upon/within any of the four bone surfaces/envelopes: periosteal, endocortical, trabecular, and intracortical. During intracortical remodeling, teams of osteoclasts and osteoblasts burrow through the matrix. This group of cells, and its associated blood vessels, is called a bone multi-cellular unit (BMU). The final product of remodeling within the cortex is an osteon, a structure composed of concentric layers of bone enclosed by a cement line with a central (haversian) canal (see Chapter 1, Fig. 1.7). New BMUs can be initiated from

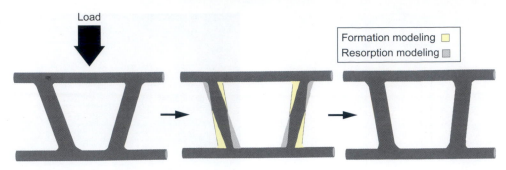

FIGURE 4.7 **Modeling occurs on trabecular surfaces in order to normalize loading-induced strains.** Areas of high strain undergo formation modeling, while regions of low strain are resorbed. These activities reorient individual trabecular structures in the direction of the principal loads.

within the bone marrow, the periosteum, or the vasculature within an existing osteon. When remodeling takes place on endocortical/cancellous and periosteal surfaces, it does not incorporate a blood vessel, and its resulting structure is called a *hemiosteon* (see Chapter 1, Fig. 1.7). Both endocortical/cancellous remodeling and intracortical remodeling are common during growth and development, as well as in adulthood, while remodeling on the periosteal surface is far less frequent at all stages of life.

Events that Signal Remodeling

Bone remodeling can be classified as targeted or stochastic. In targeted remodeling, there is a specific local signaling event that directs osteoclasts to a given location to begin remodeling. The two most accepted signaling events for targeted remodeling are microdamage and osteocyte apoptosis, although these may not be independent. Stochastic remodeling is considered a random process, with osteoclasts resorbing bone without a location-specific signaling event. Targeted remodeling serves to repair bone matrix that is mechanically compromised, while stochastic remodeling is thought to play more of a role in calcium homeostasis.

The concept of microdamage serving as a signal for remodeling was first theorized in the 1960s, when Harold Frost suggested microdamage needed to be actively remodeled by the bone in order to prevent catastrophic failure. Over 20 years later, the theory was experimentally tested using a model in which microdamage was induced in canine bone using supraphysiologic mechanical loads. Damage was imparted in one animal limb and, after one week, the same loads were applied to the contralateral limb. Histologic analyses documented that the levels of microdamage in both limbs were similar and significantly higher than

in the limbs of nonloaded animals. Most importantly, the limb initially loaded had significantly more resorption cavities than did the contralateral limb, and these cavities were spatially associated with the microdamage. These results provided the first evidence that microdamage serves as a signal for targeted bone remodeling.

Subsequent to these initial large animal studies, several experiments have advanced our understanding of the mechanism underlying microdamage-induced remodeling. These experiments have taken advantage of the fact that rats and mice do not normally exhibit intracortical remodeling. Thus, through various interventions (such as supraphysiologic mechanical loading to induce microcracks or genetic manipulations to ablate osteocytes), intracortical remodeling can be linked to specific causative factors. It is important to realize that although rodents do not normally undergo intracortical remodeling, they still have remodeling on the other bone envelopes similar to that of other animals and humans.

These rodent studies have shown that microdamage results in localized disruption to the osteocyte network via physical breakage of the cytoplasmic connections between cells. These osteocytes are then cut off from the remainder of the network and begin to undergo apoptosis. Prior to dying they begin to actively produce factors, such as tumor necrosis factor ligand superfamily member 11 [receptor activator of the NF-κB ligand (RANKL)], a key factor in osteoclast development. In addition, the cells more distant from the microdamage produce strong antiapoptotic signals [such as/tumor necrosis factor receptor superfamily member 11B (OPG)]. This pattern of signaling by the viable and dying osteocytes probably serves as a *target* for the osteoclasts to begin their remodeling activity (Fig. 4.8). When microdamage is produced and osteocyte apoptosis is inhibited through pharmacological intervention, remodeling is also inhibited.

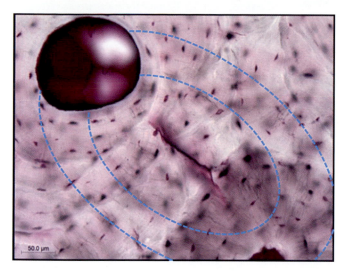

FIGURE 4.8 Regions around microdamage have both pro- and antiremodeling signals that are involved in targeted remodeling. Osteocytes near the microcracks express high levels of RANKL and low levels of OPG, thus favoring osteoclast recruitment. The osteocytes farther away express low levels of RANKL and high levels of OPG. This is believed to serve as a *target* for the osteoclasts to know which bone to remodel.

Alternatively, when the osteocyte network is disrupted in the absence of microdamage, such as through loss of estrogen, mechanical disuse, or glucocorticoid excess, intracortical remodeling is enhanced in association with osteocyte apoptosis. Collectively, these studies illustrate that although microdamage leads to targeted remodeling, osteocyte apoptosis is the critical event in the process.

Historically, calculations estimating the balance between targeted and stochastic remodeling have used the assumption that microdamage was the targeting event. Mathematical models based on experimental data have calculated that about 30% of remodeling is targeted to microdamage. The recent evidence that osteocyte apoptosis is probably the key event suggests that targeted remodeling, albeit to something other than microdamage, is likely to form an even greater percentage of total remodeling. Indeed, osteocyte apoptosis may be a key precondition for remodeling to occur on any surface. This considerably blurs the distinction between targeted and stochastic remodeling, which are probably better demarcated by the initiating event (local versus systemic) and by function (repair versus mineral homeostasis).

Remodeling Cycle

Whether or not the remodeling is targeted or stochastic, the cellular events are similar. The process of remodeling is divided into five stages: activation, resorption, reversal, formation, and quiescence. This is often referred to collectively as the remodeling cycle (Fig. 4.9). At any given time, there are thousands of remodeling cycles taking place throughout the body. These cycles are at various stages, depending on when they were initiated. The entire remodeling process normally takes roughly 4–6 months in humans, although this can be highly altered by disease (see Chapter 7).

Activation

The activation stage represents the recruitment of osteoclast precursors to the bone surface followed by their differentiation and fusion to become fully functional osteoclasts. The process of osteoclast differentiation and maturation is outlined in Chapter 2.

Resorption

Once mature osteoclasts are present, bone lining cells retract from the surface to expose the mineralized matrix to osteoclasts. This appears to be an active process and is stimulated either by the osteoclasts themselves as they approach the surface or by the same signals that initiated the remodeling. Without retraction of the bone lining cells, osteoclasts are unable to bind to the bone and begin resorption. Upon attachment, the osteoclasts actively dissolve the mineral and liberate collagen fragments. These fragments can be measured in the blood and urine, thus providing useful biomarkers for the assessment of bone remodeling.

As resorption proceeds, new osteoclasts can be recruited to the remodeling site to either support existing osteoclasts or replace those that die. There is significant variability in the size of individual remodeling sites on endocortical and trabecular surfaces. Regulation of this variability is well not understood. Intracortical radial resorption spaces, which can be quantified by measuring osteon diameter, are relatively consistent in size. Osteon length, on the other hand, ranges from several hundred micrometers to several millimeters.

Reversal

The reversal phase is characterized by the cessation of osteoclast resorption and the initiation of bone formation. The signal for reversal within BMUs is unknown, although several theories exist. Direct cell-cell interaction between osteoclasts and osteoblasts (or their precursors) may induce signaling for cessation of one cell type and activation of another. The discovery of ephrin extracellular proteins on both osteoblasts (EphB4) and osteoclasts (ephrin-B2) supports this theory, although experimental evidence of direct contact between cells is lacking. Another plausible mechanism (also theoretical) is the presence of factors released from the bone matrix during resorption [e.g. transforming growth factor beta (TGF-β) and BMPs] that stimulate osteoblast migration and differentiation.

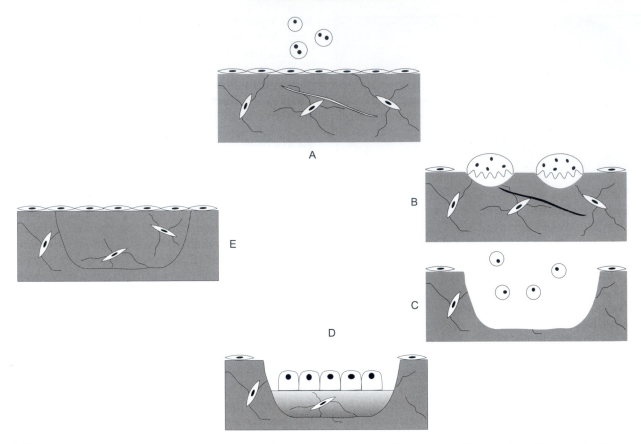

FIGURE 4.9   The remodeling cycle. The remodeling cycle involves five stages: (A) activation; (B) resorption; (C) reversal; (D) formation; and (E) quiescence. At any one time, remodeling cycles throughout the body are at various stages.

However, this theory fails to explain how osteoblasts are specifically signaled to the remodeling site, when the liberation of these molecules would probably be widespread in the nearby marrow. Recent descriptions of the *remodeling canopy*, which creates an underlying bone remodeling compartment (BRC) and provides a mechanism to localize these liberated factors, seem to address this limitation. The remodeling canopy is a physical structure, made up of bone lining cells, under which the remodeling unit exists (see Chapter 2, Fig. 2.10). The junctional complexes that link adjacent bone lining cells may allow some factors to be exchanged between the remodeling compartment and the outside environment, while maintaining appropriate molecular concentrations within the compartment.

Once osteoclasts have finished resorbing bone, the remaining collagen fragments on the exposed surface must be removed. It is currently thought that this is done by a specialized form of bone lining cell. If such fragments are not removed, then bone formation by osteoblasts does not proceed. These specialized cells are also thought to deposit a thin layer of new bone matrix (the cement or reversal line), a clear histologic feature that delineates the boundaries of osteons and hemiosteons from the surrounding, older matrix. The

cement line is rich in proteoglycans, such as osteopontin. Controversy exists regarding the composition of the cement line; specifically, whether it is highly or minimally mineralized. Regardless, it is widely accepted that cement line mineralization differs from the surrounding bone and that this plays an important role in its mechanical properties.

Formation

During the bone formation stage osteoblasts lay down an unmineralized organic matrix (osteoid), which is primarily composed of type I collagen fibers and serves as a template for inorganic hydroxyapatite crystals. Osteoid mineralization occurs in two distinct phases. Primary mineralization, the initial incorporation of calcium and phosphate ions into the collagen matrix, occurs rapidly over 2–3 weeks and accounts for roughly 70% of the final mineral content. Secondary mineralization, the final addition and maturation of mineral crystals, occurs over a much longer time frame (up to a year or more) (see Chapter 1, Fig. 1.6).

The osteoblasts participating in new bone formation undergo one of three fates. The majority (90%) die through apoptosis. These are replaced by new osteoblasts as long as formation is still necessary at the local site. Another fraction of osteoblasts is incorporated into the osteoid matrix and eventually become osteocytes. The osteoblasts remaining at the conclusion of formation remain at the bone surface as inactive bone lining cells. These cells retain the capacity to become activated and begin producing bone matrix again.

Quiescence (Resting)

At the completion of a bone remodeling cycle, the resulting bone surface is covered with bone lining cells. The matrix within the remodeling unit will continue to mineralize over time. At any given time, the majority of bone surfaces within the bone are in a state of quiescence.

Bone Remodeling Cycle Duration

Absent of pathology, a complete remodeling cycle takes about 4–6 months from the time of activation to the time that osteoblasts finish producing matrix (Fig. 4.10). Mineralization of the matrix continues for months after production. This time is not routinely taken into account when assessing remodeling cycle duration. The duration of a remodeling cycle is not evenly divided between resorption and formation. Osteoclasts typically resorb bone for 3–6 weeks (at a given site), with the remainder of the cycle comprising

bone formation. Consequently, when one looks at bone under the microscope, it is much more common to find formation sites than resorption sites (by a ratio of about 4:1). The duration of a remodeling cycle is altered in a number of diseases (some of which are detailed in Chapter 7).

Bone Remodeling Rate

The rate of bone remodeling is very high during growth and then slowly decreases until peak bone mass is attained. In adulthood, the rate is highly variable and is influenced by age and genetics, as well as a number of modifiable factors such as physical activity, nutrition, hormonal activity, and medications. In females, remodeling increases at menopause due to the loss of circulating estrogen, which normally acts to suppress remodeling through direct effects on osteoclasts and the suppression of osteoclast apoptosis. The increase in remodeling is progressive in the years following menopause. Individuals who take hormone replacement therapy can offset this increase in remodeling, as can those who take antiresorptive pharmaceutical agents (see Chapter 17). Men experience less dramatic increases in remodeling, and these typically begin to occur about a decade later than the increase observed in women. Eventually, with age (around the

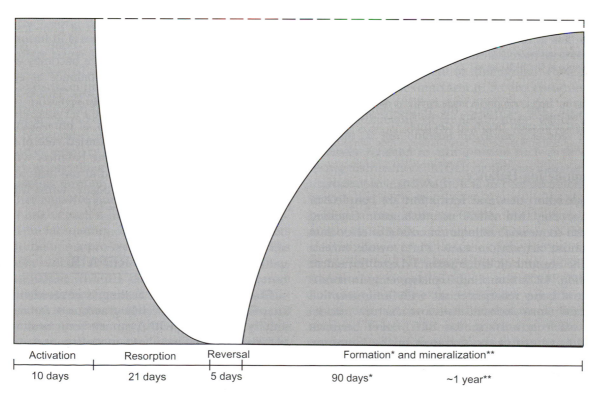

Activation	Resorption	Reversal	Formation* and mineralization**
10 days	21 days	5 days	90 days* ~1 year**

FIGURE 4.10 The stages of bone remodeling occur over different time frames, with the formation phase taking 4–5 times longer than the resorption phase. Final mineralization of the newly formed bone can take up to 1 year.

STUDY QUESTIONS

1. Compare and contrast intramembranous and endochondral ossification, making special reference to which cells are involved in each process, and how the final product of each process differs.

2. Describe the process of skeletal maturation at the growth plate. What are the major characteristics of each region, and how does this facilitate transition from the original cartilage model to mature skeletal tissue?

3. Differentiate between bone modeling and bone remodeling. List and describe the various functions of each.

4. Describe the five major stages of bone remodeling. Which cells are involved and how do they communicate with one another? What events lead to the transition from one stage to the next?

5. How does bone remodeling affect the material properties of the skeleton?

6. What methods are used to assess bone remodeling in a clinical setting, and what aspect of remodeling do they detect? List the advantages and disadvantages of each method.

Suggested Readings

Frost, H.M., 1963. Bone Remodeling Dynamics. CC Thomas, Springfield, IL.

Frost, H.M., 1986. Intermediary Organization of the Skeleton, vols. I and II. CRC press.

Hall, B., 2005. Bones and Cartilage: Developmental and Evolutionary Skeletal Biology. Academic Press.

Henriksen, K., Neutzsky-Wulff, A.V., Bonewald, L.F., Karsdal, M.A., 2009. Local communication on and within bone controls bone remodeling. Bone. 44, 1026–1033.

Parfitt, A.M., 1994. Osteonal and hemiosteonal remodeling: the spatial and temporal framework for signal traffic in adult bone. J. Cell Biochem. 55, 273–286.

Parfitt, A.M., 2002. Targeted and nontargeted bone remodeling: relationship to basic multicellular unit origination and progression. Bone. 30 (1), 5–7.

Recker, R.R., 1983. Bone Histomorphometry: Techniques and Interpretation. CRC Press.

Robling, A.G., Castillo, A.B., Turner, C.H., 2006. Biomechanical and molecular regulation of bone remodeling. Annu. Rev. Biomed. Eng. 8, 455–498.

ASSESSMENT OF BONE STRUCTURE AND FUNCTION

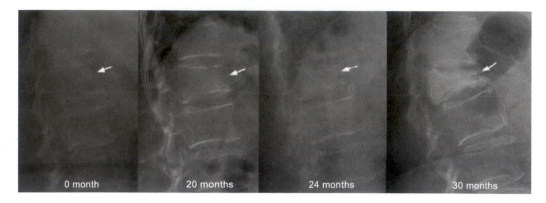

FIGURE 5.1   X-ray images represent the standard method for assessing vertebral fracture. In lateral radiographs, incremental fracture of the T12 vertebral body can be observed. No fracture is present at baseline (arrow); a mild (20—25% loss of height) vertebral fracture is present at 20 months (arrow); a moderate (25—40% loss of height) vertebral fracture is present at 24 months (arrow); and a severe (>40% loss of height) vertebral fracture is present at 30 months (arrow). *Reproduced with permission from James F. Griffith, et al., Radiologic Clinics of North America, 48, 3, 2010, 519—529.*

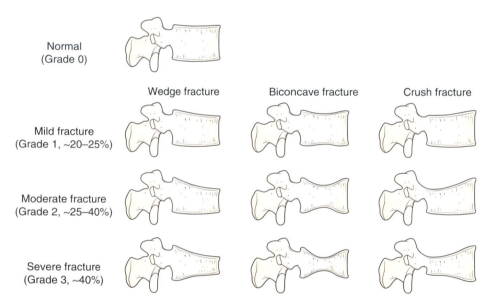

FIGURE 5.2   Genant semiquantitative method for vertebral fracture assessment. This method distinguishes vertebral fracture by comparing radiographs with a standard chart. Classification of the amount of height loss (mild, moderate, severe) as well as whether the fracture is a wedge (height loss at the anterior region), biconcave (height loss in the center), or crush (height loss at the posterior region) fracture.

radiation, the effective radiation dose from a single lateral lumbar spine scan is lower than that of a CT scan. The low radiation is due in part to the very rapid scan time. The resolution of X-ray is often low, but high-resolution radiography with spot sizes down to 10 μm can be obtained if a limited region of interest is scanned. As with all imaging methods, increasing the resolution results in a greater radiation exposure to a given region.

Lateral radiographic images remain the standard approach for diagnosing vertebral fractures. The most common method for the assessment of vertebral fracture by X-ray is the Genant semiquantitative method. This method distinguishes six grades of vertebral fractures by comparing radiographs with a standard chart (Fig. 5.2). Vertebrae are assessed for height loss using semiquantitative criteria and the fracture is then classified as either a wedge (height loss at the anterior region), biconcave (height loss in the center), or crush (height loss at the posterior region) fracture. Although this technique is useful, it has limitations for diagnostic purposes because the images represent 2D projections and the classification is only semiquantitative. In clinical research, a common approach is to have two independent readers look at the lateral spine X-rays in a blinded manner. They can use the semiquantitative method, which is purely visual assessment, or they can use measurement tools on X-rays or digital images which enable relatively accurate measurements of vertebral heights. If there is disagreement between the two readers, then a third reader is used as a tie-breaker. Having multiple readers has been valuable in diagnosing mild fractures that may often be considered as anatomical variants. However, there is little or

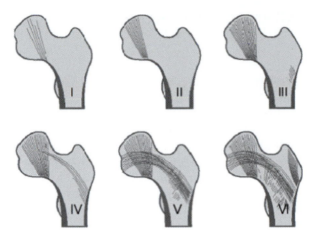

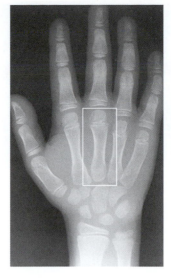

FIGURE 5.3 Scheme of changes in trabecular pattern of the proximal femur. Image shows Singh index values from I to VI. *Reproduced with permission from O. Hauschild, et al., European Journal of Radiology, 71, 1, 2009, 152−158.*

FIGURE 5.4 Radiogrammetry, the morphological quantification of bone from X-rays, is most easily applied to cortical bone of the metacarpals with measures of mid-diaphyseal periosteal and endosteal diameters (noted by black lines superimposed on the image) and calculation of cortical thickness. *Reproduced with permission from John A.M. Taylor, Tudor H. Hughes, Donald Resnick, Ch. 17, Wrist and Hand, Skeletal Imaging (Second Edition), W.B. Saunders, Saint Louis, 2010, Pages 972−1067.*

no benefit for multiple readers in diagnosing moderate or severe vertebral fractures.

Plain X-rays are commonly used to classify and track healing of long bone fractures. Some bones form significant callus at the healing site. Often a radiographic measurement of fracture healing is the presence of cortical bridging across the fracture line (see Chapter 10).

The Singh index is a grading system used to classify the pattern of trabecular bone in the femoral neck (Fig. 5.3). The index is a six-point scale ranging from VI (best) to I (worst). In grade VI, trabecular bone is present throughout the proximal femur region. Grade V has a prominent Ward triangle region, with an accentuation of trabeculae in regions of principal tensile and compressive stresses. In grade IV, the principal tensile trabeculae are reduced in number, although some still exist spanning from the lateral cortex to the upper femoral neck. Grade III is often considered to be osteoporotic and is characterized by lack of continuity in trabeculae. In stage II, only the principal compressive trabecular are prominent. Stage I has very few trabeculae. The Singh index is often considered to be too variable for an accurate diagnosis of bone loss, although it remains a useful semiquantitative tool.

Radiogrammetry, the morphological quantification of bone from X-rays, has been in use for skeletal assessment since the early 1960s. Its main use was in assessing cortical thickness and bone geometry (e.g. periosteal expansion), although it is rarely used these days because more advanced techniques now exist. This technique is most easily applied to cortical bone, often on the metacarpals. Although several morphological parameters can be assessed, the key parameters that are often measured include mid-diaphyseal

periosteal and endosteal diameters, with a calculation of cortical thickness (Fig. 5.4). Early radiogrammetric data was generated using caliper measures on the developed film but, more recently, digitized images mean that these measures can be done with computer software.

BMD can be estimated from plain X-ray images, with a 'brighter' image representing a more dense bone. Qualitatively, changes in BMD need to be in the range of 20−40% to enable visual detection of a difference in density. If a standard phantom with known densities is scanned simultaneously, then a quantitative assessment can be made which facilitates greater ability to detect differences. Still, the advent of more sophisticated methods for assessing bone density has made these measures from X-rays quite obsolete in the clinic.

Radiographic assessment of the skeleton remains a useful tool in the laboratory due in part to its speed and ease of use. Bone phenotypes, such as overall structure and density, can be quickly obtained either on whole animals or on specific bones. As with clinical scans, large differences in bone density are needed to allow detection with X-rays, although this is easily achievable in the case of some treatments or genetic manipulations (Fig. 5.5). It is not uncommon to see radiographs associated with descriptions of transgenic animal phenotypes, especially if there are large differences in animal size or morphology, or if there are dental abnormalities. In most cases, these images are

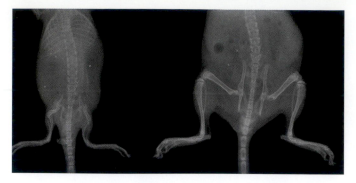

FIGURE 5.5 X-ray imaging is used in the laboratory to provide qualitative assessment of skeletal morphology of animals. This technique can often provide the necessary information to assess gross morphological differences in skeletal structure, such as differences between genetic strains of mice.

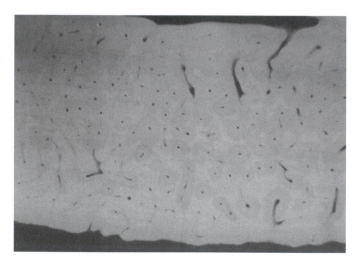

FIGURE 5.6 Microradiography, i.e. X-ray images of prepared bone sections, can be used to assess mineralization patterns of bone, with lighter regions depicting regions of high density and darker regions those of low density. Image of rabbit tibia cross-section.

not quantitatively assessed and thus other imaging methods are also employed to provide a more complete picture of the phenotype.

An adapted use of X-ray assessment, microradiography (sometimes called contact radiography), utilizes X-rays to view prepared bone sections, often with a thickness of 50–100 μm. This imaging technique can provide in-plane resolution up to 10 μm and allow detailed assessment of density throughout various regions of bone when calibrated with step-wedge phantoms. Microradiography is often used to assess bone mineralization patterns (Fig. 5.6), with lighter regions depicting regions of high density and darker regions those of low density.

ABSORPTIOMETRY (PHOTON AND X-RAY)

Photon absorptiometry was first described in the early 1960s. This imaging modality is based on the known relationship between photon attenuation and tissue density. A very dense tissue, such as bone, will absorb a significant number of protons, while soft tissue will absorb less. Because soft tissue absorbs some protons, but it was impossible to determine how many, scanning required the site of interest to be submerged in water (which similarly absorbs protons) so that the thickness of the scan could be kept constant over time (and between individuals). Site submersion limited the utility of single photon absorptiometry to peripheral sites (often the forearm) and eventually led to the development of dual photon absorptiometry. Adding a second photon source of a different energy allowed soft tissue and bone tissue to be separately assessed. Submersion in water was no longer needed and central sites such as the spine and proximal femur could be evaluated. Dual photon absorptiometry allows assessment of isolated regions in about 30 min or the whole body in about an hour. The main limitation of photon absorptiometry is that it necessitates specialized equipment that is often found only in nuclear medicine departments. The source of the gadolinium required renewal on a regular basis, which limited its use in the clinic.

In the early 1980s, the concept of photon absorptiometry was adapted to use X-rays instead of photons. The principle of X-ray absorptiometry is similar to that of photon absorptiometry: X-rays are attenuated in proportion to tissue density. The advantage of X-rays as a source was that it eliminated the need for radioactive isotopes and thus increased the ease of scanning and eliminated the need to replace the gadolinium source. The first X-ray absorption scanners, using single energy X-ray absorptiometry (SXA or SEXA), had a single energy source and detector, and were limited to peripheral sites (such as the calcaneus) due to the need to submerge tissue in water to control for soft tissue photon absorption. This technique is now mostly obsolete due to the advent of DXA, which eliminates the need for water submersion and thus allows scanning of central bone sites and/or total body (Fig. 5.7).

Early DXA scanners were equipped with a pencil beam with a single detector. Pencil beams have a narrow X-ray beam that moves in concert with the detector to scan the region of interest. Scan times with these scanners were roughly 5–10 min and had a resolution of 1 mm. Current DXA scanners employ fan beam sources with a bank of detectors. This allows a larger field of view to be scanned at one time and has

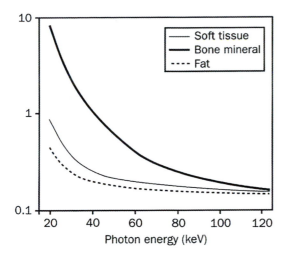

FIGURE 5.7 **Dual-energy X-ray absorptiometry can be used to assess bone density.** Dual-energy X-ray absorptiometry works by exposing the tissue to two different energy sources, one that can be attenuated by soft tissue and fat and the other by bone.

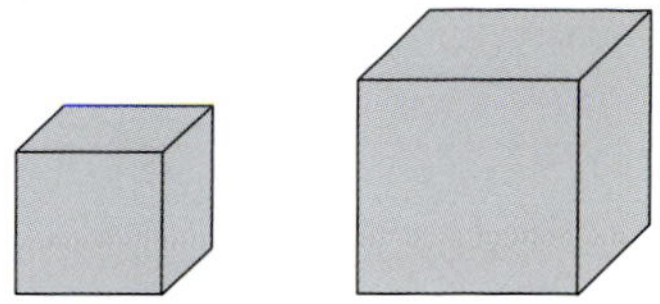

	1 cm³ cube	1.5 cm³ cube
Cube size	1 cm³	1.5 cm³
Volumetric bone mineral density, grams/cm³	1.0 g/cm³	1.0 g/cm³
DXA bone area, cm²	1 cm²	2.25 cm²
DXA bone mineral content, grams	1 g	3.375 g
DXA bone mineral density, grams/cm²	1 g/cm²	1.5 g/cm²

FIGURE 5.8 Bone density assessed by dual-energy X-ray absorptiometry (DXA) does not account for 3D bone size. As depicted, in a cube with a constant volumetric density (1.0 g/cm³), if the area is larger, the areal bone mineral density (aBMD) assessed by DXA is also larger.

resulted in scan times being reduced to about 1 min/site, with resolution of 0.5 mm. Although the resolution of DXA is inferior to that of standard radiography, this is overshadowed by the significantly lower radiation dose. Newer DXAs have improved resolution and are used to assess vertebral fracture in addition to BMD. Normal background radiation to the human body is estimated at 7 microsieverts (μSv)/day. While X-ray exposure is about 600 μSv, the dose from a typical DXA is about 3 μSv.

DXA provides a projection image of the bone and calculates BMC, bone area, and BMD based on these images. It is important to recognize that the BMD values obtained from DXA are areal density (g/cm²) as opposed to a volumetric density (g/cm³). That is, the BMD is calculated by dividing the BMC of a given bone region by its area, with no consideration of bone depth (Fig. 5.8). Take, for example, two objects that are composed of the same material (that is to say they have the same volumetric density), yet their sizes differ by 50%. Using DXA imaging, these objects would have a 50% different areal density. This is a limitation for comparing DXA-based BMD among individuals with different bone sizes, but in cohorts with similar bone sizes (or tracking an individual longitudinally when bone geometry is not expected to change), it is less of an issue.

The standard clinical sites of DXA assessment are the lumbar spine and proximal femur. The distal radius can be scanned but this is less frequent and is primarily done when either the lumbar spine or proximal femur cannot be used due to degenerative changes, fractures, or metal hardware. Thoracic spine, although clinically relevant as a fracture site, is not assessed because of artifacts caused by the overlying ribs. Proximal femur scans encompass roughly the proximal a third of the femur. Measurement values are routinely given for the total hip and the femoral neck (Fig. 5.9). The typical clinical assessment of the vertebra is the average value across L1-L4 levels, but data is presented for each vertebra and individual ones can be excluded if necessary (Fig. 5.9). Typically, at least two of the four vertebrae need to be included to obtain a valid measurement. Techniques are also available to assess vertebral fracture (using criteria similar to the Genant method, based on radiographs); although this requires a separate scan from the normal one obtained for BMD evaluation, it can be done sequentially. This technique has been shown to have a lower sensitivity for mild fractures compared to radiographs but higher sensitivity for moderate to severe vertebral fractures. The key advantage is that the patient has both their BMD and vertebral fracture assessment completed with minimal radiation and a short (approximately 10 min) examination.

Data generated from DXA are reported both as raw data (BMD, BMC, and area) and as T-scores and Z-scores (Fig. 5.10). A T-score represents the number of standard deviations of a given individual above or below the average BMD value for a young adult population. Conversely, the Z-score represents the number of standard deviations of a given individual above or below the average BMD value for an age-matched population. Because of the widespread use of DXA, there are large databases with normative data for the calculation of both T- and Z-scores. There are normative databases for both men and women. When

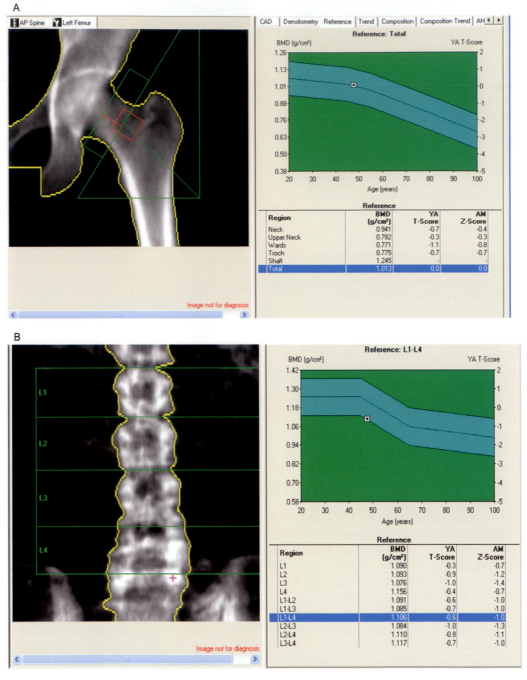

FIGURE 5.9 Dual-energy X-ray absorptiometry-based assessment of vertebral (A) and femoral neck (B) bone mineral density (BMD). Standard scanner software is used to identify various subregions and then BMD is provided for each site. Reference values, both T-scores and Z-scores, are often provided for each individual scan.

calculating a T-score for men, a male normative database is used. The World Health Organization (WHO) has created BMD criteria for the diagnosis of osteoporosis in postmenopausal women. A T-score of > -2.5 standard deviations is considered to indicate osteoporosis. BMD with T-score between -1 and -2.5 is considered to indicate low BMD or osteopenia. Although BMD has a strong correlation with fracture risk, the limitation of using BMD as the sole method to assess risk has been illustrated through several studies showing that over half of individuals who fracture do not have low BMD as measured by DXA (see Chapter 16, Fig. 16.1). Clinical diagnosis of osteoporosis would therefore include patients with classical fragility fractures even if their BMD does not meet the WHO criteria of a T-score of -2.5 or lower.

Successful DXA measures depend on a skilled technician to conduct the scan and analyze the data. Positioning the subject properly and in a standardized position in the scanner significantly affects the ability to visualize the bones during the analysis and to calculate BMD in a manner that allows comparison to the standard databases. During analysis, although the computer software will automatically define the various regions of interest, it is necessary to review these to assure accurate placement (computers do make mistakes). Improper placement of the region of interest will significantly affect the bone area and thus could either under- or overestimate BMD. It is also important to assess for any artifacts within the scan such as osteophytes, the presence of a vertebral fracture (same amount of bone compressed into a smaller area), metallic implants, or aortic calcifications in the field of view.

The low radiation and ease of use has resulted in the widespread use of DXA technology in animal research. Small animal DXA machines are routinely used to make longitudinal in vivo measures of BMD in rodents (Fig. 5.11), as well as ex vivo assessments of bones from both smaller and larger animal models. In whole body scans, individual bones (or segments of vertebrae) can be isolated in a fashion similar to that done in humans. Technical advances in other imaging modalities, such as CT, have reduced the reliance on DXA as a primary way to characterize phenotypes of genetically modified mouse models, although DXA does remain the principal technique for longitudinal assessment of human and animal bone, with the advantages of rapid scan times, low radiation, and whole body scanning capacity (compared to CT).

DXA has important limitations as an imaging modality. Cortical and cancellous bone cannot be differentiated in DXA scans. Because the images are projections, even analysis of regions of interest rich in cancellous bone (such as the trochanteric region of the femur or central vertebra) still have cortical bone incorporated into the analysis. It also does not allow assessments of architecture and geometry, two key factors that play a role in the overall assessment of bone health. The inability to account for 3D geometry can also artificially affect the determination of BMD (see Fig. 5.8).

Techniques to overcome these limitations have focused on estimating bone geometry and volumetric density from DXA images. Variables such as hip axis length and neck-shaft angle can be calculated from the projection images and these parameters have been found to be associated with fracture risk in large

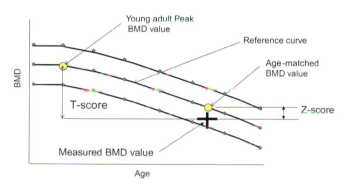

FIGURE 5.10 Graphical depiction of T- and Z-scores for bone density. See text for details. BMD, bone mineral density.

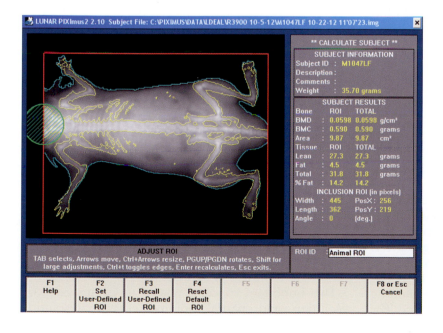

FIGURE 5.11 Dual-energy X-ray absorptiometry (DXA)-based assessment of bone mineral density (BMD) can be conducted in the laboratory for in vivo or in vitro measures. Although the limitations of human DXA remain (e.g. limits to accounting for geometry) and more sophisticated methods for assessing in-vivo properties exist, it is still a commonly applied technique for longitudinal assessment of bone properties in rodents. Whole body scans with analysis of BMD (excluding head and tail) as well as individual bone analysis (long bone and vertebral segment), can be obtained.

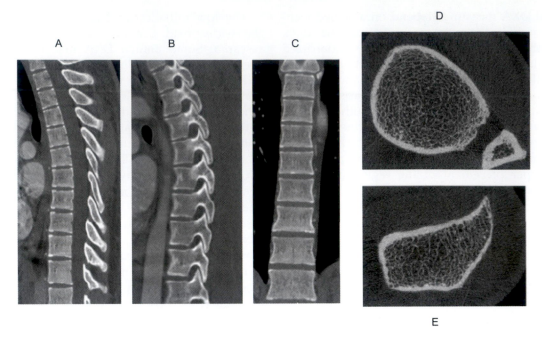

FIGURE 5.12   Computed tomography scans can be used to obtain volumetric density and geometric parameters of the vertebra (A, sagittal; B, parasagittal; C, coronal) and peripheral sites (D, distal tibia; E, distal radius). Peripheral sites can be scanned with higher resolution in order to allow visualization of individual trabeculae. *Panels A-C reproduced with permission from Mark W. Anderson, Seminars in Spine Surgery, 22, 1, 2010, 8–19. Panels D and E reproduced with permission from Jan S. Bauer, et al., European Journal of Radiology, 71, 3, 2009, 440–449.*

populations. A more comprehensive approach to estimating bone geometry is through hip structural analysis (HSA). These algorithms estimate both cross-sectional area and cross-sectional moment of inertia of the neck, trochanter, and shaft. It is important to remember that these calculations are estimates, and assume a circular geometry. Volumetric DXA (VXA or 3D-XA) is a step toward volumetric assessment of BMD from DXA scanners. In this method, two orthogonal DXA images are obtained and then compared within a database of 3D CT datasets to infer bone geometry. Volumetric bone density is then interpolated.

COMPUTED TOMOGRAPHY

First introduced in the 1970s, CT utilizes a series of 2D X-ray images to produce cross-sectional images (often called slices). These image slices have a known thickness and thus the data generated represent a volume of tissue. This volume, which results in a 3D data set, offers several advantages over 2D imaging techniques. CT scans can be used to assess bone geometry, have sufficient resolution to differentiate between cortical and cancellous regions, are not subject to superimposition of objects (such as ribs over the thoracic vertebra), and can assess tissues other than bone. For

the prediction of fracture risk, several studies have documented the superior ability of CT compared to DXA at both the vertebra and proximal femur.

Clinical whole body CT machines, used to assess vertebrae or hip, have a typical in-plane pixel size of 200–500 μm and slice thickness of around 1 mm (Fig. 5.12). Smaller CT units designed to assess peripheral sites, such as the distal tibia or radius, have typical in-plane pixel size of 100–300 μm, with slice thickness of about 1 mm. Newer generation multidetector whole body machines are able to achieve an in-plane pixel size of approximately 250 μm and slice thicknesses of 300 μm. High-resolution peripheral CT machines can achieve an in-plane pixel size of <100 μm.

The main limitation of CT is radiation exposure. The amount of radiation exposure depends on several factors, including the duration of scan (which depends on resolution and number of slices) and specifics about the machine. Radiation exposure for a typical central CT scan approximates 10 mSv, equivalent to 3–4 years of background radiation exposure and roughly 15 times higher than a typical radiograph. Radiation from the peripheral scanners is much lower. Motion artifact (from patient movement) is also a concern with CT imaging, although this can be minimized by skilled technicians administering the scans.

Two important concepts related to CT imaging are resolution and partial volume effects. Resolution

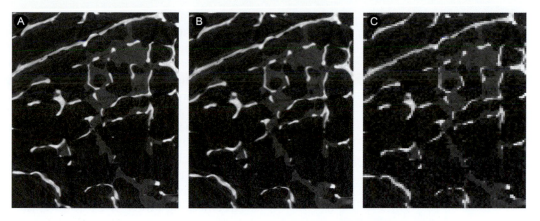

FIGURE 5.13 **Understanding the difference between pixel size and resolution.** Pixel size represents the size of the pixels during image acquisition. Resolution represents the pixel size after postimage processing. To illustrate this, an image obtained and reconstructed with a pixel size of 12 μm is shown in (A). If the original scans are reconstructed with pixel sizes 2× larger, the image in (B) is the result (which has a resolution of 24 μm). If the original image is reconstructed with a pixel size 8× larger, having a final resolution of 96 μm, the image in (C) is the result.

quantifies how close together two objects can be yet still be differentiated (resolved). The resolution capability of an imaging modality is ideally reported as its modulation transfer function (MTF), an algorithm that describes the resolution and performance of an optical system by taking into account the source, camera, and pixel size. More commonly, resolution is reported as *nominal resolution*, which is defined as the linear voxel size used to display the image. Unfortunately, pixel size is often reported in substitution for resolution. Although pixel size may be equivalent to nominal resolution, postprocessing of images can result in an image that has a large number of pixels and poor resolution. Take for example a high-resolution CT scan of a bone obtained using a pixel size of 12 μm (Fig. 5.13). Reconstruction of this image using the scanning pixel size produces an image with a nominal resolution of $12 \times 12 \times 12$ μm^3. However, if the original scans are reconstructed with larger pixel sizes, which may be done for various reasons including to reduce the overall scan file size for applications such as 3D modeling, the nominal resolution can become increasingly large. In this case, reporting the pixel size would be misleading, as the spatial resolution, which is what data analysis is based on, is much lower. Although it would be ideal to scan at the highest resolution possible on a given machine, limitations such as scan time and radiation exposure can limit the resolution that is acceptable for a given scan.

An important concept related to pixel size, and relevant for CT imaging, is partial volume averaging. This occurs when a particular pixel contains tissue of two (or more) different densities (Fig. 5.14) and the voxel is therefore assigned an average value. Partial volume averaging comes into play at the interface between the periosteal bone surface and soft tissue, the interface between the endocortical bone surface and bone marrow, and within the cancellous bone between bone trabeculae and marrow. It is important to be aware that when thresholds are used to separate the various compartments, partial volume effects will either over or underestimate areas. Smaller pixel sizes minimize partial volume effects but cannot eliminate them.

The primary sites of CT analysis include vertebra (lumbar are most common), proximal femur, radius, and tibia. CT data are generated as Houndsfield units (HU), a standard unit of measure for radiodensity that describes the linear attenuation coefficient of the X-ray. The HU of water, a standard calibrator, is 0 (air is −1000). The HU of bone matrix approximates 1000. Using standard phantoms of known density, HU data from the CT can easily be converted into BMD (mg/cm^3).

CT imaging generates data on both density and geometry. In a similar fashion to DXA, CT analysis involves identification of bone area and BMC, with a calculation of bone density. Because CT has a higher resolution than DXA, it can provide a more accurate measure of bone area; since the images have a known thickness, a volumetric density can also be obtained. The high resolution of CT scans also allows a more detailed assessment of the interface between the cortex and trabecular bone regions, which allows these compartments to be delineated (either manually or using segmentation algorithms) so that data can be collected for the individual cortical and cancellous bone compartments (Fig. 5.15). With the exception of

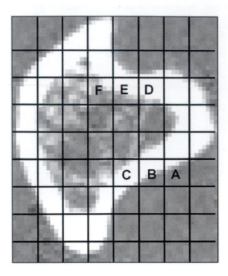

FIGURE 5.14 Partial volume averaging affects bone density and geometric measures from computed tomography (CT). Each pixel in a CT scan contains bone (depicted in white), soft tissue/marrow, or a combination of bone and soft tissue/marrow. In this example, the pixels (boxes) are quite large relative to the bone size (this makes it easier to illustrate the concept). In those pixels with two tissues having different densities, the pixel is assigned an average density (or grayscale value). For example, in pixel A, although there is a small amount of cortical bone, this pixel is mostly soft tissue and will have a low density. Therefore the bone in this pixel will be effectively ignored (making the assessed bone size lower than it actually is). Pixel B is half cortical bone and half soft tissue, and will either be counted as bone or ignored, depending on what the threshold is set at for determining the periosteal boundary. Pixel C is almost completely cortical bone. Pixel D will be considered cortical bone while, depending on the threshold for defining the endocortical boundary, pixel E will either be considered cortex or part of the cancellous region. Pixel F will probably be considered as part of the cancellous region under most thresholding criteria. If pixel E is considered cortex, then it will lower the density of the cortex, as it will have a density value that is half cortical bone and half marrow.

high-resolution peripheral CT machines (those with pixel size <100 μm), clinical CT scanners are not able to 'see' individual trabeculae and do not provide data about architecture. Furthermore, the density values for cancellous bone represent an integrated density of the trabecular bone plus the marrow. This is why cortical bone density values are about 1000–1200 mg/cm^3, whereas cancellous BMD values are often around 300 mg/cm^3—only about a third to a quarter of the cancellous region contains bone. Cortical thickness can be assessed in regions where it is greater than the pixel size (such as the tibia or radius diaphysis). A quantitative assessment of muscle area can also be made using CT scans.

The resolution of peripheral CT (<100 μm) has significantly advanced our understanding of the differential response of cortical and cancellous bone in various conditions. At the ultradistal radius, young adult men have higher cancellous bone volume, driven by higher

trabecular thickness without a difference in trabecular number, relative to women of a similar age. There is no significant gender difference in young adults in cortical thickness or cortical porosity, but overall cortical bone is significantly greater in males because their bones are larger. Age-related changes over seven decades in overall cancellous bone volume are similar in both men and women, but the loss in women is driven by a loss of trabecular number while the loss in males is due to declines in trabecular thickness. Females also lose more cortical thickness and have a greater increase in cortical porosity with age than do males.

An emerging utility of CT scans is the ability to generate finite element models (FEMs) and perform finite element analysis (FEA) that can be used to estimate bone mechanical properties including strength (Fig. 5.16). FEM is a common technique used in the engineering field to provide estimates of mechanical properties for objects with complex geometries. Objects are represented in the model as elements (nodes), with each node being given a material property value and the overall model assigned certain boundary conditions (load and displacement levels). After creating the model with each node (usually a voxel from the CT image) having an assigned material property, a virtual load is placed upon the bone. For lumbar vertebrae, the computer usually places an axial load upon the vertebra to the point of failure. For the proximal femur, there is usually a simulated sideways fall landing on the trochanter, which simulates falls in elderly patients that may result in hip fractures. Strain and stress distributions throughout the object are the main outputs from the model. A large number of studies have used FEA to assess the properties of ex vivo bone, from both humans and animals. Human data exist for both whole bone (vertebra, proximal femur, and distal radius being the main sites) and trabecular core analysis. These studies have, in general, nicely shown correlations between FEA results and traditionally assessed mechanical properties. These computer techniques can also be used to understand the individual contributions of the cortical and trabecular shell to altered mechanical properties. To do this, computational simulations can be run with the entire bone included in the model, and the secondary analysis can be run with either the cortical shell or the trabecular bone region 'virtually' removed. It is important to note that in most FEA models the voxels are assigned a single value for mechanical properties; thus, the parameters that are calculated represent the mechanical properties of the bone geometry and architecture, and not necessarily those of the composite structure (bone geometry plus the tissue-level properties). That is, if a bone has significant amounts of hypo- or hypermineralized bone tissue, the true stiffness (which would be

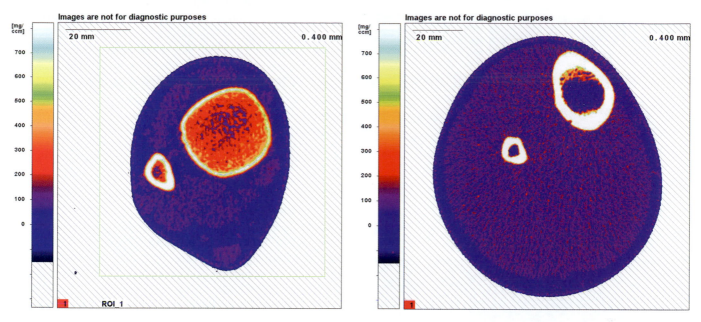

FIGURE 5.15 Peripheral quantitative computed tomography (pQCT) scanners can provide higher resolution scans compared to traditional CT systems. pQCT can distinguish between cortical and trabecular compartments at sites such as the proximal tibia. It can also provide assessments of geometry at this site as well as at a diaphyseal site. Muscle volume can also be assessed on pQCT images.

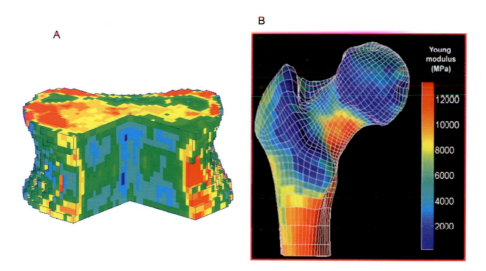

FIGURE 5.16 Finite element model analysis has emerged as a supplement to computed tomography scans for estimations of bone strength. In these models (A and B), the pixels from the scans are assigned mechanical properties and then the model can be artificially loaded (using computer simulations) to assess the whole bone mechanical properties. *Panel A reproduced with permission from R. Paul Crawford, Bone, 33, 4, 2003, 744–750. Panel B reproduced with permission from Banu B. Kalpakcioglu, Bone, 48, 6, 1 2011, 1221–1231.*

low and high, respectively) would not be accurately estimated in a model in which 'normal' material properties were assigned. However, in the absence of these extreme conditions, FEA appears to be a useful option for estimating changes in bone strength and stiffness. More recently, with the increase in clinical CT scans, studies have begun to assess in vivo FEA to measure changes associated with pharmacologic therapy, with encouraging results. This work builds on early

evidence that FEA can discriminate between women with and without a prior fracture with less overlap than is present in traditional BMD measures.

CT has become an essential tool for assessing skeletal structure and density in the laboratory. The advantage of CT over DXA or radiography is that CT offers the ability to separate cortical and cancellous bone, as well as to obtain architectural measurements. Prior to high-resolution CT, architectural parameters were routinely

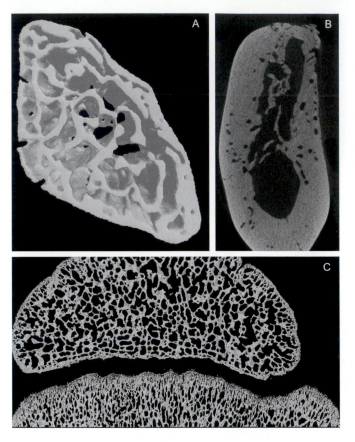

FIGURE 5.17 Micro-computed tomography systems provide high-resolution images used predominately for ex vivo assessment of bone density and architecture. Although most often used for describing phenotypes of rodent bone, e.g. (A) mouse vertebra, larger species such as dog mandible (B) or pig femoral head (C) also can be assessed for properties of trabecular and cortical bone.

Trabecular architecture and cortical geometry are the key parameters assessed by micro-CT. Trabecular bone volume (BV/TV; %) represents the amount of bone tissue within a given volume of interest. Most analyses focus on the secondary spongiosa of rodents for trabecular analysis, as this region avoids the bone directly beneath the growth plate (Fig. 5.18). A set number of slices, or a certain anatomical distance, is then marked off and all the bone within the endocortical boundary is analyzed using thresholds to determine the total volume (bone plus marrow) and bone volume. The tissue identified as bone is then assessed for architectural parameters such as trabecular number, trabecular thickness, and trabecular separation (Fig. 5.19). The structural model index (SMI) is often calculated in these software programs and provides an estimate of whether the trabecular struts are more "rod-like" (SMI = 3) or more "plate-like" (SMI1 = 0). The SMI calculation is sensitive to bone volume and if the trabecular bone region is very dense, then the SMI parameter can become negative due to the concavity of some trabecular surfaces. This can accentuate differences between experimental groups. Micro-CT assessment of cortical bone can provide data on cortical bone geometry such as bone area, cortical thickness, and cross-sectional moment of inertia (Fig. 5.19). Using calibration phantoms, volumetric BMD can also be calculated at both cortical and cancellous sites.

Capabilities now exist to perform in vivo micro-CT measures on small animals. Longitudinal assessment provides a powerful tool for assessing interventional efficacy, as it eliminates variability among animals at baseline, which can be quite large in some strains. In vivo scanning uses software to account for respiration during the scan, thus minimizing movement artifacts. This is essential if central vertebral bone is of interest, although unnecessary when assessing peripheral sites. It also utilizes registration software to allow regions of interest to be precisely copied from one scan to its follow-up. Using these methods, it is possible to understand how localized changes, at the level of an individual trabecula or pore, occur over time (Fig. 5.20). Given the significant radiation exposure related to in vivo CT scanning, some concern exists about how such exposure influences local bone responses. Although single exposures have not been shown to have a significant effect on cell viability, multiple exposures (2–3 exposures) over weeks have been shown to cause osteoblast death and to affect dynamic bone remodeling.

High resolution CT is also permitting exciting advances in quantitating other aspects of bone tissue. Recent work has focused on using CT to assess osteocyte lacunar density, vasculature, and microdamage. Lacunar density can be assessed by applying a reverse threshold to the scan and quantifying void spaces

determined using histology. CT offers the advantage of providing a 3D assessment of the entire region, in contrast to histology, which provides information on a single section and assumes that it is representative of the entire bone. Interestingly though, most studies that have examined relationships between trabecular architecture from 2D histology to 3D CT have found strong correlations.

Ex vivo micro-CT (or μCT) scanners typically produce scans with pixel sizes in the range of 1–30 μm. This provides sufficient resolution to accurately detect individual trabecular structures in small rodents, such as rats and mice, as well as in larger species (Fig. 5.17). Higher resolution scans, on a nanometer scale, can be obtained with nano-CT and synchrotron-CT machines. Depending on several parameters, such as the specific machine, size of the scanning region, pixel size, and rotation step, scan times can range from 5 min to 31 min for a given bone region. Postprocessing protocols are vital to analysis integrity and recommendations provide guidance on the use of these protocols (see suggested readings).

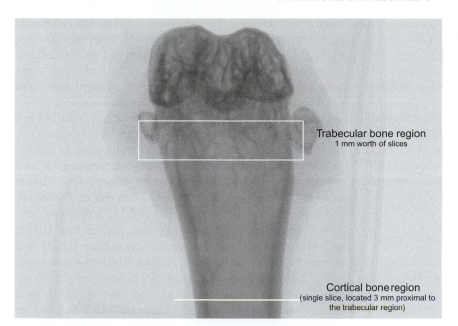

FIGURE 5.18 Micro-computed tomography imaging of excised bone involves collection of a series of projection images. Based on these images, regions rich in trabecular bone, or those containing just cortical bone, can be chosen for assessment.

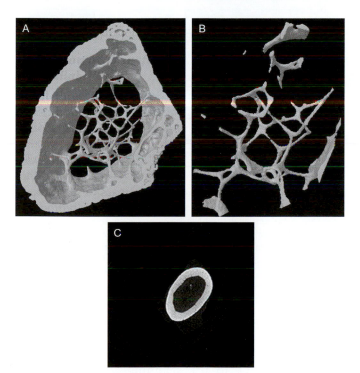

FIGURE 5.19 Micro-computed tomography assessment is routinely done on both trabecular and cortical bone sites. Metaphyseal bone, such as the distal femur, can be assessed as a whole (including cortical bone) (A) or the cancellous bone can be isolated and assessed separately from the cortex (this is the most common method) (B). Properties such as bone volume/tissue volume (BV/TV; %) and trabecular thickness, number, and separation can be calculated. Cortical bone sites can be assessed for geometric measures of bone area, cortical thickness, and cross-sectional moment of inertia (C).

(Fig. 5.21). Given the size of lacunae, the resolution of these scans is of the order of only a few micrometers. Bone vasculature can be assessed by a similar method,

although this makes the assumption that the void spaces within the bone (osteonal canals) contain a patent vessel. A more accurate assessment of the osseous vasculature is done using a contrast agent. Once the bone is perfused, the mineral can be removed through decalcification and the contrast agent-filled vasculature can be easily quantified using CT (Fig. 5.22). In fact, many of the CT analysis algorithms used to describe trabecular architecture can be applied to the vasculature to describe its dimensions and orientation. In a similar way, contrast agents have been used to assess microdamage in bone using CT (Fig. 5.23). One advantage of this technique, compared to traditional morphological assessment with histology, is that it provides a 3D view of damaged regions. Contrast agents, targeted to proteoglycans, also have been used to image cartilage with CT (Fig. 5.24). This has advantages for assessing changes to the cartilage and subchondral bone associated with osteoarthritis, which currently is limited to 2D histologic analysis.

MAGNETIC RESONANCE IMAGING

MRI represents an alternative method to CT for the assessment of trabecular architecture since it does not involve ionizing radiation. The technique is based on the distinct magnetic susceptibility and signal decay (the outcome measure of which is called T2*) of bone and marrow. The fat and water within the marrow have free protons that generate a magnetic signal, while the bone has no free protons and thus generates no signal. Images generated from MRI therefore depict a pattern of the bone marrow and "voids" where bone

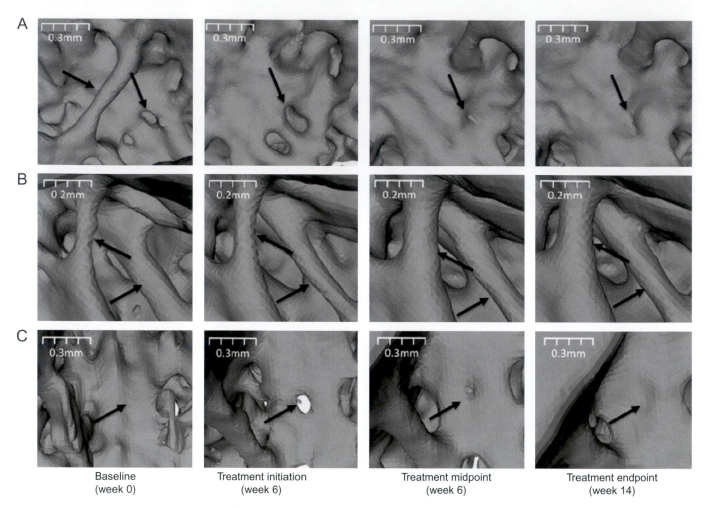

Baseline   Treatment initiation   Treatment midpoint   Treatment endpoint
(week 0)   (week 6)   (week 6)   (week 14)

FIGURE 5.20 In vivo micro-computed tomography provides an emerging tool for longitudinal assessment of changes in cancellous bone structure. Local changes in the cancellous structure of a rat given combined parathyroid hormone and alendronate treatment. (A) Trabecular elements that are lost with bilateral ovariectomy (OVX) do not reform. A trabecular element at baseline (week 0; left arrow) is lost after 6 weeks of OVX, and is not replaced with treatment. Perforations in the plate-like trabeculae (right arrow) are enlarged with OVX, and filled in after 10 weeks of treatment. (B) Trabecular elements that become thin but stay intact with OVX thicken with treatment. A trabecular element aligned with the loading direction (left arrow) and another not aligned to the loading direction (right arrow) are shown. The element on the right becomes thin after OVX, while the left element does not. Both elements increase in thickness after combined therapy. (C) Plate-like trabeculae become perforated after OVX but can fully recover. The arrow indicates a hole that develops 6 weeks after OVX in a plate-like element and fills in 8 weeks after treatment. *Reproduced from Campbell GM, et al., Bone 2011;49:225−232.*

exists (Fig. 5.25). These voids produce a negative image of the trabecular network in two dimensions. MRI images are often referred to as T1- or T2-weighted, with the specific type of image being dependent on the pulse sequence used and the tissue of interest. In T1-weighted images, fat and calcifications appears bright, while water appears dark. Conversely, in T2-weighted images, water appears light, while calcifications and fat are dark. Similar to CT, slices are obtained with MRI and datasets can be built as 3D volumes.

The clinical application of MRI for bone began in the late 1970s. The widespread availability of MRI instruments in medical facilities has now made this a viable alternative to CT scanning for bone assessment. MRI instruments generate magnetic resonance pulses that are classified as either *spin echo* or *gradient echo*. Spin echo pulses are considered superior for assessing cancellous bone, despite being slower to acquire compared to gradient echo pulses. The speed of acquisition and the high signal-to-noise ratio of gradient echo make it more attractive for in vivo scanning.

Signal-to-noise ratio is a major consideration for MRI scanning. In general, the signal-to-noise ratio is proportional to the strength of the magnetic field and

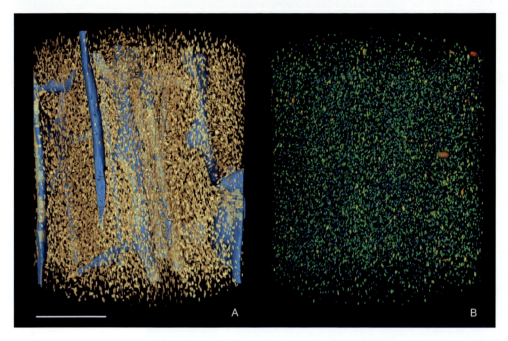

FIGURE 5.21 **High resolution computed tomography can be used to assess bone voids, including both haversian canal and lacunae spaces.** (A) 3D render of single anterior region of interest 1 × 1 mm showing 19,140 lacunae in gold with the vascular canals in blue. (B) Same region of interest with ellipse fit colored for the degree of equancy, with red being the most equant and blue the least. Scale bar, 300 μm. *Reproduced with permission from Carter Y, et al., Bone 2010;52:1−7.*

is inversely related to the spatial resolution of the scan. Strength of the magnetic field is determined by the size of the scanner magnet, with clinical magnets currently ranging from 1.5 to 11.7 Tesla (T; 1.5T and 3T scanners are most common in clinical settings). Higher strength magnets can yield higher resolutions (Fig. 5.25). High-resolution MRI (hrMRI) and micro-MRI (μMRI) machines now exist and these can achieve much higher resolution than the clinical scanners. hrMRI using a 1.5-T coil has yielded in plane resolution of 300 μm with a slice thickness of around 0.5 mm at peripheral sites. At this resolution, individual trabeculae cannot be delineated clearly, although apparent structural parameters can still be determined (Fig. 5.26) using postimage processing techniques. hrMRI can be applied in vitro (and in vivo with small animals) and can achieve a spatial resolution of >100 μm. Although hrMRI can achieve resolutions approaching those needed to image individual trabeculae, its restriction to ex vivo sample analysis and animals, scans for which radiation is less of a concern, have limited its widespread use; CT remains the standard.

Challenges exist to imaging the axial skeleton because of the low signal-to-noise ratio and low resolution achievable by the large coils. This is due in part to the larger amount of hematopoietic marrow (which does not contrast as well as fatty marrow) in the axial skeleton compared to the appendicular skeleton. Due to these challenges, most MRI data has been collected on peripheral sites such as the calcaneus, distal radius, and distal tibia as these can be scanned by smaller coils with higher resolution. The proximal femur has also been assessed in vivo using MRI with reasonable image resolution. Most properties related to trabecular architecture using MRI are similar to those assessed with CT. MRI datasets can also be used to develop FEMs. Thus far, hrMRI has primarily been a research modality based in academic radiology departments. Attempt to perform multicenter clinical trials using hrMRI has been challenged by the complexities of using peripheral MRI coils with the standard MRI units used for clinical imaging. The advantage of using MRI in human clinical trials is the complete lack of radiation exposure for the patient.

Although much of the focus of MRI has been on the mineralized matrix structure, MRI has the ability to assess other aspects of bone. Ultrashort echo times allow water assessment, including water that is free within the bone marrow and cortical pores, as well as water bound to the matrix. Specialized coil pulse sequences (called *WASPI*; water and fat suppressed proton projection MRI) combined with specially designed phantoms have been developed to assess collagen matrix components. MRI can image cartilage

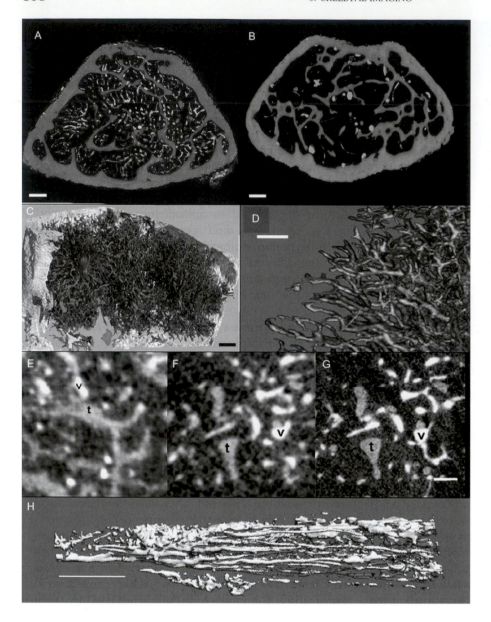

FIGURE 5.22 **Micro-computed tomography applications for imaging bone vessels.** (A-D) Synchrotron radiation micro-computed tomography (μCT). (A and B) Stack of thirty 1.5 μm-thick slices of femoral distal metaphysis of mice infused with either barium sulfate (left) or silicon containing lead chromate (right). Scale bar, 100 μm. (C and D) 3D rendering of barium sulfate infused vessels (D, zoom on microvessels). Scale bar, 100 μm. (E-H) Conventional μCT. (E-G) 2D slices of undecalcified trabecular bone; voxel size at 10 μm, 6 μm, and 3 μm, respectively. t, bone trabecula; v, vessel. Scale bar, 50 μm. (H) 3D image of decalcified femur, at 10 μm voxel size. Scale bar, 1 mm. *Reproduced with permission from Roche B, et al., Bone 2012;50:390−399.*

and soft tissue and is advantageous in combination with CT for assessing cartilage using multimodality imaging with registration.

POSITRON EMISSION TOMOGRAPHY

The imaging modalities described above provide a snapshot of bone density and structure, but do not provide any information about the dynamic processes responsible for those values. For example, two patients could have similar DXA values, with one patient having high bone remodeling and the other having low bone remodeling. Furthermore, DXA and CT are often focused on discrete skeletal sites that are prone to fracture. Positron emission tomography (PET) imaging addresses both of these limitations by providing a whole bone scan aimed at detecting regions of high bone metabolic activity. It can also be used to quantify rates of blood flow, as well as protein expression, thus making it potentially useful for assessing the dynamic activities of osteoblasts and osteoclasts in the living skeleton. Although most widely used for detection of benign and metastatic involvement of the skeleton, it has applications outside of the cancer setting, including as a fracture diagnosis tool.

PET, a nuclear medicine technique, detects gamma (γ) rays emitted by a radionuclide tracing agent. Once a tracing agent is administered, it becomes preferentially deposited on bone surfaces that are remodeling

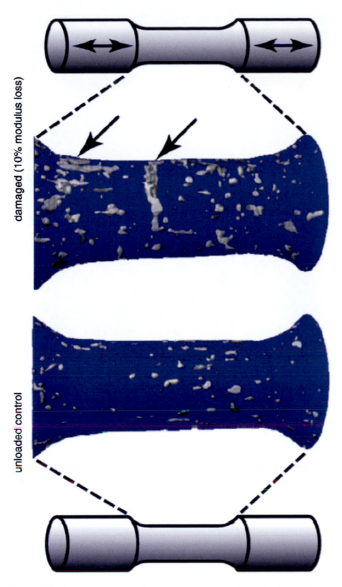

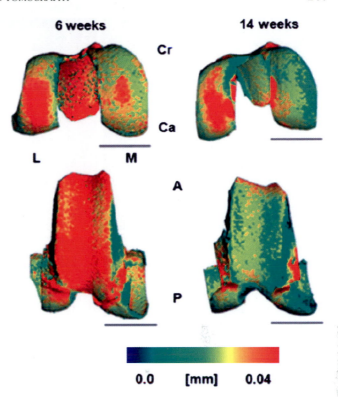

FIGURE 5.24 **Computed tomography-based methods are being developed for assessing cartilage.** 3D thickness maps of the morphology of the articular cartilage of 6- and 14-week-old mice show thicker cartilage in the younger mice. Scale bars, 1 mm. A, anterior; Ca, caudal; Cr, cranial; L, lateral condyle; M, medial condyle; P, posterior. *Reproduced with permission from Kotwal N, et al., Osteoarthritis and Cartilage 2012;20:887–895.*

FIGURE 5.23 **Computed tomography-based methods are being developed for assessing microdamage.** Segmented, 3D micro-computed tomography (μCT) reconstructions of the entire gauge section (2.5 mm in diameter by 5 mm in length) for an unloaded control specimen compared to a specimen loaded in cyclic uniaxial tension to a 10% reduction in secant modulus, showing the ability of μCT to detect spatial variation in damage accumulation. Arrows highlight regions of concentrated BaS04 staining characteristic of fatigue damage and/or propagating microcracks. *Reproduced with permission from Landrigan M, et al., Bone 2010.*

rapidly. The concentration of tracer in the tissue subsequently undergoes decay, which is tracked by the scanner (Fig. 5.27). Modern machines combine PET scans with CT in order to enable the tracer information to be superimposed on detailed CT-image structures. The disadvantage of this technique is that the combined CT and PET scan involves significant ionizing radiation exposure. PET scans can also be combined

with MRI imaging. Regional PET scans, with regions of interest on the scale of 10–20 cm, can take as little as 5–11 min, while whole body scans are possible but take roughly 41 min. Adding CT or MRI to the scan increases the acquisition time.

Fluorine-18-labeled sodium fluoride (^{18}F-NaF) and technetium-99 m methylene diphosphonate (^{99m}Tc-MDP) are commonly used tracers for assessing bone metabolism. ^{18}F-NaF was one of the first nuclear medicine skeletal imaging tracers; it combines with hydroxyapatite crystals to form fluoroapatite. One advantage to ^{18}F-NaF is that it has minimal binding to plasma proteins and so protein binding does not need to be measured or estimated. This increases the accuracy of determining how much tracer is going to a particular site of interest. An alternative tracer is ^{99m}Tc-MDP, an analog of etidronate which is an early generation bisphosphonate used for the treatment of metabolic bone diseases (see Chapter 17). The use of ^{99m}Tc-MDP for tracing is complicated by the fact that it has significant plasma protein binding, which is maximal about 4 h after administration (50% is bound

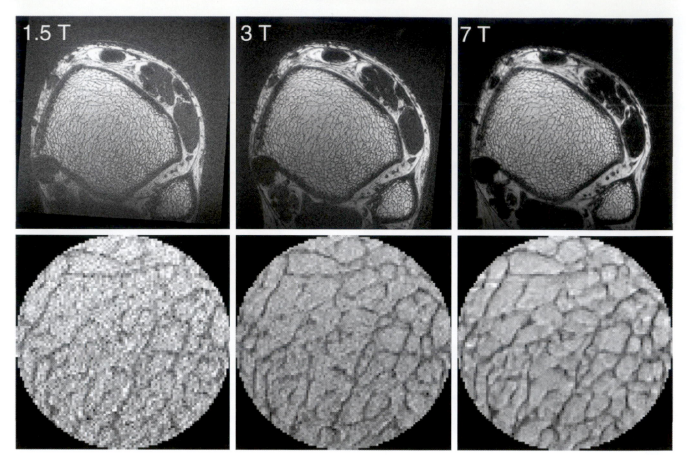

FIGURE 5.25 In vivo MR images of the distal tibia acquired at 1.5T, 3T, and 7T using the same pulse sequence and identical parameters. Image data acquired at 1.5T and at 3T were registered to the data acquired at 7T using trabecular and cortical bone of the tibia. Lower panels, zoomed views of a central region in the tibia. *Reproduced with permission from Wright AC, et al., Journal of Magnetic Resonance 2011;210:113–122.*

to these proteins). Because of this plasma binding, ^{99m}Tc-MDP is often measured as 24-h uptake under the assumption that everything is cleared by then except what is bound to bone (usually about 30% of the injected compound). Renal function plays a significant role in ^{99m}Tc-MDP imaging, as reduced renal clearance allows more time for the agent to bind to the skeleton. These compounds are adsorbed onto hydroxyapatite crystals and therefore provide an index of bone formation. They are also highly influenced by blood flow.

PET can provide important supplemental information to other imaging modalities in various instances. For example, observation of a vertebral crush fracture using X-ray, DXA, or CT would provide no information on the time of occurrence (whether recent or long ago) of that fracture. Supplementation with a PET scan would show high tracer activity if a fracture were recent. PET scanning is also useful in Paget disease of bone (also called osteitis deformans;

see Chapter 16), as the assessment of high bone metabolic activity and extent of skeletal involvement associated with Paget disease makes it more sensitive than the more traditional radiographic skeletal survey.

ULTRASOUND

Although not technically an imaging modality, quantitative ultrasound (QUS) is used clinically as a method for noninvasively assessing bone. QUS measures the distance and speed at which sound waves travel from the source to the detector. Ultrasound waves are reflected and almost completely attenuated by air, so ultrasound cannot be conducted on the axial skeleton such as the vertebrae due to air in the lungs and bowel. Rather, sites with limited soft tissue such as the calcaneus (the most frequently used site), tibia, or radius are used. Even at these peripheral sites,

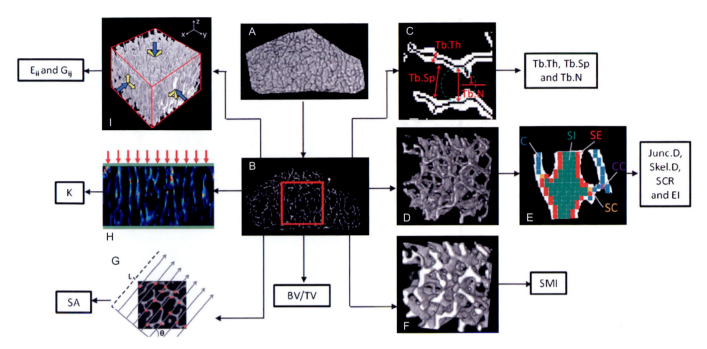

FIGURE 5.26 **Process for extraction of structural and mechanical parameters.** Parameters are derived from both whole trabecular bone (TB) volume and subvolume unless specified. (A) Center slice of a micro-magnetic resonance image of the whole TB volume region of interest after retrospective registration and segmentation of region common to baseline and follow-up images. (B) Bone volume fraction (BVF) image of (A) yielding BV/TV. Region in box indicates analysis subvolume. (C) Fuzzy distance transform based computation of trabecular thickness (Tb.Th), spacing (Tb.Sp), and number (Tb.N). (D) 3D-rendered binarized skeleton after sinc interpolation. (E) Digital topological analysis classification of skeleton voxels. (F) 3D rendering of a subvoxel processed and binarized image yielding structure model index. (G) Computation of mean intercept length yielding structural anisotropy (SA; subvolume only). (H) Finite element analysis (FEA) providing estimation of axial stiffness, K (whole TB volume only). (I) FEA of subvolume compressive and shear moduli (Eii and Gij, where i, j = {x, y, z}). *Reproduced with permission from Lam SCB, et al., Bone 2011;49:895−903.*

submersion of the site in a water bath, or use of a coupling gel, enhances the signal-to-noise ratio of the technique. QUS has several advantages as an assessment tool, including that it is relatively inexpensive, scans times are quick, there is no ionizing radiation, it does not require technical expertise, and machines are highly portable.

Ultrasounds are attenuated by both scattering and absorption. Scattering sends the waves in directions that differ from the principal one, while absorption is the transformation of energy into heat. Bone is acoustically inhomogeneous because of its architecture and the marrow cavity, and the attenuation or scattering of bone is dependent on the bone mass, architecture, and material properties.

QUS generates two primary variables: speed of sound (SOS) and broadband attenuation (BUA). SOS is a measure of the velocity of the waves through the tissue; BUA assesses how much the waves are attenuated, reported in decibels per megahertz (dB/MHz) (Fig. 5.28). This represents the oscillation of the wave and the assessment of its decline over time. The higher the bone density, the more energy is lost per unit time, although it has been suggested that bone architecture

can affect BUA independently of bone density. In addition to the main outcomes, some machines generate composite variables such as stiffness index or QUS index, both of which are a combination of SOS and BUA. The commercially available heel ultrasound machines have a normative database and will generate a T-score. These T-scores are not interchangeable with central DXA T-scores and should not be used with the WHO BMD criteria for the diagnosis of osteoporosis. Given the fairly large standard error of heel ultrasound results, they should not be used to follow individual patients or to determine whether a pharmacologic intervention has been efficacious. The role of heel ultrasound in the community has been primarily used for population screening and subsequent referral of appropriate patients for central DXA.

In summary, imaging modalities continue to offer the researcher and the clinician methods to assess bone health noninvasively. Techniques continue to evolve and the quantification of the objective data acquired from the imaging continues to give basic and clinical scientists and clinicians options to assist in fracture risk assessment and following individual patients with and without pharmacological intervention.

Skeletal Hard Tissue Biomechanics

Joseph M. Wallace

Department of Biomedical Engineering, Indiana University-Purdue University, Indianapolis, Indiana, USA

INTRODUCTION TO BASIC BONE STRUCTURE AND FUNCTION

Main Functions of Bone in the Body

As an educational exercise, approach a random person on the street and ask them what they know about bone. Chances are that the majority would tell you that bone is a structure in our bodies that grows until we reach our adult size, and then becomes static and inert. As the other chapters in this book have demonstrated, this statement is quite incorrect. Instead, our skeleton is constantly changing and adapting in response to the dynamic needs of our bodies.

In general, bone serves four main roles in the body. Two of these roles (hematopoiesis and mineral ion homeostasis) are discussed in Chapters 1 and 13. The two remaining roles are structural in nature. Bone serves to protect the vital organs in the body. The skull protects the brain and four of the major sense organs (eyes, ears, nose, and mouth). The thoracic cavity encloses the heart and lungs, and the vertebral column protects the spinal cord. In addition, bone also supports the weight of our soft tissues and acts as a lever system for muscular activity. In general, bone structure is directly related to these functional needs.

During normal daily activity, bone constantly bears dynamic mechanical stimuli (e.g. the cyclic compressive loading of walking or static bending loads induced by carrying a heavy bag). The ability of bone to bear these loads is dependent on the type and magnitude of the applied load, as well as the structural properties of the bone that is loaded. When loads exceed the structural strength of the bone, damage and eventually failure (fracture) will occur. As discussed elsewhere in this book factors such as trauma, age, and disease can compromise bone's ability to effectively perform its structural load-bearing roles in the body.

Bone is a Hierarchical Structure

As a material and structure, bone is composed of elements that themselves have structure on smaller scales (see Chapter 1). This hierarchy plays a major role in determining the bulk mechanical properties of the overall bone. Structural hierarchy is employed because hierarchical structures typically contain less material to achieve a desired strength. In addition, one can achieve a material that is simultaneously strong and tough, for which there is normally a trade-off (i.e. materials with high strength are often more brittle and have lower toughness compared to more ductile materials, which have higher toughness but lower strength).

An obvious example of a real world object with structural hierarchy is the Eiffel Tower (see Chapter 1, Fig. 1.12). As a whole, the tower is composed of pig iron and stands just over 300 m tall, with a square base of 100 m per side. A closer look reveals that the larger structure is actually composed of a latticework at each of the four corners where diagonal girders connect elements together. These units, with dimensions on the order of tens of meters, are composed of girders on the order of meters in length. Each individual girder has cross-sectional dimensions of 10^{-2} to 10^{-1} meters composed of an iron material with an atomic structure made up primarily of iron, carbon, silicon and manganese ions in a lattice structure on the Angstrom scale (i.e. 10^{-10} m!). In bone, structural hierarchy exists over 9–10 orders of magnitude in scale (see Chapter 1, Fig. 1.1). In general, there are seven discrete hierarchical levels represented, although different subgroups and naming conventions are used.

As previously noted, when loads on a bone exceed the overall strength of the bone, fracture will occur. Osteoporosis is a significant medical and economic burden facing our society and is characterized by low bone mass that leads to fracture. Each year, an estimated 1.5 million Americans suffer a bone disease-related fracture resulting in direct care expenditure of up to US$ 18 billion dollars a year. To understand and prevent these failures, a grasp of the major contributors of bone strength is needed. In general, there are three factors that contribute to bone strength: bone mass, geometry (the size, shape, and distribution of material), and the material properties of the bone tissue itself. These factors will be discussed individually, but first it is important to introduce some fundamental concepts of solid mechanics.

FUNDAMENTALS OF SOLID MECHANICS

In the laboratory, loads can be applied in many ways to derive the various structural properties of bone (Fig. 6.1). When the two ends of a bone are pulled apart and the bone length increases, the bone is said to be under tension. If, instead, the ends of the bone are pushed together causing the length of the bone to decrease, then the bone is under compressive loading. If load is applied perpendicular to the bone's long axis, sliding the ends of the bone opposite each other, then the bone experiences shear. If the ends of the bone are twisted relative to one another, then the bone is under torsion. Finally, if the bone is bent about its long axis, then part of the sample experiences compression while other parts experience tension.

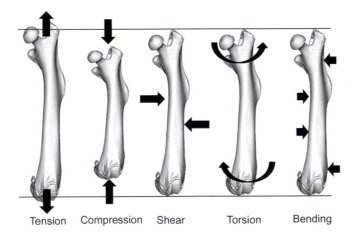

| Tension | Compression | Shear | Torsion | Bending |

FIGURE 6.1 Mechanical testing modes in bone. When the ends of a bone are pulled apart, length increases and the bone is under tension. If the ends are pushed together, length decreases under compressive loading. Loading across the long axis of the bone induces shear. If the ends of the bone are twisted relative to one another (e.g. the top is twisted counter clockwise and the bottom is twisted clockwise), then the bone is under torsion. Finally, when a bone is bent about its long axis, the initial straight shape of the bone is bent into a curve. Part of the sample experiences compression (left side) while other parts experience tension (right side).

The Force-Displacement Curve

When a force is applied to a bone (regardless of the loading modality), deformation of the bone will occur. When the load is examined as a function of this deformation, a characteristic curve called the *force-displacement curve* is produced (Fig. 6.2). Several important extrinsic characteristics of the structure can be determined by analyzing this curve. There is an initial linear portion of the curve; a secondary nonlinear portion of the curve where the maximum force is defined; and finally a location where the bone can no longer carry a load and fails. The initial linear portion of the curve is known as the *elastic region*. Within this region, loading and unloading will follow the same curve, indicating that no energy is lost during the loading cycle. Like a spring, deformation will return to zero upon unloading and no permanent deformation of the structure will exist. The slope of the curve within this elastic zone is a measure of the structural stiffness of the bone, indicating the resistance of the entire structure to deformation under a given applied load. If loading continues past this linear elastic region, the bone will yield and begin to behave in a nonlinear manner. Past this yield point, permanent (or *plastic*) deformation will occur and unloading to zero force will result in a structure that has sustained permanent deformation (damage). The yield point is difficult to define using observation alone and is typically calculated using an offset method described in the section below on the stress-strain curve. A maximum level of force is reached before the internal structure of the bone begins to fail, the load level falls, and catastrophic failure eventually occurs. In bone, the maximum load and failure load are generally close to each other. The maximum load level achieved is defined as the ultimate or structural strength, while the force separating the elastic and plastic regions is called the *yield force* (with corresponding yield or elastic deformation). The total deformation to failure minus this elastic deformation defines the plastic, or post-yield, deformation. In general, the amount of post-yield deformation a structure exhibits is a measure of ductility. The opposite of ductility is brittleness, meaning that little to no post-yield deformation occurs before failure. Finally, the area under the loading curve is a measure of work, or the energy absorbed by the structure during the loading cycle. Overall, structural strength, stiffness, deformation, and energy dissipation can be calculated from the force-displacement curve.

Stress and Strain in Axial Loading

The force-displacement curve depicted in Fig. 6.2 describes only what happens to a whole structure

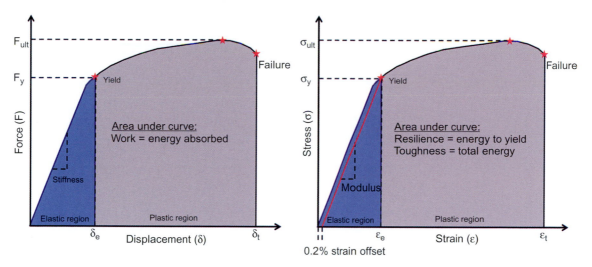

FIGURE 6.2   **Force-displacement curve and stress-strain curve.** When a force is applied to a bone (or any structure), deformation of the structure occurs. When the load level is plotted as a function deformation, the force-displacement curve is produced (left graph). Several important characteristics of the structure can be determined by analyzing this curve, including strength, stiffness, deformation, and energy dissipation parameters of the structure as a whole. If force and displacement are normalized for the geometry of the structure, a complimentary stress-strain curve results (right graph). This curve provides similar information (e.g. strength, stiffness, deformation, and toughness), but the properties are characteristics of the material that are independent of the size and shape of the structure. For both curves, properties can be calculated at a variety of points, including to the yield point (the blue region), to the ultimate/maximum force/stress, to failure, or for the region of the curve past the yield point (the gray region). The yield point is usually calculated in the stress/strain domain using the 0.2% offset method (see text), and then transposed to the force/displacement curve.

when a load is applied. This curve is dependent on the mass, geometry, and material properties of the structure being tested. To understand what these structural properties mean, one must introduce the concepts of *stress* and *strain*. Stress has the units of force per unit area (e.g. pressure) and represents a way to normalize force to the geometry of the specimen being tested. Depending on the direction of the applied load, two types of stress are possible. *Normal* stresses occur on a plane perpendicular to the applied loads when loads are along the axis of a member, and can be compressive (if the material is shortened) or tensile (if the material is lengthened). In comparison, *shear* stress occurs when loading is transverse to the axis of the specimen. The equation for normal stress (σ) and shear stress (τ) follows:

$$stress = \frac{force}{area} = \frac{P}{A} \qquad (6.1)$$

In these equations, P is the applied load and A is the cross-sectional area of a plane perpendicular to the member axis. Under simple tension and compression, only normal stresses exist on this plane (defined as 0°). However, on other planes within the material, a combination of normal and shear stresses exist (Fig. 6.3A). For instance, for a plane oriented 45° from the member axis, shear stress is maximized and has the same magnitude as the normal stress at this

orientation. Fig. 6.3A depicts how normal and shear stress varying with orientation under axial loading.

As noted above, when a force is applied to a sample such as a bone, the sample will deform. Within the material itself, strain is defined as the relative deformation induced by the load (Fig. 6.4). For a sample under normal loading, load is applied along the axis of the sample (perpendicular to a plane cross section), causing the material to either shorten (compression) or lengthen (tension). The resulting normal strain is calculated as the ratio of the change in length to the original length (Fig. 6.4A).

$$normal\ strain = \varepsilon = \frac{Change\ in\ length}{Original\ length} = \frac{\Delta L}{L} \qquad (6.2)$$

Because normal strain is a ratio of lengths, it is a dimensionless quantity. Strains in bone are typically small and are often reported in units of microstrain ($\mu\varepsilon$): 1% strain is the same as $0.01\,\varepsilon$ or $10,000\,\mu\varepsilon$. Conversely, $1\,\mu\varepsilon$ is equal to $1 \times 10^{-6}\,\varepsilon$ or 0.0001%. Strains in human bones are generally less than $1000\,\mu\varepsilon$ in tension and $2000\,\mu\varepsilon$ in compression. Under shear, load is applied perpendicular to the axis of the sample (parallel to a plane cross section) causing the surface of the section to slide past adjacent parts of the samples (Fig. 6.4B). For the resulting shear strain, the distortion ratio $\Delta L/L$ is calculated as the angle, in radians, through which the sliding deformation occurs.

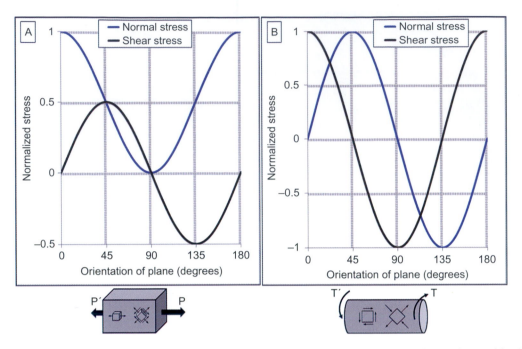

FIGURE 6.3 **Normal and shear stress as a function of orientation in axial and torsional loading.** (A) Under axial loading, only normal stresses exist on a plane that is perpendicular to the loading direction (defined as 0°). However, at other orientations within the material, a combination of normal and shear stresses exist (Figure 6.4A). For a plane oriented 45° from the member axis, shear stress is maximized and has the same magnitude as the normal stress. (B) In torsion, a portion of the surface with a plane oriented either perpendicular or parallel to the axis of the sample (i.e. 0° and 90°) is under maximum shear stress, while normal stress is equal to zero. As one rotates the plane from 0° to 90°, a combination of normal and shearing stresses exists. At 45° and 135°, the shear stress falls to zero while the normal stress is maximized (with a magnitude equal to the magnitude of maximum shear stress seen at 0°). In torsion, these are called the principal directions.

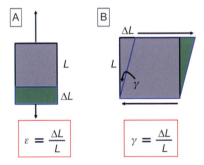

FIGURE 6.4 **Normal and shear strain.** (A) Under normal loading, load is applied along the axis of the sample, causing the material to either shorten or lengthen. The resulting normal strain is the ratio of the change in length (ΔL) to the original length (L). (B) Under shear, load is applied perpendicular to the axis of the sample causing the surface of the section to slide past adjacent parts of the sample. Shear strain is the angle γ through which the sliding deformation occurs, and is calculated using the distortion ratio $\Delta L/L$ for small strains.

For every point on the force-displacement curve, the corresponding stress and strain can be calculated and plotted on a stress-strain curve (Fig. 6.2). As opposed to the force-displacement curve, which depends on the mass and geometry of the sample being tested, this stress-strain curve has been normalized to geometry

and mass, and therefore represents the material properties of the sample. From the stress-strain curve, properties analogous to those on the force-displacement curve can be calculated. For instance, the stress and strain at the yield point define the elastic region, and the area under the curve to this point is the resilience of the material or the energy to yield. The yield point defines where the stress-strain curve begins to become nonlinear but, similar to the load-displacement curve, this point is seldom well defined. Most often, an offset method is used, where a line that is parallel to the initial linear portion of the curve, but offset by 0.2% strain, is produced. The point where this offset line intersects the stress-strain curve is then defined as the *yield point*. If loading remains below this yield point, then all energy absorbed during loading will be returned upon unloading to zero stress, and the strain will return to zero. The maximum stress is the ultimate strength of the material and the total area under the stress-strain curve is the toughness or total energy absorbed before failure.

An important point must be addressed regarding the derivation of properties from the post-yield region of the stress-strain curve. The equations used to calculate stress and strain make some important assumptions about the material being tested. Among these is the assumption that the material behaves in a linearly elastic manner and that loading does not exceed the proportional limit (yielding). Before yielding, the strain-strain curve for bone does remain linear. However, in bone and many other materials, these assumptions are not adhered to during the entire loading cycle and, therefore, properties derived from the post-yield portion of the stress-strain curve need to be considered with caution.

Modulus of Elasticity and Poisson's Ratio

The slope of the stress-strain curve represents the stiffness that, as opposed to the structural stiffness defined above, is intrinsic to the material being tested. Most structures are designed to undergo little deformation and remain in the elastic region. This statement is true for bone under normal loading conditions. In this elastic region, stress is directly proportional to strain:

$$\sigma = E\varepsilon \tag{6.3}$$

Eqn 6.3 is known as Hooke's Law and is the equation of the linear portion of the stress-strain curve where E is the slope of the curve. Hooke's Law defines the direct proportionality of stress and strain. Since strain is a dimensionless parameter, E has the units of stress [pounds per square inch or pascals (Pa)]. E is called the elastic modulus, modulus of elasticity, or Young's modulus and is an intrinsic property of materials. The stiffer the material is, the steeper the slope and the larger the modulus. A larger modulus means that a larger stress is necessary to lead to a given strain level. Although the modulus of bone depends on many conditions, typical values for cortical bone range between 10 and 15 GPa. In comparison, the modulus of rubber is around 0.01−0.1 GPa, while that of diamond is greater than 1200 GPa. Analogous to Young's modulus for axial loading, a modulus also exists for shear:

$$\tau = G\gamma \tag{6.4}$$

Here, G is known as the modulus of rigidity and is the slope of the linear portion of the stress-strain curve in shear. Because the shear strain, γ, is dimensionless, G has the same units as shear stress.

Within the elastic limit (i.e. prior to yielding), stress and strain satisfy Hooke's Law (Eqn 6.3). Therefore, for loading in one direction only (e.g. along the axis of the sample, the x direction), Eqn 6.3 can be rearranged as:

$$\varepsilon_x = \frac{\sigma_x}{E} \tag{6.5}$$

In Eqn 6.5, the subscript x indicates the stress and strain on a plane whose normal (perpendicular) points in the x direction. At the same time, normal stresses exist only in the direction of the applied load (i.e. the x direction) and, therefore

$$\sigma_y = \sigma_z = 0 \tag{6.6}$$

It would be tempting to say that strains in the y and z directions are also equal to zero, but this statement is not true. Siméon Poisson, a nineteenth century French mathematician, noted that axial elongation (ε_x) is always accompanied by lateral contraction ($-\varepsilon_y$ and $-\varepsilon_z$). Stated in simple terms, when a material is stretched in one direction, the cross-sectional length in the other two directions gets smaller. For a given material, the ratio of lateral strain to axial strain is constant, and is given the name *Poisson's ratio*:

$$\nu = -\frac{lateral\ strain}{axial\ strain} = -\frac{\varepsilon_y}{\varepsilon_x} = -\frac{\varepsilon_z}{\varepsilon_x} \tag{6.7}$$

Poisson's ratio ranges from 0 for a perfectly incompressible material (e.g. cork) to 0.5 for a perfectly compressible material (e.g. rubber). Poisson's ratio for bone is often said to be between 0.3 and 0.35. The negative sign in Eqn 6.7 is needed since the axial and lateral strains have different signs. Rewriting Eqn 6.7 yields:

$$\varepsilon_y = \varepsilon_z = -\nu\varepsilon_x \tag{6.8}$$

Combining Eqn 6.5 with Eqn 6.8 yields expressions for the lateral strains in terms of an applied axial stress:

$$\varepsilon_y = \varepsilon_z = -\frac{\nu\sigma_x}{E} \tag{6.9}$$

Eqns 6.5 and 6.9 define the strains that exist in a material subjected to uniaxial loading in the x direction only. A uniaxial load applied in the y or z direction yields Eqn 6.10 and Eqn 6.11, respectively:

$$\varepsilon_x = \varepsilon_z = -\frac{\nu\sigma_y}{E}; \varepsilon_y = \frac{\sigma_y}{E} \tag{6.10}$$

$$\varepsilon_x = \varepsilon_y = -\frac{\nu\sigma_z}{E}; \varepsilon_z = \frac{\sigma_z}{E} \tag{6.11}$$

In the case of combined general loading applied in the x, y and z directions simultaneously, the equations for the generalized Hooke's Law under multiaxial loading are:

$$\varepsilon_x = \frac{\sigma_x}{E} - \frac{\nu\sigma_y}{E} - \frac{\nu\sigma_z}{E} \tag{6.12}$$

$$\varepsilon_y = -\frac{\nu\sigma_x}{E} + \frac{\sigma_y}{E} - \frac{\nu\sigma_z}{E} \tag{6.13}$$

$$\varepsilon_z = -\frac{\nu\sigma_x}{E} - \frac{\nu\sigma_y}{E} + \frac{\sigma_z}{E} \tag{6.14}$$

bone. In this case, the solid bone would have a 19% larger periosteal diameter compared with the original solid bone and a 42% increase in cortical area.

Bone Tissue Material Properties

As noted at the beginning of this chapter, bone is hierarchical and the structural elements that exist at various levels of the hierarchy contribute to the properties of the material. At the tissue level (10^{-2} to 10^{-3} m), bone is distinguished as cortical or cancellous, most often classified as a function of porosity. Differences in mechanical properties between cortical and cancellous bone have been studied and do exist, but detailed comparisons using traditional mechanical approaches are challenging. Cancellous bone is better suited to compression testing, while cortical bone is better in tension and bending. Since mechanical properties can vary based on the type of loading, it is challenging to directly compare cortical and cancellous samples. For a given volume of bone, cancellous tissue is weaker and more compliant than cortical tissue because it is a porous structure. When a cancellous sample is tested, there is a yield and ultimate point similar to that found in cortical bone. However, the sample does not undergo a catastrophic failure. Instead, the void spaces between trabeculae collapse and the individual trabeculae push against one another causing stress to rise past the ultimate point. Further discussion of the mechanical differences between cortical and cancellous bone is provided elsewhere (see Cowin, 2001, Evans, 1973, and Yamada and Evans, 1970 in the "Suggested Readings").

As one explores levels in the tissue hierarchy beyond the microstructure and submicrostructure that make up cortical and cancellous bone, one arrives at the nanostructure (10^{-7} to 10^{-9} m), where differences between cortical and cancellous tissues are blurred. The properties of bone at this length scale are often lumped together into a single term, *tissue quality*, which refers to the inherent chemical and physical properties of bone independent of mass and distribution. At this level and below, bone is predominantly a two-phase composite material composed of a relatively flexible organic matrix (approximately 90% type I collagen) impregnated with and surrounded by a stiffer reinforcing carbonated apatite mineral phase. Collagen makes up anywhere from 20% to 25% of the tissue (by weight) and provides tensile strength and ductility. Hydroxyapatite accounts for about 65% of the tissue (by weight) and provides compressive strength and stiffness. The balance of strength, stiffness, and toughness that are mechanical characteristics of bone as a tissue are initially derived from the intimate interaction between these nanoscale constituents of vastly differing mechanical properties. Consider the schematic representation of a mechanical test in Fig. 6.9.

On its own, a fully demineralized bone will easily deform without much increase in stress, and often without yielding (similar to a rubber band). On the other hand, a fully deproteinated bone is stiff and has high strength, but behaves in a brittle manner (similar to chalk). Bone balances these two extremes by demonstrating moderate stiffness, strength, and toughness. A real world example of this balance is concrete (strong in compression) that has been reinforced with steel rods or mesh (strong in tension).

The Bone Extracellular Matrix: Type I Collagen

Little is known about the direct impacts that collagen structure and organization have on bone mechanical properties. As discussed in Chapter 1, the structural organization of mineralized collagen fibrils in bone is important. In woven bone, collagen is rapidly deposited and mineralized without the need for an existing bone substrate. This type of temporary bone is formed during periods of rapid growth or injury. Collagen fibrils are randomly packed and organized, and there are widely varying degrees of mineralization where crystals are not necessarily directly associated with the collagen template. Since woven bone has no specific organization, the material is isotropic and mechanical properties are similar in all directions. Overall, woven bone has a lower tissue modulus in comparison to other more organized types of bone. In lamellar bone (whether in primary bone or in secondary osteons or trabecular packets), mineralized collagen fibrils require a preexisting

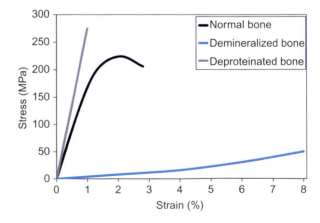

FIGURE 6.9 **Contributions of collagen and mineral to the mechanical behavior of bone.** On its own, a fully demineralized bone easily deforms without much increase in stress, and often without yielding. A fully deproteinated bone is stiff and has high strength, but behaves in a brittle manner. Bone balances these two extremes by demonstrating moderate stiffness, strength, and toughness.

cartilage or bone substrate, are formed slowly, and are much more organized than in woven bone. Within a single 3−7 μm-thick lamellae, fibrils are all arranged in one direction and there is a consistent level of mineralization. The tissue within single lamellae is transversely isotropic, meaning that the elastic modulus is higher in the direction of the fibrils (the grain axis) and lower but similar in the two perpendicular directions. There are competing theories for explaining how lamellae vary from layer to layer; these are discussed in Chapter 1. Regardless of the specific theory, there are variations of fibril orientation and density from lamellae to lamellae. This organization leads to orthotropic behavior, where the elastic modulus and other mechanical properties are different in each of the three perpendicular directions. Overall, the complex organization of the extracellular matrix means that the stress-strain behavior of bone is strongly dependent on the loading direction. As an example, cortical bone is very stiff and strong when loaded in the longitudinal direction (i.e. in the direction of the osteons). However, bone loaded perpendicular to the longitudinal axis behaves more like a brittle material. The important point here is that for bone, the direction of loading matters.

Collagen cross-link quantity and maturity are important to the mechanical stability of the tissue, affecting tissue strength, stiffness, and deformability. However, the specific impacts that cross-links have are unknown. In general, the mechanical stiffness and strength of bone and other collagenous tissues increases with tissue age, and much of this increase has been associated with changes and maturation in the cross-linking profile within the tissue. When cross-linking in bone is inhibited, mechanical integrity of the tissue suffers. Lathyrism is a pathologic condition in which cross-linking is inhibited through the inhibition of lysyl oxidase. When lathyrism is induced in animals using β-aminopropionitrile (or BAPN), significant changes in strength, deformation, and toughness are noted with no changes in mineralization levels or stiffness. The formation of cross links through the actions of lysyl oxidase requires the presence of vitamin B_6 (pyridoxine). Animals fed a diet deficient in pyridoxine demonstrate increased levels of divalent cross-links coupled with decreased strength. These types of studies reinforce the notion that proper collagen cross-linking plays a pivotal role in determining the quality and mechanical integrity of bone.

A variety of reducing sugars, including glucose, can react with free amino groups in proteins, lipids, and nucleic acids in a nonenzymatic process to create a heterogeneous group of molecules known as advanced glycation end products (AGEs). This process is not specific to collagen, but extensive investigations of AGEs in collagenous tissues have been undertaken, partially due to the prevalence of AGEs in type 1 or type 2 diabetes mellitus. The presence of AGEs can modify proteins such as collagen through the accumulation of permanent and dysfunctional intra- and inter-fibrillar cross-links. These cross-links accumulate with tissue age and negatively impact the strength, stiffness, and toughness of affected tissues.

The Bone Extracellular Matrix: Inorganic Hydroxyapatite

The physicochemical characteristics of bone mineral, along with total mineral content, density, and the orientation of the mineral relative to collagen have

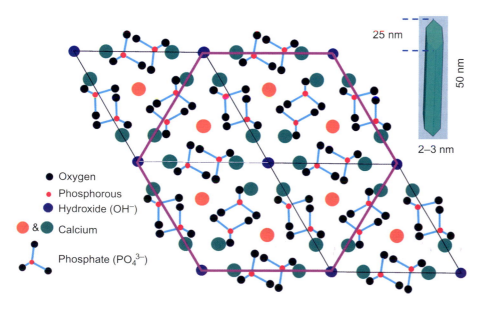

- ● Oxygen
- • Phosphorous
- ● Hydroxide (OH⁻)
- ● & ● Calcium

Phosphate (PO_4^{3-})

25 nm

50 nm

2−3 nm

FIGURE 6.10 **The crystal lattice of hydroxyapatite.** About two-thirds of the weight of bone is a mineral phase generically described as a form of hydroxyapatite [$Ca_{10}(PO_4)_6(OH)_2$]. The ions initially form a hexagonal crystal lattice, but as the crystals mature they take on a plate shape, with an average size of 50 nm × 25 nm and an average thickness of about 2−3 nm. Other ionic species such as carbonate can be easily substituted into each space in the crystalline lattice.

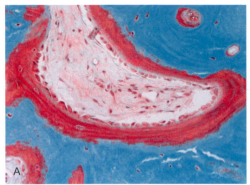
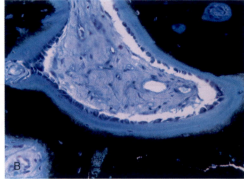

FIGURE 7.5   Plastic-embedded tissue stained with Goldner's trichrome (A) or von Kossa and MacNeal (B). Using Goldner's trichrome the mineralized bone stains green while the osteoid stains red. Von Kossa and McNeal staining results in black mineralized bone and blue osteoid.

endogenous proteins than those substances used for processing calcified bone. One limitation of paraffin embedding with bone is that complete decalcification takes considerable time (weeks to months, depending on the size of the specimen). The time can be shortened by increasing the acid concentration, but this can impact the integrity of the specimen. A second significant limitation, especially if one wishes to use the histologic sections for quantification, is that paraffin-embedded sections can distort and shrink up to 15% (pers. comm. George Costanza), compared to only 1−2% for plastic-embedded specimens. This can make a big difference in histologic measurements, and potentially obscure real differences between groups. Another limitation of paraffin embedding is that decalcification is usually incomplete. Thus, tissue sectioning is very challenging, resulting in suboptimal sections for analysis.

An alternative processing method is to embed sections without decalcification into a hard plastic. Methyl methacrylate is the most commonly used plastic, although others can and have been used. The advantage of processing a bone in this manner is that it allows the assessment of fluorochrome labeling. It is also faster than decalcification if the specimens are large. Plastic embedded cancellous bone sites are routinely sectioned using a microtome with tissue thickness between 4 μm and 8 μm. Cortical bone can be thin sectioned using a microtome but is more routinely sectioned at a thickness of 80−100 μm using a wafer or wire saw. Bone structure and cellular analyses can be conducted on these sections, as with paraffin, but fluorochrome labeling can be assessed. Immunohistochemistry techniques can also be applied to sections embedded in plastic (though protocols must often be adapted from the more common paraffin-based methods of immunohistochemistry). Undecalcified sections can also be obtained using cryotomy. In this technique, bones are flash frozen in liquid nitrogen, embedded in a special medium (optical cutting temperature compound, commonly referred to as OCT),

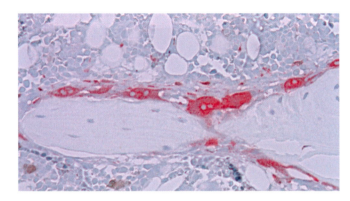

FIGURE 7.6   Histologic assessment of osteoclasts is routinely conducted on sections stained with TRAP. This stain provides clear distinction of the multinucleate cells and is often combined with a counterstain to allow visualization of the bone surface.

and cut with a cryostat. This technique is emerging as an optimal method for immunohistochemistry and fluorescent cell markers [such as green fluorescent protein (GFP)], although it is more technically rigorous than plastic embedding.

The most commonly used stains for examining tissue and cellular properties on plastic embedded sections are Goldner's trichrome and von Kossa tetrachrome. In Goldner's trichrome, the mineralized bone is green and the osteoid is red, whereas in von Kossa tetrachrome the mineralized tissue is black and osteoid is blue (Fig. 7.5). While osteoclasts can be assessed using these stains, it is more common to stain a separate section with tartrate-resistant acid phosphatase (TRAP), which specifically labels the osteoclasts (Fig. 7.6). TRAP staining can be carried out on either plastic or paraffin sections using slightly different techniques. Several other special stains can be used to assess specific features of the tissue. Because it binds to large proteoglycans (such as aggrecan), Safranin O is used to assess cartilage. It is often applied to studies of fracture repair or growth

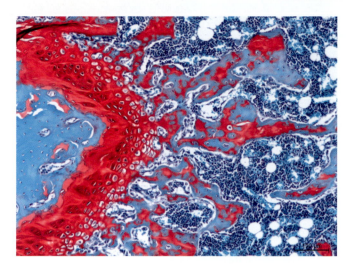

FIGURE 7.7   Safranin O is a useful stain for regions with cartilage such as the growth plate. Combined with a counterstain such as fast green, the cartilage regions of the growth plate stain red while the mineralized bone stains blue/green. This stain is also useful for studying fracture repair where classification of the different tissue types (cartilage and bone) is important.

plate dynamics (Fig. 7.7). Toluidine blue provides a stain for visualizing the growth plate and cement lines surrounding osteons and hemiosteons (Fig. 7.8).

HISTOMORPHOMETRIC ANALYSIS

Histomorphometric analysis of bone can range from simple assessments of bone structure to more detailed analyses of cell numbers and function. An essential reference for anyone interested in histomorphometry is the 1987 publication by A. Michael Parfitt and colleagues. This article provides standardization of the nomenclature, describes concepts related to histologic primary measurements and referents, and details essential information to be collected and reported for histologic methods in papers. The document was updated in 2013, although the key aspects such as nomenclature and standardization remain unchanged from the original.

Static versus Dynamic Measurements

Static measurements are those that measure bone structure without regard to rates of change or dynamic bone remodeling processes, such as resorption or formation. Examples of these would be measurements that characterize trabecular bone structure—trabecular thickness, number, and separation—or describe the amount of tissue—bone volume, cortical area, and porosity. They describe the result of all of the growth, modeling, and remodeling processes that have occurred without any reference to the time over, or rate at, which those

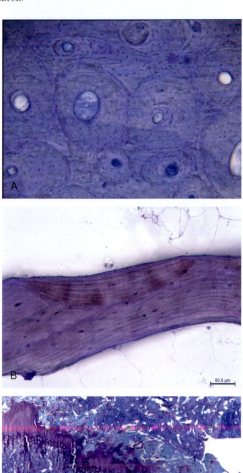

FIGURE 7.8   Toluidine blue is most commonly used to identify cement lines in cortical (A) and trabecular (B) bone sections. It can also be used to study the growth plate (C), as calcified cartilage stains more intensely than do noncalcified cartilage and bone.

structures may have been produced. Parameters such as osteoblast, osteoid, and osteoclast surface are also considered static measures as they provide a single snapshot of the tissue at the time it is viewed. Dynamic measurements employ fluorochrome labels to assess the rates and magnitudes of change in bone tissue, either at the time the tissue was taken or at various points in the past, depending on when the fluorochrome labels were given. Thus, dynamic measurements can be used to assess the consequences of a single treatment or intervention over time and, therefore, can be used to interpret the specific effects of that intervention. They can also be used to determine whether there were variations or aberrations in the normal physiologic processes of bone modeling or

TABLE 7.1　Common Histomorphometric Variables[a]

Variable	Abbreviation	Units	Definition
PRIMARY			
Tissue area	Tt.Ar	mm^2	Total tissue area within the ROIs
Bone area	B.Ar.	mm^2	Total area of trabecular bone within ROI
Bone perimeter	B.Pm	mm	Total length of bone surface examined
Single label perimeter	sL.Pm	mm	Total length of single label surface examined
Double label perimeter	dL.Pm	mm	Total length of double label surface examined
Osteoid perimeter	O.Pm	mm	Total length of osteoid surface examined
Osteoblast perimeter	Ob.Pm	mm	Total length of surface occupied by osteoblasts
Osteoclast perimeter	Oc.Pm	mm	Total length of surface occupied by osteoclasts
Osteoid width	O.Wi	μm	Average width of osteoid seams
Interlabel width	iL.Wi	μm	Average width between double labels
Wall width	W.Wi	μm	Average width of completed BMU
Erosion depth	E.De	μm	Average depth of resorption lacunae
DERIVED			
Bone volume	BV/TV	%	Percentage of ROI occupied by bone = (B.Ar/Tt.Ar) $\times$ 100
Mineralizing surface/bone surface	MS/BS	%	Percentage of bone surface undergoing active formation = [(dL.Pm + 0.5 sL.Pm)/B.Pm] $\times$ 100
Osteoid surface/bone surface	OS/BS	%	Percentage of bone surface covered by osteoid = (O.Pm/B.Pm) $\times$ 100
Osteoblast surface/bone surface	Ob.S/BS	%	Percentage of bone surface covered by osteoblasts = (Ob.Pm/B.Pm) $\times$ 100
Osteoclast surface/bone surface	Oc.S/BS	%	Percentage of bone surface covered by osteoclasts = (Oc.Pm/B.Pm) $\times$ 100
Mineral apposition rate	MAR	μm/day	Average rate of osteoblast activity at each BMU = iL.Wi/days between label administration
Mineralization lag time[b]	Mlt	day	Average time between deposition of osteoid and initiation of mineralization = O.Wi/MAR
Bone formation rate/bone surface	BFR/BS	μm^3/μm^2/year	Rate of bone formation, surface referent = MAR*MS/BS $\times$ 365
Activation frequency	Ac.f.	number/year	Frequency of appearance of new remodeling units at a given location = (BFR/BS)/W.Wi

[a]This table represents some of the most common 2D histologic measures and calculated variables. For a more complete list of variables please see works by Parfitt, Recker, and Dempster in the suggested readings.
[b]The equation O.Wi/MAR is more traditionally referred to as osteoid maturation time (Omt), while mineralization lag time is calculated as O.Wi/Aj.AR. Aj.AR (adjusted apposition rate) is calculated as MAR $\times$ (MS/OS) and represents the mineral apposition rate averaged over the entire osteoid surface. Aj.AR is seldom calculated; thus MAR is presented in the calculation of Mlt. More details regarding these variables can be found in several of the suggested readings.
BMU, basic multicellular unit; ROI, region of interest.

remodeling during the period of labeling and separate these from processes that may have been preexisting.

Primary versus Derived Variables

Two types of variables are used in histomorphometric analysis. *Primary* variables are those that are directly measured from a histologic section (Table 7.1). These can be either static parameters (total bone volume) or dynamic parameters from labeled sections (mineralizing surface). Variables that require some calculation are called *derived*. For instance, when bone volume is normalized with a referent (bone volume divided by total tissue volume), it becomes derived from the original measurements. In fact, the most useful variables are those that are derived in some form. Specifically, they allow (1) comparison across groups by normalizing the measurements to a common referent (see below) that also may differ between the groups and (2) calculation of rates and remodeling processes that cannot be directly visualized or measured through the microscope. One difficulty with derived measures is that each time a calculation is performed it increases the variability of that particular parameter because each measurement is associated with its own standard deviation. Consequently, some derived bone measures, such as activation frequency (Ac.f), can be wildly variable, thereby minimizing their utility in some cases. However, the importance of derived variables for the purpose of referents outweighs this limitation.

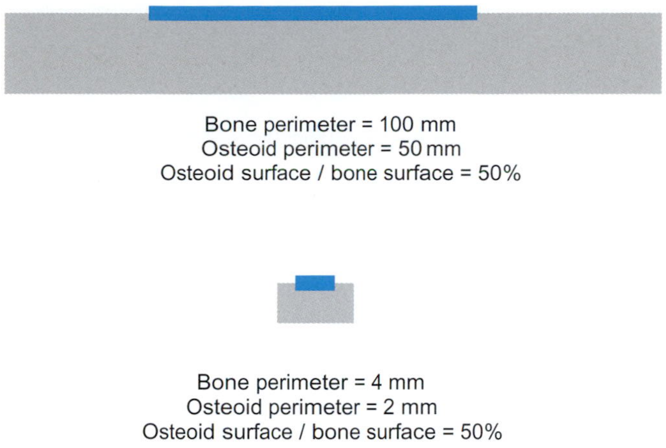

Referents

Quantification of cells, osteoid, and fluorochrome label should be normalized by a referent, i.e. some standard measure across all samples. Common referents include tissue volume, bone volume, and bone surface. The importance of referents can be illustrated with the following example (Fig. 7.9). Imagine two bones with dramatically different amounts of bone (for example, one with 100 mm of surface compared to one with 4 mm of bone surface). If in both cases the osteoid surface is half of the total bone surface, then it will constitute 50 mm of surface in the first case but only 2 mm of surface in the second. In absolute terms (mm of surface), the osteoid surface is lower in the bone with less surface and so, without accounting for bone surface, the conclusion would be that less bone formation is taking place in this bone. One might even conclude that there is a problem with osteoblast activity. Yet, if the data were normalized to the bone surface, in each case the osteoid surface would constitute 50% of the existing surface upon which it is possible to deposit bone. Having normalized the measurement, one would then conclude that the amount of bone formation at the time of tissue collection was the same in each case. Thus, normalizing the measurements leads to quite different conclusions than when no referent is utilized.

Bone Architecture and Geometry

Cancellous bone volume can be measured on histologic sections. A more detailed assessment of trabecular architecture, such as thickness, number, and separation can be calculated from the primary measures of bone area and surface, but the equations employed make a number of assumptions about the structures (such as whether they are rod-like or plate-like). Cortical geometry, such as bone area and periosteal and endosteal perimeters can be directly

Bone perimeter = 100 mm
Osteoid perimeter = 50 mm
Osteoid surface / bone surface = 50%

Bone perimeter = 4 mm
Osteoid perimeter = 2 mm
Osteoid surface / bone surface = 50%

FIGURE 7.9 Histologic assessment involves measuring primary variables and then adjusting those by referents. Whether one examines the primary variables or the referent-adjusted variables can significantly affect data interpretation (see text for explanation). Gray box represents bone and blue line represents osteoid.

measured. Due to the advancement of imaging techniques [such as micro-computed tomography (micro-CT)], bone structure analyses are now rarely derived from histologic sections. Several studies have compared bone volume measured by CT and histology and found strong correlations. Thus, despite the three-dimensional (3D) analysis enabled by CT, 2D sampling by histology provides a valid and representative indication of the structural parameters. The added value of a CT analysis is that it provides details on the morphology of the cancellous network (avoiding assumptions of trabecular morphology) and more detailed information on cortical bone geometry.

Tissue Types

Differentiating between woven and lamellar bone tissue can be useful for determining whether bone formation is occurring in a normal fashion. Assessment of lamellar and woven bone is accomplished using polarized light microscopy on unstained sections, although some stains allow collagen orientation to be visualized. As described in Chapter 1, lamellar bone is characterized by a series of parallel laminar sheets, while woven bone is rapidly formed and highly disorganized. When viewed using crossed-polarized light, collagen fibers in bone are birefringent (light and dark patterns; Fig. 7.10). Lamellar bone patterns can be easily observed under polarized light, whereas woven bone is unorganized. The majority of histologic assessments of woven and lamellar bone are qualitative, with papers reporting simple statements such as "all bone was lamellar in nature" or "no woven bone was observed." Alternatively, in pathologic conditions such

FIGURE 7.10 The varying collagen orientation of lamellar bone results in the tissue being birefringent when viewed under polarized light. Polarized light images can be used to distinguish woven versus lamellar bone. In addition, properties of lamellae such as thickness and number can be quantified from these types of images.

as Paget disease of bone (also called osteitis deformans), the presence of woven bone provides a key diagnostic criterion. Lamellar bone, viewed under polarized light, can be assessed in more detail to elucidate features such as the number of lamellae within a given basic multicellular unit (BMU), thickness of lamellae, or the type of lamellar organization (alternating or homogeneous).

Using stains for osteoid, the examination of mineralized versus nonmineralized bone can provide information about changes in the mineralization process. Analysis of osteoid involves measuring the extent of the bone surface covered by osteoid (and then normalizing it by the total bone surface examined) and either the width or volume of osteoid. Although called osteoid volume in the literature this is actually an area (given that it is a 2D assessment). If osteoid width is normal, increased osteoid surface is indicative of greater bone formation. Increased width of osteoid is indicative of a mineralization defect. While it is possible to measure osteoid seams in rats, measuring them in mice is quite challenging due to short mineralization lag time (the time from formation of osteoid to its mineralization), resulting in few osteoid seams being present in normal mouse tissue.

Cell Number and Activity

The extent of surfaces covered with osteoblasts and osteoclasts provides a primary index of how bone formation and/or resorption are altered under various conditions. Osteoblasts can be identified using morphological characteristics (as outlined in Chapter 2) on sections stained with Goldner's trichrome, von Kossa and McNeal's, or even hematoxylin and eosin. Primary outcomes related to osteoblasts include osteoblast surface and their number, both typically normalized to bone surface. Identification of active osteoblasts can be challenging, and their assessment is sometimes restricted to regions with osteoid. In these situations, osteoid is first identified and then partitioned into two categories: osteoid with osteoblasts (active formation surfaces) and osteoid without osteoblasts (inactive formation surfaces). The extent of surface with osteoblasts and the number of osteoblasts are then normalized either to total surface or to osteoid surface. An alternative approach to quantifying osteoblasts, although not commonly utilized, is to perform immunohistochemistry (such as with alkaline phosphatase) and then assess positive cells adjacent to the bone surface.

Osteoclast assessment is often conducted on TRAP-stained sections, although this is not necessary as morphological features can be used to see osteoclasts in most staining preparations. The most commonly used osteoclast outcomes are osteoclast surface and number, with both parameters normalized to total bone surface within the region of analysis. The number of nuclei per osteoclast is less frequently measured but can sometimes be helpful in assessing osteoclast activity. The presence of osteoclasts on the bone surface does not necessarily reflect their activity and thus there has long been an interest in a dynamic method for assessing osteoclast activity. The most commonly employed technique for assessing activity resorptive is to measure eroded (or resorption) surfaces or erosion depth. Erosion surfaces with osteoclasts on them are considered active resorption sites, whereas erosion surfaces without evident osteoclasts are considered inactive. The latter can occur either because the remodeling process was in the reversal phase at the end of the experiment or because the histologic section simply did not pass through an osteoclast that is actively eroding that surface or one adjacent to it (thinking three-dimensionally). Morphologically, eroded surfaces are defined as scalloped surfaces and, on human bone, are fairly easy to delineate because the majority of surfaces are relatively smooth (Fig. 7.14A). Rodent bone presents more of a challenge for assessing resorption surfaces since the majority of surfaces are not smooth (Fig. 7.14B). In cross-section, active resorption sites are visible as scalloped resorption holes (the cross-section of a cutting cone) within cortical bone, or divots on trabecular or endocortical surfaces. Measures of erosion depth are feasible in clinical biopsies and in larger animals, but difficult in rodents. In cortical and cancellous bone, erosion depth is estimated by projecting the preexisting bone surface and measuring maximum depth down to the base of the resorption lacuna (Fig. 7.14C). Various techniques based in stereology exist for

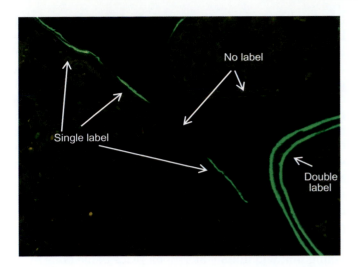

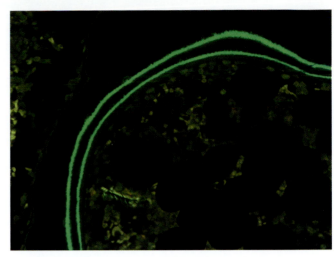

FIGURE 7.11 A primary measurement of fluorochrome-labeled bone is the extent of surfaces having single and double label. All bone surfaces within a region of interest are classified as single, double, or no label. Total bone surface (BS) is calculated as the sum of these three classifications. Mineralizing surface (MS) is calculated as the total double label surface plus one half of the single label surface. The resulting parameter is mineralizing surface/bone surface (MS/BS; %) and is generally considered an index of the extent of osteoblast activity such that interventions that stimulate or inhibit osteoblast proliferation and/or differentiation would be expected to have increased or decreased MS/BS, respectively.

FIGURE 7.12 Along with mineralizing surface/bone surface (MS/BS), a second primary outcome variable in fluorochrome-labeled bone is the distance between the two labels on surfaces with double label. The distance between the two labels (interlabel distance) is measured and averaged across all double labels. Dividing the interlabel distance by the number of days between label administration results in the mineral apposition rate (MAR), reported in μm/day. MAR is often considered to be an index of osteoblast vigor such that interventions that stimulate or suppress cell activity would have higher or lower MAR, respectively.

determining erosion depth (see works by Eriksen and Parfitt). A variable related to erosion depth from previous remodeling activity is average wall width (W.Wi), a measure of the amount of bone formed at a given BMU (Fig. 7.14C). The balance between W.Wi and erosion depth determines BMU balance. In cortical bone W.Wi is the radius of the osteon minus the radius of the central canal. In cancellous bone, W.Wi is the distance from the cement line to the BMU surface. In order to assure accurate measures of W.Wi, the analyzed BMU must have completed bone formation activity. There is another assumption inherent in this measurement if one is attempting to assess the results of an intervention; W.Wi is a measure of erosion depth at the time that the BMU was formed, which may have preceded the initiation of the intervention or treatment that is being evaluated. In this case, it is important to only measure W.Wi in BMUs that had a fluorochrome label given during the period of interest, but which do not have active osteoblasts laying down osteoid on the bone surface. These can often be difficult to find experimentally, and so most investigators measure W.Wi without regard to when the BMU was formed.

Dynamic Histomorphometry

Although osteoblast function can be inferred through measures of osteoid, the most commonly used method is the assessment of fluorochrome labels (*dynamic histomorphometry*), because it allows the calculation of rates of modeling and remodeling. On unstained tissue sections, the extent of surface having a single label, double label, or no label is measured (Fig. 7.11). From these primary measures, the mineralizing surface can be calculated as the total double label plus half the single label, which is equivalent to the mean of the separately measured first label length and second label length. This follows the scientific procedure of taking the mean of two separate observations when they are available. Mineralizing surface is reported per unit bone surface (MS/BS; %) by dividing the mineralizing surface by the total bone surface measured. MS/BS is often considered an index of osteoblast activity such that interventions that impact osteoblast proliferation and/or differentiation would be expected to change MS/BS.

In regions with double label, the distance between these two labels is measured—often from the midpoint of one label to the midpoint of the second (Fig. 7.12). This primary distance variable, i.e. interlabel width, is divided by the number of days between label administrations to derive mineral apposition rate (MAR; μm/day). MAR is often considered to be an index of osteoblast vigor at the individual BMU level such that interventions that stimulate or suppress cell activity would have higher or lower MAR, respectively.

The product of MS/BS and MAR is bone formation rate (BFR), which represents the cumulative bone formation activity, including both the number of sites undergoing active formation and the rate at each site. Since this is a composite variable, it is important to recognize that the same BFR value can be achieved through several combinations of MAR and MS/BS (Fig. 7.13). For example, doubling of BFR can occur by increasing the vigor of osteoblasts with no change in

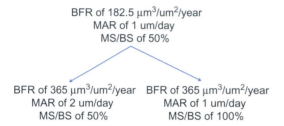

FIGURE 7.13   Bone formation rate (BFR) is calculated from the measures of mineralizing surface/bone surface (MS/BS) and mineral apposition rate (MAR) in fluorochrome-labeled analyses. BFR is often reported in the literature without MS/BS and MAR, as it provides a single value to index the overall formation activity. However, it is important to understand that similar BFRs can exist with varying combinations of MAR and MS/BS. In this example, the baseline rate of BFR (182.5 µm³/um²/year) is doubled through one of two ways. In the case of increasing MAR, the tissue-level mechanism of increased formation activity is through increased vigor of osteoblasts—with no change in the number of sites actively forming bone. In the case of increasing MS/BS, the tissue-level mechanism of increased formation is through increased number of active sites—with no change in the individual osteoblast vigor.

the number of sites actively forming bone. This would be reflected by an increased MAR. Alternatively, doubling of BFR can occur by increasing the number of active remodeling sites. This would be reflected by an increased MS/BS with no change in the individual osteoblast vigor. It is also possible for a BFR designation of *normal* to be caused by abnormal increases in either MS/BS or MAR and decreases in the other. Given the unique and important information of each variable, it is recommended that all three parameters are calculated and reported.

Another parameter often calculated from these dynamic variables is Ac.f. This is often assumed to represent the birth rate of new remodeling units, but it is actually a measure of the probability that a new remodeling unit will initiate at a given place in the bone over time (Ac.f does not measure the birth rate because it also depends on the rate and duration of activity of the BMU remodeling unit). Ac.f is derived by combining BFR with the amount of tissue remodeled at the BMU level. Activation frequency is often reported in units of time (typically years) and provides the best tissue-level assessment of the turnover rate. However, because it is a derived variable, it is usually more variable than BFR, which can make it difficult to detect differences between groups subjected to an intervention.

Dynamic bone properties can be assessed on trabecular, periosteal, and endocortical surfaces using the methods described above, in which surfaces are measured and normalized to total surface. In humans, and in animals that undergo intracortical remodeling, it is

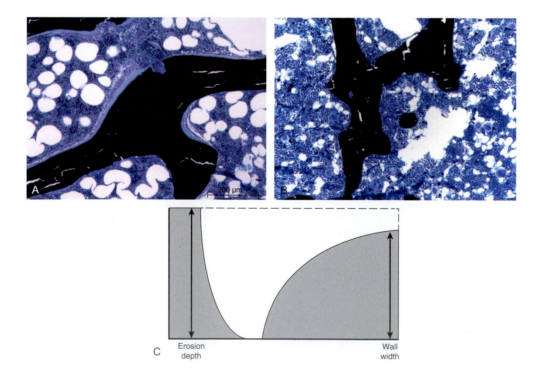

FIGURE 7.14   Assessment of eroded surfaces provides an index of resorption activity. Morphologically, eroded surfaces are defined as scalloped surfaces that may or may not have osteoclasts within them. In human bone, eroded surfaces tend to be fairly easy to delineate because the majority of surfaces are relatively smooth (A). In rats and mice these measures are much more subjective since the majority of surfaces are *not* smooth (B). (C) Erosion depth is defined as the distance from the preexisting bone surface to the base of the resorption lacuna. Wall width is a measure of the amount of bone formed at a given basic multicellular unit (BMU). The balance between wall width and erosion depth determines BMU balance (see Chapter 4 for more details).

also possible to measure BFR within this envelope. This is accomplished by assessing labeled osteons, measuring the extent of single and double label, and then normalizing parameters to total bone area. In this case, BFR is calculated as a percentage of the total area, often reported as %/year.

An emerging and important question with respect to dynamic histomorphometry is how to interpret assessment when double label is not present. A lack of double label can occur when both labels are not properly administered or, more commonly, when the rate of bone formation is such that it is not captured by the labeling schedule. The best illustration of this is in the case of low BFRs, either due to pathologic conditions or remodeling suppression by drugs. In conditions of normal remodeling, a region of bone is likely to have a relatively large number of remodeling events, and thus the probability of catching one or more of these formation events during the 2—3-week period of labeling tends to be high. If the number of sites undergoing remodeling is reduced by 70—90%, then this significantly reduces the probability of capturing a region that is forming bone during this time frame. While the duration between successive labels can be adjusted, this is not commonly done, as the degree of suppression is not typically known in advance. In practice, if no double label is present in a histologic section (either only single label exists or there is no label at all), then there is no way to calculate MAR. Thus, BFR is either 0 (if there is no label at all) or a missing value (if single label is present but no double labels exist). In cases where there is single label without double label, MS/BS can be reported as measured, while MAR can be handled in one of two ways. Early work from human biopsies revealed the lower range for accurate measurement of MAR was 0.3 μm/day. Hence, several papers have imputed a value of 0.3 for MAR when single but no double label is present. This method of imputing a MAR value allows the calculation of BFR. Alternatively, MAR has been considered to be a missing value, and thus BFR calculation is not possible, making it a missing value.

An alternative means of assessing resorption surface utilizes fluorochrome labels in a slightly different fashion than for assessing formation. In this technique, a fluorochrome label is given at time zero and a baseline group of animals is then euthanized and the extent of surface label is measured. A number of days later, perhaps after a treatment suspected to affect resorption, a second group of animals is euthanized and the extent of label is assessed. The difference between the label in the baseline animals and the experimental animals is assumed to occur by resorption. Although not routinely used, this approach does afford a means of estimating resorption "activity" in rodents.

Osteocytes

Histologic assessment of osteocytes is primarily focused on cell number and status. Although lacunar number is often used for this, the number of lacunae does not necessarily reflect the number of viable osteocytes. Living osteocytes produce lactate dehydrogenase (LDH), and stains for LDH can determine whether the cell is viable or not. The challenge in this is that staining for LDH has to be done on relatively fresh bone specimens, as either fixation or freezing will kill the cells. The other side of the coin—cell death by apoptosis—is a more common measure of osteocyte health. Terminal deoxynucleotidyl transferase dUTP nick end labeling (TUNEL) staining can be used on paraffin- or plastic-embedded tissue to differentiate osteocytes undergoing apoptosis from those that are viable. A variety of stains for markers of apoptosis (such as caspase-3 or -8) can also be used to assess bone cell health. The number of apoptotic osteocytes relative to total osteocytes is the most commonly reported outcome.

Microdamage

Despite emerging techniques that allow quantification of microdamage using CT imaging (see Chapter 5), histologic quantification remains the gold standard. Bone is stained en bloc with basic fuchsin, which penetrates the bone tissue and fills all of the voids including Haversian canals, lacunae, and microcracks. Alternatively, bones can be stained en bloc with fluorochromes such as calcein; these do not fill the voids but rather bind to the surfaces of the cracks. Bones are then embedded in plastic and sectioned. This prevents the staining of damage created by histologic processing, but is limited by the possibility that the stain does not penetrate all voids or cracks and probably underestimates the amount of damage. Fuchsin-stained cracks can be visualized with brightfield or ultraviolet (UV) microscopy (fuchsin has fluorescent properties); fluorochrome-stained cracks are visualized using UV microscopy (see Chapter 1, Fig. 1.21). Key outcome parameters include crack number and crack length, with calculations of crack density (crack number/bone area) and crack surface density (crack number × length/bone area).

Marrow

The two most commonly assessed features of the bone marrow are fibrosis and adiposity. In some pathologic conditions, such as hyperparathyroidism, the marrow becomes highly fibrotic (Fig. 7.15). Assessment of marrow fibrosis is routinely made on a percentage

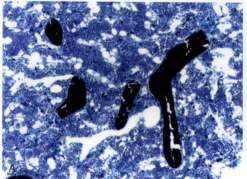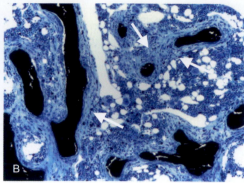

FIGURE 7.15 Marrow fibrosis can be quantified on sections stained for bone analysis. Normal marrow is highly cellular (A), while in some conditions fibrous tissue (shown by arrows) is produced and accumulates (B). Fibrotic areas are typically measured and reported relative to total marrow area.

basis, by measuring the relative amount of marrow that is cellular versus fibrotic. Marrow adipocyte number and area are commonly reported and are expressed relative to total marrow area.

ASSUMPTIONS AND TECHNICAL ASPECTS

It is important to understand that a number of assumptions are associated with bone histomorphometry. The greatest of these is the assumption that bone turnover is in a steady state at the time of analysis. This means that for a given period (typically the time it takes for one complete remodeling cycle) there has been no alteration in the signals that govern osteoblast or osteoclast development and function. For example, if a biopsy is taken too soon after an intervention that stimulates bone turnover, there would probably be an increase in osteoclast surface without a concomitant increase in osteoblast surface or fluorochrome-labeled surface (as there has been insufficient time for formation to increase). This would represent a transient deficit in bone volume that might not reflect the underlying long-term bone balance. The interpretation also assumes that the 2D data can be translated into the 3D structure, i.e. what you measure in one section is representative of the whole region. This assumption can be strengthened by assessing more than one section (ideally not adjacent sections) or, in some cases, by mathematical transformation of the measured data.

There are a number of important technical and operational aspects of bone histomorphometry. For example, it is important to make measures on high quality sections. If the tissue sample is cracked or incomplete, the marrow adjacent to the trabeculae is pulled away, there

are significant folds in the section, the staining is incomplete or too heavy, or fluorochrome labels are too weak, then the quality of the data could be reduced. It is also important to ensure that sufficient tissue is sampled, especially for parameters that have low percentages (e.g. osteoclast number). In some cases, this means sampling a large area on a single slide, while in other cases this necessitates measuring several slides. For human biopsies, the target tissue area to be sampled is 30 mm^2 and the target bone perimeter is 60 mm. In rats, the target tissue area is approximately 6–8 mm^2, while the target bone perimeter is approximately 25 mm. Target tissue area and perimeter in mice is approximately 3–4 mm^2 and 12 mm, respectively.

HISTOLOGIC FEATURES OF DISEASE AND TREATMENT

Osteoporosis

The clinical criterion for diagnosing osteoporosis (and osteopenia) is based on low bone mineral density from dual-energy X-ray absorptiometry (DXA) measures (see Chapters 5 and 16). There are a number of pathophysiologic mechanisms that result in low bone mass but, histologically, they all result in a loss of cancellous and cortical bone mass. Cancellous bone volume (BV/TV) is reduced and those trabeculae that remain tend to be thinner and less connected (Fig. 7.16). Cortical bone becomes more porous due to a greater number of resorption cavities that either do not fill in or fill in less than their original amount.

At a cellular level, osteoporosis can have several characteristic combinations of osteoblast and osteoclast activity. These features can be inferred through serum levels of biomarkers, but direct assessment can only be

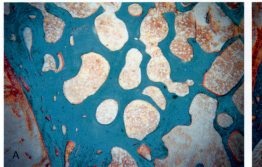
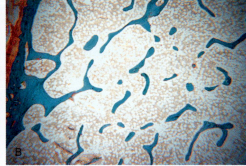

FIGURE 7.16 The hallmark of osteoporosis is low bone mass. Biopsies from healthy (A) and osteoporotic (B) individuals show the clear effects of the condition. In osteoporosis, cancellous bone volume is lower, individual trabeculae tend to be thinner, and there is less connectively among trabecular struts. Cortical bone is also thinner.

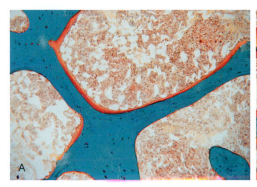
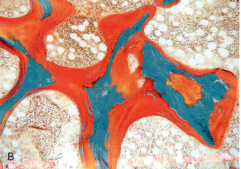

FIGURE 7.17 Osteoid accumulation is an important characteristic used for diagnosing osteomalacia. Individual seams in normal bone are relatively thin (A), while in osteomalacia they are significantly thicker due to a deficiency in mineralization (B). There is also an increase in the number of surfaces covered with osteoid.

achieved through histologic measures. The combination of effects includes:

1. *Low formation with normal or high resorption.* Bone volume is lower and accompanied by reduced osteoid surface or labeled surface, while the number of osteoclasts is either normal or high.
2. *Normal formation with high resorption.* Bone volume is low despite normal bone formation. It would be expected that osteoclast surface is increased, but in some cases, it may be normal (if osteoclasts are digging deeper).
3. *Low formation and low resorption with greater relative reduction in formation.* Bone volume is low and accompanied by both low formation and low resorption. Detecting the relative difference between formation and resorption (i.e. to confirm that one is changing more than the other) may be challenging due to limitations in measuring dynamic bone resorption.
4. *High formation and high resorption with greater relative increase in resorption.* Bone volume is low and accompanied by high formation and high osteoclast surface. As in (3), detecting the relative difference

between formation and resorption may be challenging.

It is possible that differences between formation and resorption may not be entirely consistent, due in part to the fact that osteoclast surface does not really represent the best index of osteoclast function. That is, the number of osteoclasts may be unchanged but the activity of the cells is increased or decreased (in the latter case as in bone treated with antiresorptive agents)—this would be impossible to capture with the static measure of osteoclast surface.

Other key histologic features of osteoporosis include a reduction in W.Wi (negative bone balance), driven by either increased activity of osteoclasts or reduced activity of osteoblasts (or a combination of both).

Osteomalacia

Osteomalacia exists when newly formed bone matrix does not mineralize in a timely manner. This leads to accumulation of thick osteoid seams and, in some cases, a larger than normal osteoid volume (Fig. 7.17). In extreme cases, osteoid can represent up to 40% of the total bone volume. A related, but distinct, condition to osteomalacia is defective mineralization, a condition characterized by increased volumes of osteoid due primarily to increased amounts of osteoid surface (osteoid width tends to be normal). Histology plays an essential role in diagnosing osteomalacia because imaging assessment shows low bone mass when in fact bone volume may be normal, although a significant portion of it is not mineralized. Because treatments for correcting deficiencies in mineralization are distinctly different from those used for treating osteoporosis, distinguishing osteomalacia from low bone mass is important for making treatment decisions.

Osteoid surface, osteoid volume, and osteoid width are each important to help identify osteomalacic bone. For example, large amounts of osteoid surface do not necessarily indicate osteomalacia, as increased rates of bone formation accompanied with normal mineralization rates would result in greater osteoid volume, but not in thicker osteoid seams. Rather, osteomalacia (sometimes called *active osteomalacia*) occurs when osteoid volume is increased due to both an increase in osteoid surfaces and osteoid width. Osteoid width of >12 μm is generally considered to be an indication of osteomalacia. Bone remodeling may be high, shown by greater levels of osteoblast and osteoclast surface, although this is not always the case. Fluorochrome labels tend to be diffuse (as opposed to the normal crisp label appearing in Fig. 7.12) due to the slow rate of mineralization at formation sites, thus making them difficult to analyze. It is not uncommon to see woven bone or marrow fibrosis in these conditions.

Mineralization lag time (Mlt), i.e. the mean time between deposition and mineralization of a volume of matrix, is sometimes calculated to confirm osteomalacia (Table 7.1). Mlt is calculated as osteoid width divided by the adjusted appositional rate (Aj.AR), where Aj.AR is the product of MAR and labeled surface over osteoid surface. In situations where osteoid width is elevated with no change in osteoid surface or MAR, Mlt is increased.

Hyperparathyroidism

Primary Hyperparathyroidism

Increased parathyroid hormone secretion due to direct effects of the parathyroid gland elevates bone remodeling well above normal rates. There is an increase in the number and size of both osteoblasts and osteoclasts. Osteoid surface, MS/BS, BFR, and Ac.f are all higher in these patients and marrow fibrosis develops in some cases. Although osteoclast surface is higher, erosion depth is actually lower, indicating reduced osteoclast activity. The combination of lower erosion depth

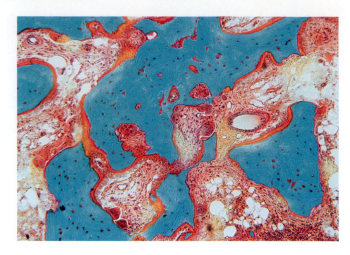

FIGURE 7.18 Paget disease typically presents with a histologic feature of a *moth-eaten* morphology, where the surfaces are highly irregular due to the rampant bone resorption and disorganized bone formation.

and normal or high formation results in a positive BMU bone balance and an increase in W.Wi (in cancellous bone). This helps explain why patients with primary hyperparathyroidism have normal or even elevated cancellous BV/TV: despite the higher turnover, there is a positive BMU balance to help preserve the structure. Cortical bone is not as fortunate and is the primary site of destruction in primary hyperparathyroidism. There is an increase in cortical porosity and a reduction in cortical width. This loss of cortical bone puts these patients at higher risk for hip fracture.

Secondary Hyperparathyroidism

Increased parathyroid hormone levels in the absence of parathyroid gland pathology can have numerous etiologies including renal failure, hypocalcemia, parathyroid hormone (PTH) resistance, low levels of 1,25-dihydroxyvitamin D_3 [1,25(OH)$_2$D$_3$], and low phosphate. Secondary hyperparathyroidism tends to present with high bone remodeling, along with increases in both osteoblasts and osteoclasts. The bone formed is predominantly woven and there can be significant marrow fibrosis. Cancellous BV/TV tends to be high but this is due to the significant accumulation of woven bone.

Paget Disease of Bone (Osteitis Deformans)

Paget disease is characterized as a defect in bone remodeling that is often (but not always) localized to a given skeletal site (see Chapter 16). A key histologic feature of Paget disease is an increase in the number and size of osteoclasts. The number of nuclei is also typically greater, with reported instances of over 100 nuclei in some cells. Resorption lacunae are increased

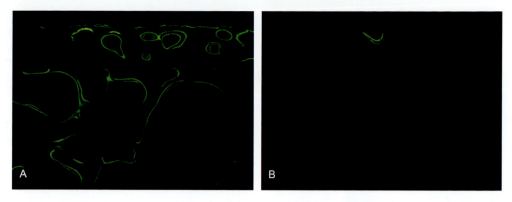

FIGURE 7.19 The goal of antiresorptive treatment is to reduce the number of active remodeling units. This can be visualized histologically by assessing fluorochrome labels. Under normal conditions (A), there is bone remodeling activity on trabecular, endocortical, intracortical, and periosteal surfaces. In response to antiresorptive treatment (B), this activity is greatly reduced. The degree of reduction is dependent on several factors, including the duration of treatment and the potency of the antiremodeling agent.

in number and depth is greater (as assessed by estimations of W.Wi). This results in cancellous bone often having an erratic *moth-eaten* morphology (Fig. 7.18). The enhancement of osteoclasts triggers a compensatory stimulation of osteoblast number and activity. Contrary to the abnormal osteoclasts, the osteoblasts in Paget disease are intrinsically normal, yet the drive to produce sufficient matrix to keep pace with osteoclast resorption results in the formation of woven bone. The increased bone turnover occurs in many different areas, but can be quite local, resulting in the radiologic appearance of patchiness, with areas of dense bone interspersed with bone that appears normal. Even so, overall bone mass tends to be higher due to the rampant bone formation, although a significant portion of this bone is woven, and thus of inferior quality. The presence of this woven bone matrix is a key histologic feature of Paget disease. Fluorochrome labels are diffuse in Paget disease due to the irregularly formed woven bone. Osteoid volume may be higher because of the rapid turnover but the seams tend to be of normal width and thus the bone is not osteomalacic. It is not uncommon to find significant marrow fibrosis in patients with active Paget disease. Over time, regions with high activity may become less active, yet the mosaic pattern of woven and lamellar bone remains making it possible to diagnose Paget disease even if it is not active. Treatment of Paget disease involves suppression of the rampant osteoclast activity using antiresorptive agents. The efficacy of these treatments supports the primary role of osteoclasts in the disease.

Antiresorptive Therapy

Treatments that reduce resorption are collectively known as *antiresorptive agents* (also referred to as antiremodeling or anticatabolic). These include estrogen (hormone replacement therapy), selective estrogen receptor modulators, calcitonin, bisphosphonates, and denosumab (see Chapter 17). They function to either slow osteoclast activity or slow the development and maturation of osteoclasts.

The main histologic characteristic of antiresorptive treatment is a reduction in the dynamic variables, specifically MS/BS, BFR, and Ac.f (Fig. 7.19). As outlined in Chapter 4, bone remodeling is coupled with osteoblast activity subsequent to osteoclast activity at the level of the BMU. Since dynamic assessment of bone resorption is challenging, assessment of dynamic bone formation provides a surrogate for overall activity. Treatment with antiremodeling agents produces a robust reduction in Ac.f, i.e. the number of active remodeling sites, on trabecular, endocortical, and intracortical surfaces (Fig. 7.19). This is reflected, and most often reported in the literature, as a reduction in MS/BS and BFR (recall that Ac.f. is calculated using BFR, which is in turn calculated using MS/BS).

The degree of remodeling suppression is drug dependent, even within a given class of agents such as the bisphosphonates (see Chapter 17, Fig. 17.7). For the most potent suppressing agents, the challenge becomes having sufficient double label to enable calculation of MAR and subsequently of BFR and Ac.F. Prior to the publication of recommendations in 2011, there was no standard for how to address these problems and many papers differed in their handling. Some papers simply excluded subjects in whom two double labels were not found, while others used an imputed value of 0.3. Both these methods of assessment have significant effects on interpretation (Fig. 7.20). The current recommendation is to present the data both with and without the imputed values, and also to present the total number of single and double labels.

Alterations in osteoclast parameters are mixed in the setting of antiresorptive treatment. Agents that

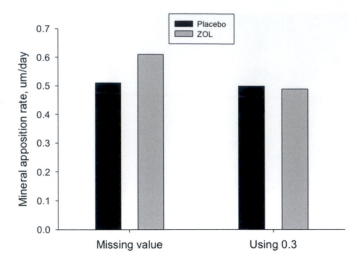

FIGURE 7.20 The approach to assessing specimens lacking double label can have a significant effect on conclusions. A study on the effects of zoledronic acid (ZOL), a bisphosphonate, on dynamic properties of iliac crest cancellous bone studied 59 zoledronate-treated patients and 52 placebo-treated patients. Of these, four placebo-treated patients had issues that precluded assessment of mineral apposition rate (MAR; two patients had double label only in cortical bone, one had three labels, and one had insufficient double label surface to reliably measure). In the zoledronic acid group, 16 patients had double label only in cortical bone, three had only single label and two had evidence of previous tetracycline label. The authors chose to exclude all of these patients from the assessment of MAR, entering the value as missing data and drew the conclusion that patients treated with zoledronic acid had significantly higher MARs. In all cases, there was evidence of some bone formation activity (either single or double label) and thus it would be acceptable to impute a value of 0.3 for those patients to indicate that formation was occurring but at a low level. When imputed values are used, the zoledronic effect on MAR disappears. This illustrates the dramatic impact that such decisions make on conclusions and has resulted in recent recommendations that the way in which data is handled should be explicitly outlined in papers and, ideally, data be analyzed with and without imputing values. *Data from Recker R, et al. Journal of Bone and Mineral Research 2008;23:6—16.*

suppress development or maturation of osteoclasts (e.g. denosumab) have lower osteoclast surface and erosion surfaces. However, agents that simply inhibit osteoclast activity may result in no change or even in greater osteoclast surface compared to untreated individuals. This is a good example of why simply measuring osteoclast surface by itself is not sufficient to understand what is happening with bone remodeling, as the number could be high but activity low. Antiremodeling treatments do result in reduced W.Wi and, more generally, in reduced BMU size either because osteoclast activity is reduced or because there are fewer of them within a given BMU.

The effects of antiremodeling treatments on bone modeling are less clear, although they do differ among the various agents. Estrogen inhibits bone modeling, most notably at the periosteal bone surface. This is evident through experiments in which animals undergo ovariectomy (to remove endogenous estrogen), which results in enhanced periosteal BFR due to the loss of estrogen inhibition. When these animals are treated with estrogen, periosteal BFR is returned to normal. Similar results have been observed in humans, although the histologic data are less convincing because they are measured on the periosteal surface of the iliac crest, where there tends to be little modeling. Reports on the effects of bisphosphonates on modeling are conflicting: some animal studies have shown inhibition of periosteal MAR with bisphosphonate treatment, while others have shown no effect.

Reductions in bone resorption translate into increased cancellous and cortical bone volume. Most of the gain in bone volume occurs in the early phase of treatment. BMUs that were previously initiated finish the remodeling cycle, while few new BMUs are initiated. This results in a tissue-level balance of more bone formation than bone resorption. Trabecular bone volume is increased (or, at a minimum, any reduction is averted), whereas cortical porosity is reduced.

One consequence of reduced remodeling is the accumulation of microdamage. This has been clearly documented in animal experiments, in which bisphosphonate-induced reductions in remodeling are associated with increased microdamage. Data in humans are limited and, while some studies have found results consistent with the animal experiments, others have concluded that bisphosphonates are not associated with more damage.

Anabolic Therapy

In conditions of severe osteoporosis, in which a patient's DXA score is below −3.5 to −4.0 and the patient may have had several fractures, slowing bone loss is insufficient as a treatment goal. Teriparatide, i.e. recombinant human PTH(1—34) [rhPTH(1—34)], is the only currently approved anabolic treatment, although several others are in development (see Chapter 17). The histologic characteristics of teriparatide are dependent upon the duration of treatment (weeks to months), differ between cortical and cancellous bone, and are site specific (i.e. because the effects on cortical and cancellous bone are somewhat different, the hip and the spine may show different responses, at least initially).

One of the earliest effects of teriparatide is to stimulate the direct apposition of bone to preexisting surfaces through modeling. This occurs by both prolonging the lifespan of osteoblasts that are currently engaged in bone formation, as well as recruiting bone lining cells to reactivate into active bone-forming osteoblasts. Suppression of osteoblast apoptosis has been observed in biopsies from teriparatide-treated individuals, while conversion of lining cells to active osteoblasts has been documented in animals. These events can be observed as increases in osteoid surface and in MAR and MS/BS on trabecular surfaces in just a few days (in animals) or weeks (in humans), a time period inconsistent with the effects being mediated through enhancing remodeling. Modeling activity also may be stimulated on the periosteal bone surface in animal models, but whether this effect occurs in humans is still controversial and difficult to measure. These early effects on bone formation create an *anabolic window*, during which teriparatide produces its greatest increase in bone mass.

Over time, teriparatide also stimulates osteoclast activity, resulting in enhanced bone remodeling. Increases in MAR, MS/BS, and BFR occur, as do increases in osteoclast surface and resorption spaces. The enhanced bone remodeling activity with teriparatide is unusual compared to other conditions and treatments, as osteoblast activity is enhanced relative to osteoclast activity. At an individual BMU level, this results in overfilling and a positive BMU balance, which can be measured histologically as an increase in W.Wi of trabecular hemiosteons. This appears to be the result of an increased thickness of individual lamellae.

A somewhat unique characteristic associated with teriparatide treatment is trabecular tunneling, in which BMUs tunnel through individual trabeculae, thus producing two trabeculae instead of one (Fig. 7.21). Tunneling can be observed in untreated bone, but it is much more common with teriparatide treatment. The mechanism underlying this effect has not been determined, although it is probably related to trabecular thickness. Once a certain thickness is achieved, the central osteocytes can no longer obtain nutrients through diffusion; trabecular size is reduced and osteocyte viability maintained by splitting the structure into two. This effect is consistent with the common finding of increased trabecular number and modest changes in trabecular thickness with teriparatide treatment. This effectively increases trabecular connectivity (which may have an effect on bone strength that is independent of the increased mass) and also provides additional surfaces for bone formation, which can accelerate the process of adding bone mass and volume.

Increases in bone remodeling with teriparatide have been shown to be associated with reduced levels of trabecular microdamage, even in patients with prior bisphosphonate treatment and damage accumulation. Biopsies were obtained from patients who had previously been treated with bisphosphonates and then again after 2 years of teriparatide treatment. Baseline biopsies in these patients showed greater accumulation of damage compared to treatment-naive patients. Levels of damage after 2 years of teriparatide

FIGURE 7.21 Trabecular tunneling is commonly observed in association with treatment with the anabolic agent teriparatide. Basic multicellular units tunnel through individual trabeculae (star), effectively producing two trabeculae where one originally existed, thus serving to increase trabecular number and connectivity and reduce mean trabecular thickness.

treatment were significantly lower than baseline and equivalent to damage levels in treatment-naive biopsies. These effects were associated with significantly higher BFRs in teriparatide-treated patient biopsies.

STUDY QUESTIONS

1. Compare and contrast histological processing of bone by paraffin and plastic embedding. What are the strengths and weaknesses of each approach?
2. Describe the difference between static and dynamic histomorphometry and give examples of parameters measured in each type of analysis. How does fluorochrome labeling provide information about bone remodeling activity? What are the most often assessed dynamic variables, how are they calculated, and what information do they provide about remodeling?
3. Describe the difference between primary and derived measurements. What are the most common referents and why are they important for comparing different specimens?
4. How can histomorphometrists address the problems associated with the suppression of bone remodeling? What are the advantages and disadvantages of removing samples from the study or assigning imputed values?
5. What are the main histomorphometric changes associated with antiremodeling and anabolic therapies?

Suggested Readings

An, Y.H., Martin, K.L. (Eds.), 2003. Handbook of histology methods for bone and cartilage. Humana Press, Totowa, NJ.

Boyce, R.W., Paddock, C.L., Gleason, J.R., Sletsema, W.K., Eriksen., E.F., 2009. The effects of risedronate on canine cancellous bone remodeling: three-dimensional kinetic reconstruction of the remodeling site. J. Bone Miner. Res. 10, 211–221.

Cohen-Solal, M.E., Shih, M.S., Lundy, M.W., Parfitt, A.M., 1991. A new method for measuring cancellous bone erosion depth: application to the cellular mechanisms of bone loss in postmenopausal osteoporosis. J. Bone Miner. Res. 6, 1331–1338.

Dempster, D.W., Compston, J.E., Drezner, M.K., Glorieux, F.H., Kanis, J.A., Mallulche, H., et al., 2013. Standardized nomenclature, symbols, and units for bone histomorphometry: A 2012 update on the rerport of the ASBMR Histomorphometry Nomenclature Committee. J. Bone Miner. Res. 28, 1–16.

Eriksen, E.F., Axelrod, D.W., Melsen, F., 1994. Bone Histomorphometry. Raven Press, New York.

Eriksen, E.F., Melsen, F., Mosekilde., L., 1984. Reconstruction of the resorptive site in iliac trabecular bone: a kinetic model for bone resorption in 20 normal individuals. Metab. Bone Dis. Relat. Res. 5, 235–242.

Malluche, H.H., Faugere, M.C., 1986. Atlas of Mineralized Bone Histology. S Karger Pub.

Parfitt, A.M., Drezner, M.K., Glorieux, F.H., Kanis, J.A., Malluche, H., Meunier, P.J., et al., 1987. Bone histomorphometry: standardization of nomenclature, symbols, and units: report of the ASBMR Histomorphometry Nomenclature Committee. J. Bone Miner. Res. 2, 595–610.

Recker, R.R., Kimmel, D.B., Dempster, D., Weinstein, R.S., Wronski, T.J., Burr., D.B., 2011. Issues in modern bone histomorphometry. Bone. 49, 955–964.

Recker, R.R., 1983. Bone Histomorphometry: Techniques and Interpretation. CRC Press, Boca Raton, FL.

Skeletal Genetics
From Gene Identification to Murine Models of Disease

Kenneth E. White[1], *Daniel L. Koller*[1] *and Tim Corbin*[2]

[1]Department of Medical and Molecular Genetics, Indiana University School of Medicine, Indianapolis, Indiana, USA
[2]Department of Therapeutic Discovery, Transgenic Division, Amgen Inc., Thousand Oaks, California, USA

Many disorders of the human skeleton result from DNA variants that can be transmitted from parent to child via the normal meiotic processes. We refer to these particular transmissible traits as genetic disorders. Two dichotomies should be kept in mind when considering genetic disorders. First, we contrast rare disorders inherited in Mendelian (single-gene) fashion via defined modes of inheritance with those that are much more common in the human population and typically contain variants in many genes. Second, and strongly related to the first point, is the relative strength of the effect of a given variant on the bone phenotype in question. In bone disorders (and in nearly all human diseases), variants contributing to rare Mendelian traits typically have very large effects on gene function, often with a single variant being solely responsible for the observed phenotype. For more common diseases containing a heritable component, however, each genetic variant typically has a small effect, with the sum of these effects producing the observed phenotype. Once a variant or gene is found that links a particular phenotype to bone structure or function, or to a Mendelian bone disorder, these variants can be studied in vivo through the use of mouse models. The use of these models then becomes critical to the understanding of bone function and for targeted therapeutics; this will be discussed in the second portion of this chapter.

SIMPLE AND COMPLEX TRAITS

The mapping of rare Mendelian skeletal traits involves localizing a disease gene and variant to a fairly large chromosomal segment. This is achieved using the principles of linkage mapping in extended human pedigrees that demonstrate transmission of the skeletal trait from one generation to the next. It is important to distinguish between simple and complex traits that lead to genetic disorders. The two types of traits differ in their *genetic architecture*; this is a composite of several properties defining the genetic contribution to disease. These include the frequency of the genetic variant(s) involved, the *effect size* of the variants implicated, and the methods employed to localize and identify the genes involved. The genetic architecture of a particular genetic trait also strongly affects the public health impact of the disease and its genetic findings, as well as the interpretation of subsequent biological studies.

Linkage Mapping of Bone Traits

The mapping of a simple Mendelian genetic trait exploits a few fundamental biologic processes, notably chromosomal inheritance, meiosis and the resulting *linkage* observed between nearby loci on the same chromosome. Fig. 8.1 illustrates the meiotic recombination process that results in the linkage phenomenon. During gamete (sperm or egg) formation, specifically the stage referred to as meiosis I, homologous chromosomes pair and *crossing over* occurs at a random position along the chromosomal length. In the particular meiosis shown in Fig. 8.1, this crossover occurs between the adjacent (inner) chromatids roughly two-thirds of the way down the long arm from the centromere. This results in the production of gametes containing different combinations of genetic material that existed in either parent, as

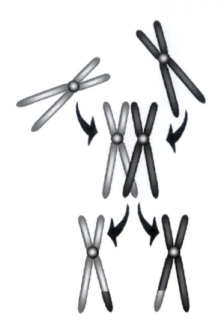

FIGURE 8.1 **The process of meiosis I, which results in the linkage phenomenon and increased genetic variability in diploid organisms.** For simplicity, only one pair of homologous chromosomes is shown, with shading distinguishing material of maternal (light gray) and paternal (dark gray) origin. Further division at meiosis II will result in four gametes, two consisting of material entirely from one parent, and two with recombined material.

the lower third of this chromatid arm has been swapped between the maternal and paternal chromosomes. Particular alleles of genes located above and below this breakpoint may now be passed on to offspring in combinations that would not occur without recombination. Importantly for linkage mapping purposes, genes that are physically close together are more likely to be coinherited; conversely, genes that are physically far apart on the chromosome are less likely to be coinherited. Genes on different chromosomes are always inherited independently.

Linkage mapping of an autosomal dominant trait observed in a very small number of pedigrees can be illustrated using the case of a high bone mass (HBM) phenotype. The phenotype is a strongly elevated bone mass [i.e. hip and spine bone mineral density (BMD) of four standard deviations above the population mean without other clinical sequelae]. The trait was fortuitously discovered on X-ray evaluation of a mother and daughter following a traffic accident. The phenotype was found to segregate in a single large pedigree (Fig. 8.2) and is considered to be autosomal dominant due to direct parent-child transmission of the trait, and both male and female offspring being affected in approximately equal numbers.

Mapping of this type of trait is normally achieved via a genome screening approach. This method seeks to identify, in families, chromosomal regions that are

consistently transmitted to affected individuals. Typically, the expectation is that only one such region in the genome (containing the disease gene or variant) will be identified. The method consists of analyzing markers located at regular intervals throughout the genome to evaluate the chromosomal regions. These genetic markers can be of several varieties [e.g. single nucleotide polymorphisms (SNPs) and microsatellite markers]. These markers are useful in that they occur frequently across the entire genome, thus enabling genome-wide screens with a single type of marker and corresponding genotyping technology. Another critical property of useful markers for linkage studies is that they are polymorphic, i.e. they have two or more distinguishable forms or *alleles* across the population.

Mapping the HBM phenotype pinpointed the trait to chromosome 11q (the short arm of human chromosome 11, near the position of microsatellite marker D11S1313; Fig. 8.2). Among all regions of the genome that were tested (several hundred DNA markers were involved), the only region demonstrating consistent transmission from affected parents to children was the region of chromosome 11 indicated by the black-shaded bar. The disease can be localized specifically to the upper part of the region depicted as the shorter black region (denoting a recombination event between the second and third markers shown; see legend). The logarithm of the odds (LOD) score, an important measure of the evidence in favor of presence of the disease locus and a marker, corresponds to the odds in favor of linkage based on the pedigree structure, phenotypes, and marker data shown. The pedigree and marker data shown correspond to odds of nearly 1 million to one in favor of linkage to this region, as the LOD score is derived from a base 10 logarithm. A LOD score of 3 (indicating 1000:1 odds in favor of linkage) is traditionally used as the threshold for significant linkage between a marker and disease locus.

Linkage mapping can also be used to identify regions harboring autosomal recessive disease loci. As an example, a rare osteoporosis-pseudoglioma syndrome exists in several pedigrees (Fig. 8.3). As is often the case, this rare disorder is most commonly observed in pedigrees with some degree of consanguinity or inbreeding. The disease phenotype for osteoporosis-pseudoglioma syndrome is quite severe and includes severe juvenile-onset osteoporosis and congenital or juvenile-onset blindness; it has been observed that unaffected carrier parents with one copy of the disease allele have lower BMD on average than the general population. As outlined above for assessment of the HBM trait, a linkage screen was performed, this time across several smaller pedigrees. Unexpectedly, linkage was observed with markers on 11q that overlapped the HBM linkage region. The shared region exists in homozygous form (two copies of the

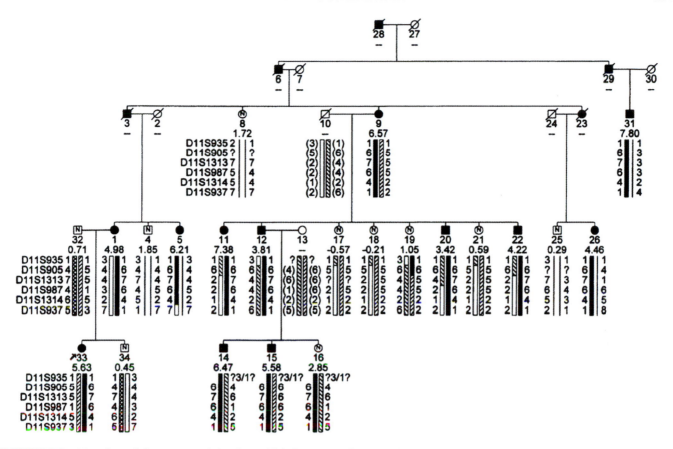

FIGURE 8.2 **Mapping of the autosomal dominant high bone mass locus to chromosome 11q.** Pedigree structure and disease status of each individual in the pedigree is shown, with dark shading of affected individuals. Males are shown as square symbols and females as circles. These individuals' haplotypes for a segment of chromosome 11 surrounding microsatellite marker D11S1313 is indicated below their symbol. Ancestral origin of each haplotype or recombinant segment is indicated by the shading of the bars. For example, a recombination event can be inferred in individual 5 between the last two markers, which places the disease locus above the breakpoint (and therefore above D11S937). *Reproduced with permission from Johnson et al., AJHG 60:1326–32.*

disease allele), again uniquely in the genome, as shown in the figure. Note that different alleles are linked to the disease locus on chromosome 11q in each family (termed *allelic heterogeneity*), but that the same chromosomal region provides evidence for linkage in each family. This differentiates a linked locus from an associated allele; association is observed when a particular allele at a locus (not a locus itself) correlates strongly with the presence of disease. In the examples shown for both autosomal dominant and recessive gene mapping, a gene at the same locus is causative of disease; however, each family can have different alleles for nearby marker loci due to recombination events between the loci across the population's history. However, in a linkage study, the fact that a particular allele segregates with disease in a particular family is what is of interest, and evidence for linkage can be combined across families (via summation of the LOD scores) even when the linked marker allele varies between pedigrees.

An extensive review and sequencing of genes within this common chromosome 11q region for the two diseases surprisingly identified that different, rare mutations in the *LRP5* gene [encoding low-density lipoprotein receptor-related protein 5 (LRP5)] result in both low and high BMD Mendelian phenotypes. Osteoporosis pseudoglioma (autosomal recessive) is caused by inactivating mutations in the *LRP5* gene, while the HBM (autosomal dominant) trait is caused by an activating mutation in the *LRP5* gene. The LRP5 lipoprotein receptor is active in the Wnt-frizzled signaling pathway and has an important role in skeletal homeostasis (see Chapters 2, 3, and 9). This is an excellent example of how the discovery of human bone-relevant mutations in a gene previously not known to impact bone has provided insight via genetic means into skeletal biology.

Genetics and Mapping of Complex Phenotypes

Osteoporotic fracture and its underlying *endophenotypes*, such as the variation between people in BMD, are two of a large number of common or (alternately) complex genetic disorders. The hypothesized causal factors and their action and interaction are described in more detail below. These disorders are typically of great importance from a public health point of view, and clearly have a genetic contribution. These common disorders include autism, epilepsy, cardiovascular disease, and dementia. All of these disorders are observed to cluster in families and particular ethnic groups, thus providing evidence of genetic contribution to disease, but examination of typical pedigrees shows they are not inherited in a simple Mendelian fashion.

Complex disorders are typically assumed to be due to the involvement and action of dozens or more genes, along with multiple environmental factors and the interactions among all these factors. Indeed, the possible involvement of these many factors and their interactions leads to the term *complex inheritance*. It is hypothesized that (possibly multiple different alleles

in) multiple genes contribute to disease susceptibility and that the effect of a gene or genotype may vary depending on the environment in which the individual is placed. These genetic and environmental factors would each be expected to have a modest impact on disease risk, and the individual factors (genetic or environmental) could interact with one other to influence phenotype only when certain combinations of the factors were present.

Several methods have been employed by researchers attempting to dissect the potentially complex networks of factors underlying these types of conditions. As an example, consider the condition of osteoporotic fracture. These fractures typically occur late in life after a large number of environmental factors (e.g. diet and exercise) may have influenced the bones at the hip and spine where such fractures typically occur. Falls as a cause or contributor to fracture would also create a more complex genetic phenotype, with a possible

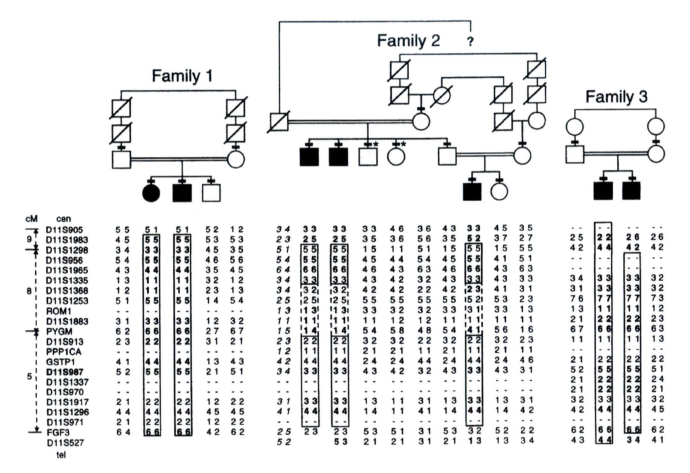

FIGURE 8.3 Mapping of the autosomal recessive osteoporosis-pseudoglioma locus to chromosome 11q. Pedigree structure and disease status of each individual in the pedigree is shown, with dark shading of affected individuals. Males are shown as square symbols and females as circles. These individuals' haplotypes for a segment of chromosome 11 surrounding microsatellite marker D11S1883 is indicated below their symbol. The marker haplotypes indicated by the box around a haplotypic segment in the affected individuals are preserved without ancestral recombination in that particular family. Overlap of these boxes indicates that the disease gene lies below D11S1298 (evidenced by an ancestral recombination and unshared segment above D11S956 in the affected individuals in family 3) and above *FGF3* (with evidence of recombination in family 2). *Reproduced with permission from Gong et al., AJHG 59:146–51.*

genetic component of their own. One way to "simplify" the phenotype in such a case might be to use a clinical or modified clinical definition of "osteoporosis" per se (deviation below mean BMD for young normal) as the phenotype. A further step that is now possible with improved analytic methods would be to consider the quantitative BMD measure itself as the phenotype. Several research groups have performed genetic studies on BMD in younger individuals (premenopausal women or children) under the hypothesis that fewer genes and environmental factors might come into play in determining BMD in these subjects.

Linkage mapping strategies and candidate gene association studies have been employed in the past to evaluate known genes and identify new genetic effects on BMD-related phenotypes, with only modest success. Particularly for linkage studies of complex disease, typical family and pedigree samples of a few hundred individuals lack the statistical power to detect the effect sizes that now seem to be realistic for complex disease-related variants. However, recent advances in genotyping technology have enabled complex disease research groups to employ a genome-wide association study (GWAS) strategy, in which several hundred thousand markers can be typed on (typically) a thousand or more subjects with phenotypic measures. The relatively common SNPs typed on arrays for GWAS studies capture or *tag* nearly all of the common genetic variation throughout the genome, which can then be tested for association with phenotypes of interest such as fracture, BMD, or other

measures. Results from many GWASs across a wide range of phenotypes (bone and many other traits and disorders) have shown that the effect sizes typical of complex disease-associated common variants are difficult to detect (owing to modest statistical power) in typical single research cohorts of a few thousand individuals. For this reason, results of several or many cohort GWASs are typically combined in a meta-analytic approach to take best advantage of a higher sample size. Fig. 8.4 presents the results of such a meta-analysis. Note the SNPs on chromosome 3 (in pink) detected in a GWAS of osteoarthritis, located within the gene encoding nucleostemin (GNL3). Levels of GNL3's transcript are raised in chondrocytes from patients with osteoarthritis. Such studies often detect genes of established skeletal function, as well as novel genes that might provide predictive function or pharmaceutical targets in the future.

The complexity of the genetic architecture of traits such as BMD and fracture risk is illustrated by the *missing heritability* problem that exists for nearly all common complex disorders. Peak BMD has heritability (i.e. a proportion of variation due to genetic factors) of 70–80%; however, GWAS studies detect common variants explaining 10% or less of total variation in the trait. This raises the question of which factors (genetic or otherwise) explain the missing heritability. Hypotheses currently being actively pursued include the unexplained factors consisting of rare variants (investigated using next-generation sequencing), gene × gene interactions (epistasis), and gene × environment interactions.

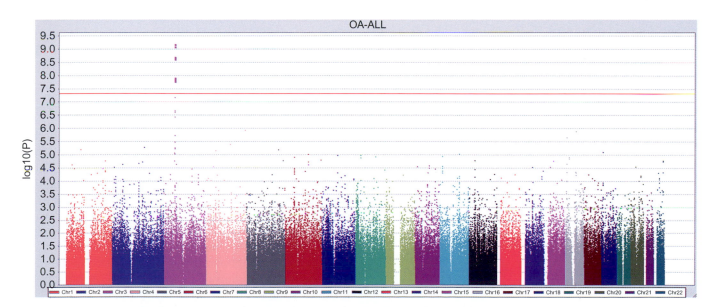

FIGURE 8.4 Results from a genome-wide association study for osteoarthritis. The vertical axis shows the significance level for association of osteoarthritis with approximately 1 million SNPs spread across the human autosomes (1–22), whose chromosomal positions are indicated on the horizontal axis. The line in red corresponds to the commonly used threshold statistical significance for a genome-wide association study, a P value of 5×10^{-8}. *Reproduced with permission from the arcOGEN Consortium, Lancet 380:815–823.*

In future studies, large increases in sample size or alternative analysis methods will be required to begin to identify factors explaining the missing heritability phenomenon. In the short term, next-generation sequencing and the identification of rare bone-related variants may provide complementary information on the common variants studied by the GWAS approach and provide us with a more complete picture of the genetic architecture of these complex bone traits and conditions.

TRANSGENIC TECHNOLOGY AND THE CREATION OF NOVEL RESEARCH MODELS

Once disease loci have been identified, the next step in terms of research may be the creation of an animal model that over- or underexpresses the gene of interest, or expresses a particular mutation. Molecular biology and the use of transgenic animals have played a crucial role in the understanding of modern genetics and the genetic manipulation of a wide variety of organisms. Transgenic animals can be defined as animals in which new or altered genes have been experimentally inserted into their genome by genetic engineering techniques. The new genetic material becomes integrated into the chromosomal DNA of the host organism, and is transmitted as an heritable trait to succeeding generations of progeny. While genetic engineering has been used in a variety of organisms, rodents (mainly mice and rats), are the model of choice when creating transgenic animals due to our understanding of their genomes and the high reproductive vigor of the species. The foreign or manipulated genes are constructed using recombinant DNA techniques in which gene fragments are amplified and purified for insertion into the genome of a host species. The rapidly growing number of transgenic animal models and the increasing applications for their use has allowed transgenic technology to span the industrial and academic research community. Applications of transgenic animals range from medical research and drug production to the study of mammalian development. The use of transgenic animals in medical research enables the identification and study of the functions of specific factors in complex systems.

Production of Transgenic Rodents

Transgenic animals can be generated by several different methodologies, including lentiviral infection of single-cell zygotes; pronuclear microinjection of foreign DNA into the pronucleus of the fertilized egg; and genetic modification of embryonic stem cells (ESCs) followed by microinjection into 8-cell or blastocyst embryos. Several complete and comprehensive transgenic manuals provide extremely detailed descriptions of transgenic methodologies and protocols (see suggested readings for more information). For the purpose of this chapter, we will concentrate on the two most common and successful methods of transgenic production:

Animals that Arise from Pronuclear Injection

Genetically modified animals that arise from the injection of the pronucleus of fertilized eggs are commonly referred to as *transgenic* animals. In this case, a cDNA or gene of interest is fused to gene promoter sequences that drive the expression of the gene cassette either ubiquitously or in a cell- and tissue-specific manner. The cDNA or gene then enters the host genome through nonhomologous recombination.

Animals Derived from Homologous Recombination in Embryonic Stem Cells

This technology is used to derive traditional *knock-out* (or KO) and *knock-in* (or KI) mice, in which critical exons of the native gene have been replaced with a drug resistance gene cassette (such as neomycin resistance) or a marker of gene activity [β-galactosidase (encoded by *LacZ*) or green fluorescent protein (GFP)]. The replacement cassette is flanked by sequences that are identical to portions of the targeted gene so that during cell division the targeted exons are replaced through homologous recombination. This replacement interrupts the normal exon sequences and 'knocks out' the function of the targeted gene. In a second step, the recombined ESCs are injected into mouse blastocysts, thereby incorporating the genetically modified cells into the transgenic animal.

Pronuclear Microinjection

By far the most popular and widely used method of transgenic mouse and rat production is microinjection of exogenous DNA into the pronucleus of a single-cell donor embryo at the 0.5 day postcoitus (dpc) stage of development. The background strain of the transgenic model is an important factor in production and must be considered carefully due to differences in genetic make-up, parental suitability, fecundity, and response to administered gonadotropins, as well as possible future experiments involving breeding to specific background strains. In transgenic mouse production the most common and popular strains are FVB/N, C57BL/6 × C3H, C57BL/6 × DBA/2, and C57BL/6 mice. For transgenic rats, the most commonly used strain is Sprague Dawley, although other strains, such

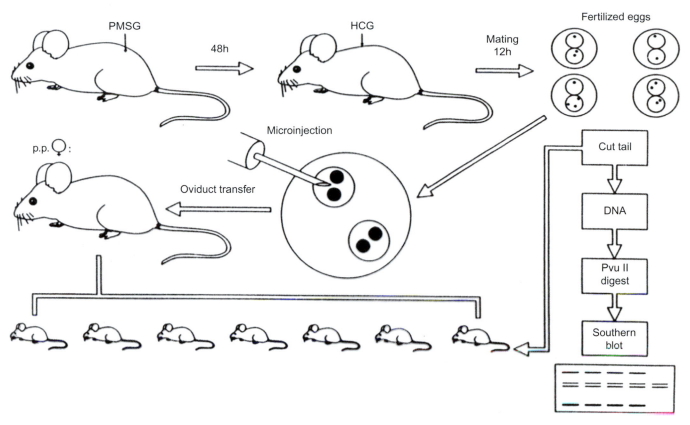

FIGURE 8.5 **Diagram of the steps involved in the production of transgenic mice (or rats) via pronuclear microinjection.** The first step in transgenic production is the superovulation of donor females. This is initiated by the administration of the pregnant mare serum gonadotropin followed 48 h later by administration of human chorionic gonadotropin. This process stimulates the female's reproductive system significantly increasing the egg yield. Donor females are subsequently mated to stud males and the fertilized embryos are collected from the donor female approximately 12 h after mating. The fertilized embryos are microinjected with the DNA construct of interest and the injected embryos are then surgically implanted into a pseudopregnant female via oviduct transfer. The offspring are born 19 days following oviduct transfer. A tissue sample is collected (cut tail) from the potential transgenic founders and DNA is isolated and purified from the sample. DNA is digested with the appropriate restriction enzymes (e.g. *Pvu*11) and assayed by Southern blot or PCR for genotypic determination of transgenic founders.

as Lewis and F344, are increasing utilized as more transgenic rat models are produced. The steps in the overall production of transgenic animals by the microinjection of DNA are shown in Fig. 8.5.

Isolation of Embryos for Microinjection

The first step in the process of pronuclear microinjection is the isolation of newly fertilized embryos from the reproductive tract of the donor female.

Superovulation

In order to obtain the maximum number of embryos for microinjection from the fewest number of donor females, the technique of hormonal superovulation is employed. Superovulation is the result of administrating exogenous gonadotropic hormones in order to synchronize and stimulate natural ovulation. Typically, nonstimulated females naturally mated will produce

approximately 8−12 oocytes per ovulation. This number can be increased to 15−40 oocytes per superovulated female, depending on the strain of mouse or rat utilized. Superovulation in mice is achieved by administration of pregnant mare serum gonadotropin (PMSG) followed days later with human chorionic gonadotropin (hCG). The superovulatory response of rats is far more variable than that of mice, and hormone response appears to be much more strain dependent. The importance of this hormone treatment is twofold: to increase the number of ovulatory follicles for each female and to control the timing and synchronize ovulation independent of the natural estrous cycle. It must be noted that the administration of hormones to elicit a superovulatory response increases the rate of chromosomal errors in the embryos obtained and can result in a large number of abnormal eggs. However, the beneficial effects and the reduction in the numbers of mice or rats needed outweigh the deleterious effects of superovulation. Following superovulation, the female mice (or rats) are mated with proven male breeders of the same strain. Fertilized embryos are then collected from successfully mated females at the appropriate developmental stage for microinjection.

Collection of Single-Cell Embryos

To obtain embryos at the correct stage of development, the embryos are harvested from the reproductive tract of the donor females the morning following mating resulting in embryos at 0.5 dpc. At this developmental stage, the pronuclei from both gametes will be visible for several hours following embryo collection and will provide the necessary time window for microinjection of prepared DNA before the pronuclei fuse and nuclear membranes are no longer visible.

To collect the newly fertilized eggs, the oviduct of the superovulated female must be isolated and dissected from the reproductive tract and placed in the appropriate culture medium. At this stage, the oviduct has a noticeably swollen area called the ampulla, which contains clusters of fertilized eggs surrounded by sticky follicular cumulus cells (Fig. 8.6A). The oviduct is visualized under a dissecting microscope and fine watchmaker forceps are used to gently tear open the swollen ampulla, allowing the fertilized eggs and cumulus cell mass to be released (Fig. 8.6B). The eggs are then treated with a 1% hyaluronidase solution, which enzymatically removes and disassociates the adherent cumulus cell from the eggs. The eggs are then put through a series of washes in appropriate culture medium to remove any excess debris. The prepared eggs for microinjection can now be placed into the appropriate medium for incubation at 37°C under 5% CO_2 to allow the embryos to stabilize before manipulation.

DNA Preparation for Microinjection

The next step is to obtain a high quality DNA preparation for microinjection. The technique of molecular cloning is used to generate and amplify the DNA fragment of interest. Typically a bacterial cell (such as *Escherichia coli*) and plasmid cloning vector are utilized to replicate and amplify the recombinant DNA molecules. The cloning vector contains multiple restriction sites that enable the foreign DNA to be inserted into the plasmid, as well as a gene for antibiotic resistance that enables selection of bacteria that have successfully incorporated the vector sequences. The plasmid is then cut with a restriction enzyme (such as *Eco*RI) to create a cleavage site where the foreign DNA will be inserted.

The DNA fragment of interest can be extracted from virtually any tissue and, if necessary, be amplified using the polymerase chain reaction (PCR) or, if a tissue sample is not available, the fragment can be designed by the researcher and chemically synthesized. The foreign DNA is then cut with the same restriction enzyme used to cut the cloning plasmid creating fragments with "sticky" ends that are compatible with the plasmid. The recombinant DNA is created by ligation of the DNA fragment into the plasmid DNA. This is accomplished by combining the cut plasmid cloning vector with the prepared foreign DNA fragments in the presence of DNA ligase, an enzyme that covalently binds the sticky complimentary ends of the plasmid and foreign DNA together. The products of the ligation reaction are then translocated to the host bacteria by transformation, a process by which exogenous DNA is taken up through the cell membrane of host bacteria that has been chemically induced to increase permeability to the recombinant DNA. Once the recombinant DNA is incorporated into the bacterial host, the bacteria are allowed to replicate and clones are subsequently selected for recombination by means

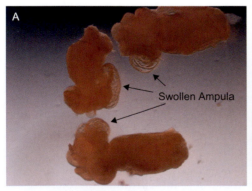
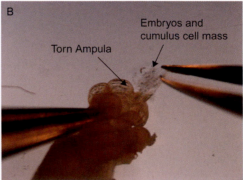

FIGURE 8.6   **Isolated oviducts from female mouse reproductive tract.** (A) Small section of uterus and complete oviduct dissected from reproductive tract of donor female mouse. Arrows show the swollen ampullas of the oviduct containing fertilized embryos. (B) The swollen ampulla of the oviduct is gently torn open with watchmaker forceps allowing the embryos and associated cumulus cell mass to be released for subsequent collection. Magnification, 10×.

of the antibiotic resistance incorporated into the plasmid vector. By growing the bacterial host on a medium containing antibiotic, only bacteria containing the recombinant DNA with antibiotic resistance can grow. The selected bacterial clones that have been successfully transformed can then be grown in a large volume culture flask to greatly amplify the recombinant DNA. At this point, the bacteria are lysed under alkaline conditions and plasmid DNA is separated and purified. Purification can be performed based on density gradients using cesium chloride (CsCl) and high speed centrifugation. Alternatively, plasmid DNA can be purified using one of several commercially available kits. Plasmid vector sequences can significantly alter the expression of the transgene and must be separated from the insert. The DNA insert is cut from the plasmid by restriction enzyme digest and isolated by gel electrophoresis. Microinjection-quality DNA preparations must be free of any contamination that could be cytotoxic or detrimental to the cell's development, as well as any particulate debris that may clog the fine injection needle. The DNA fragment is therefore extensively purified before introduction to the pronucleus of the fertilized egg.

Pronuclear Microinjection of Embryos

Microinjection is performed by observing the recently fertilized single-cell embryo through an inverted microscope. Differential interference contrast (DIC) optics are necessary to enhance the contrast and provide clear visualization of the embryo's pronuclear structure. Microinjection of single-cell embryos is typically performed at a magnification of $200\times$. Manipulation of the embryo is accomplished using two micromanipulators and joystick controllers located on either side of the microscope. One micromanipulator controls the movement of the holding pipette, while the other controls the injection needle, both of which are connected to finely controlled pressure systems to allow fine control of the embryo and injection needle.

Microinjection of foreign DNA into the pronucleus of a single-cell mouse or rat embryo is accomplished by first picking up and securing the embryo with the holding pipette. It is generally accepted that microinjection into the male pronucleus (contributed from the sperm) results in a higher rate of success in terms of transgenic offspring produced. The male pronucleus is generally the larger of the two pronuclei present and is typically located at the periphery of the cell. The micromanipulator controlling the injection needle is then used to bring the needle into sharp focus in the same focal plane as the pronucleus (Fig. 8.7A). The injection needle then carefully punctures the outer zona pellucid and cell membrane. Injection is continued until the needle penetrates the pronuclear membrane and approximately 1–5 pL of the DNA solution is introduced to the pronucleus. Successful delivery of the DNA preparation is made evident by the swelling of the pronucleus (Fig. 8.7B). The cell membrane of a rat embryo is far more elastic and resistant to the penetration of the injection needle, compared to a mouse embryo. Rat embryos require sharper needles and deeper penetration before the cellular and pronuclear membranes are successfully punctured. The pronucleus of the rat embryo is also much more difficult to visualize and the injection needle usually has to be prepositioned several times before proper alignment for injection is achieved (Fig. 8.7C–E). These issues make rat embryos far more difficult and time consuming to microinject; however, they survive the process of microinjection better than do typical mouse embryos.

Only a percentage of embryos will survive the microinjection process; others will be irreversibly damaged and lyse (Fig. 8.8A). Following microinjection, surviving embryos are placed in appropriate culture medium and placed in an incubator at 37°C under 5% CO_2 overnight. Most injected embryos will divide to the two-cell stage following overnight incubation, while some will arrest at the one-cell stage and others will develop incorrectly and show a fragmented appearance (Fig. 8.8B). Embryos that have divided to the two-cell stage are considered viable and are transferred to the oviduct of a pseudopregnant (false pregnant) recipient female. Injected embryos can also be transferred to the recipient female immediately following injection at the one-cell stage; however, in this case more embryos should be transferred, as viable embryos are not selected based on overnight development to two cells.

Transfer of Injected Embryos to Recipient Females

As the one-cell embryos were harvested from the oviduct of the donor females, they must be transferred, or reimplanted into the oviduct of a recipient surrogate female. For this process, pseudopregnant females are produced by mating receptive females with vasectomized males the night before the embryos are transferred to the female. As the male is sterile, the embryos present within the female will not be fertilized. However, the act of mating will induce the female to become hormonally receptive to the transferred embryos (at either the one- or two-cell stage), allowing the surrogate female to carry the pups to term. Injected embryos for implantation are loaded into a finely drawn glass pasture pipette, or transfer

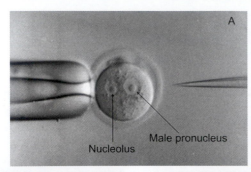
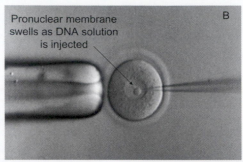
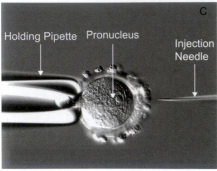

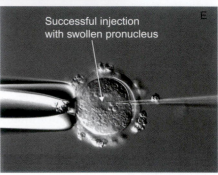

FIGURE 8.7   Microinjection of mouse (A,B) and rat (C-E) embryos. The fertilized embryo can be seen being held in place by the holding pipette on the left. The DNA injection needle can be seen on the right. (A) Mouse embryo prior to microinjection showing the larger male pronucleus and nucleolus. (B) Successful microinjection of the mouse embryo showing swelling of the pronucleus as the DNA is injected. (C) Rat embryo prior to microinjection. The pronuclei are much more difficult to visualize as compared to the mouse. Also note the cumulus cell still attached to the rat embryo. (D) Insertion of injection needle without successful penetration of the cell membrane. Arrows show the tip of the injection needle not yet within the embryo. (E) Successful penetration of the cell membrane and pronuclear membrane showing swelling of the pronucleus as the DNA is injected. Magnification, 200×.

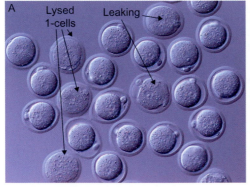
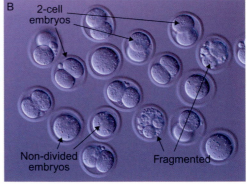

FIGURE 8.8   **Mouse embryos following microinjection.** (A) Embryos immediately after microinjection. While most embryos survive the microinjection process, some are irreversibly damaged and do not survive. The arrows show both completely lysed one-cell embryos and those that have been lysed and are slowly leaking. (B) Mouse embryos following overnight incubation. Photograph shows healthy embryos that have divided to the two-cell stage: nondivided embryos arrested at the one-cell stage, and fragmented nondeveloping embryos. Only the two-cell embryos are viable and suitable for transplantation. Magnification, 80×.

pipette. The embryos are taken up in the smallest amount of medium possible, packed as tightly as possible, with several air bubbles in the transfer pipette to control the flow of the medium (Fig. 8.9). Successfully mated females are identified by the presence of a copulatory plug in the vaginal opening of the mated female. The pseudopregnant surrogate female is anesthetized and a small lateral incision is made through the skin and body wall to identify the fat pad that is attached to the ovary. The fat pad is gently pulled from the body cavity, exposing the associated ovary, the small coils of the oviduct, and the distal portion of the left uterine horn (Fig. 8.10). The oviduct region is visualized under a dissecting microscope and the thin transparent bursa membrane surrounding the ovary and oviduct is gently torn open near the infundibulum. Once the

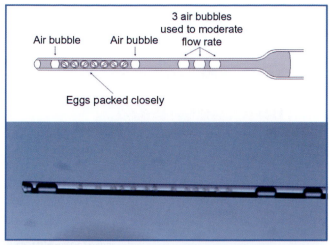

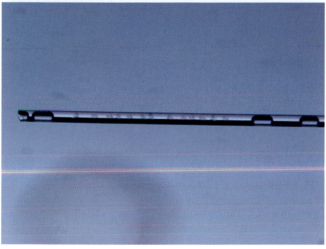

FIGURE 8.9 **Implant pipette with embryos loaded for transplant surgery.** (Above) Schematic representation of implant pipette illustrating air bubbles to moderate and control the flow rate of the medium into the pipette and arrangement of embryos in the transfer pipette. (Below) An actual photograph of the loaded transfer pipette. Embryos can be visualized in the pipette. Magnification, 20×.

infundibulum is clearly visualized, the transfer pipette is inserted and the injected embryos are gently expelled into the oviduct (Fig. 8.10). The fat pad, ovary, and oviduct are carefully replaced into the body cavity and the incision is closed. For mice, approximately 15–20 two-cell embryos or 20–25 one-cell embryos can be implanted into each recipient female mouse. The number of rat embryos should be increased to 20–25 two-cells and up to 30 one-cells per implanted female rat. Approximately 40–60% of implanted females will become pregnant and carry the pups to term.

Analysis of Potential Transgenic Founders

Successfully implanted embryos will develop to term in approximately 20 days. A typical litter size for a surgically implanted mouse is 3–6 pups, and 2–5 pups for a rat. The potentially transgenic pups, or *founders*, are weaned between 21–28 days and a small tissue biopsy is collected for genomic DNA analysis. The most common methods of genomic DNA analysis utilized by most laboratories are Southern blot and PCR analyses. PCR is currently the most widely used method for identification of transgenic animals, in part because it is far less time consuming and requires much smaller amounts of less purified DNA than does Southern blot analysis. However, Southern blot analysis is still utilized to confirm the founder and can be used to reveal the integrity of the integrated transgene sequences, determine copy number of the transgene, and reveal the number of integration sites.

The transgenic rate, or number of transgenic animals produced per total number of pups born, varies dramatically depending on the genetic construct and strain used. The average transgenic rate for mice ranges from 15–50%, and 1–15% for rats. Efficiencies for the production of transgenic mice and rats are shown in Table 8.1. As incorporation of the injected gene is a completely random event, each positive animal, or founder, produced will be genetically unique. Each founder line is typically bred to a wild-type animal of the same strain and the line is maintained in a heterozygous state.

Microinjection of Genetically Modified Embryonic Stem Cells

The second common method for generating transgenic animals involves microinjection of genetically altered ESCs into the cavity of a 3.5-day developed embryo, or blastocysts. The process of DNA homologous recombination is used to alter the genome in pluripotent ESCs. Pluripotent ESCs are capable of differentiating into any cell type in the body, including germ cells, and have the ability of replicating and propagating themselves indefinitely. Homologous recombination in ESCs is utilized to create knockout and knock-in mouse models.

To date, the creation of knockout or knock-in rats via ESC modification has been greatly limited due to the lack of availability of a reliable rat ESC line. Knockout technology involves the elimination of an endogenous gene or the deletion of the functional domain of the translated protein. This allows the researcher to completely remove one or more exons from a gene, resulting in the production of a mutated or truncated (nonfunctional) protein or no protein at all. Researchers can then determine the role of the particular gene by observing the phenotypic characteristic of the animals that lack the gene completely. Knock-in technology is the targeted insertion of a transgene at a

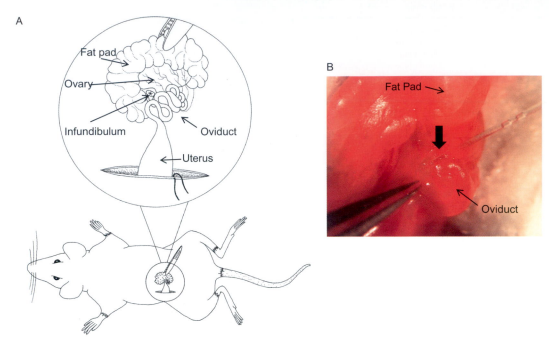

FIGURE 8.10 **Schematic representation of a mouse in the correct orientation for oviduct implant surgery.** (A) Location of incision site and close up of the exteriorized fat pad, ovary, and oviduct with infundibulum (opening). (B) Photographic insert shows actual surgery. Arrow shows the implant pipette inserted into the infundibulum of the oviduct. Magnification, $20\times$.

TABLE 8.1 Typical Microinjection Parameters for Mouse and Rat

Parameter	Mouse	Typical Mouse Injection	Rat	Typical Rat Injection
Embryo yield (embryos/female)	20–40	300	15–30	225
Embryos fertilized from total collected	70–80%	220	60–70%	150
Embryos that survive microinjection	50–70%	140	80–90%	125
Embryos that develop to 2-cell stage following incubation	80–90%	110 implanted into 6–7 females	80–90%	100[a] implanted into 4–5 females
Number pups born per implant		3–6		2–5
Transgenic rate (mice expressing transgene from total born)	15–50%		1–5%	

[a]Rat embryos are typically implanted at the one-cell stage.
Typical microinjection percentages and actual numbers for an average day of microinjection in mouse and rat embryos. Mouse embryo numbers are the average obtained from 10 superovulated C57BL/6 mice (3–4 weeks of age). Rat embryo numbers are the average obtained from 10 superovulated Sprague Dawley rats (4–5 weeks of age).

selected locus. This avoids the problems of traditional transgenic models such as random integration, multiple integration sites, and variations in copy number. With this approach, the researcher has complete control of the genetic environment surrounding the gene of interest and the DNA is not incorporated into multiple locations, which results in a more consistent level of expression from generation to generation. Additionally, the resulting phenotype is more likely to be due to the exogenous expression of the protein because the targeted transgene does not interfere with a critical locus in another location in the genome. Knock-in technology can also be used to introduce reporter molecules such as fluorescent proteins

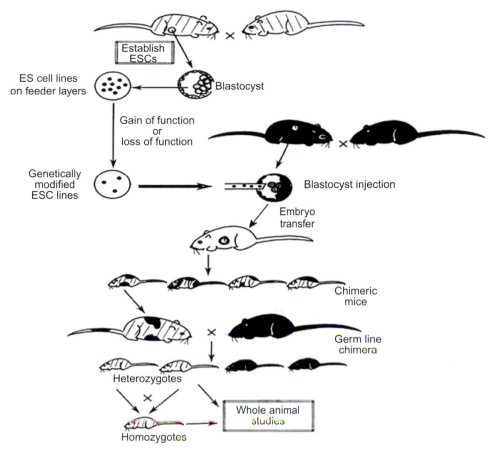

FIGURE 8.11 **Steps involved in the production of knockout and knock-in rodents via microinjection of genetically altered embryonic stem cells into donor blastocyst.** The process begins with the establishment of embryonic stem cells (ESCs) from the inner cell mass of collected blastocysts from donor females. Blastocysts are harvested from the reproductive tract of donor females 3.5 days after mating. The inner cell mass of the blastocyst is dissected and ESCs are cultured on feeder cell layers. Gain or loss of function is accomplished by homologous recombination via electroporation, and genetically altered ESCs are grown on selection medium. Genetically modified ESCs are then microinjected into donor blastocysts and transferred into 2.5-day pseudopregnant recipient females via uterine transfer. The offspring are born 17 days following uterine transfer and the genetically modified founders are easily identified by their chimeric coat color. For germline determination, the chimeric mice are mated back to the same strain as the original blastocyst donor, and germline transmission is determined by coat color. Heterozygous animals can then be crossed to produce a true homozygous knockout or knock-in for whole animal studies. See text for complete details.

(e.g. GFP) or *LacZ* to study gene expression. Researchers can very elegantly modify mouse genes to reproduce genetic defects associated with human diseases. The overall steps in the creation of knockout and knock-in models are shown in (Fig. 8.11).

Generation of Embryonic Stem Cell Lines

The first step in this process is the acquisition and genetic alteration of an appropriate ESC line. Historically the most popular ESC lines have been derived from the 129/Sv mouse strain. However, recent advances in ESC culture have led to the commercial availability of a multitude of ESC lines from a variety of mouse strains, including C57BL/6. ESCs are derived from the inner cell mass of preimplantation blastocysts. Blastocysts are 3.5-day developed embryos containing approximately 68–128 cells. The blastocyst is a hollow

structure consisting of the blastoceal cavity, the outer trophectoderm layer, and the inner cell mass (Fig. 8.12). Blastocysts are harvested from the reproductive tract of donor females. The inner cell mass, containing the uncommitted pluripotent ESCs, is gently teased away from the blastocyst and cultured in defined medium containing factors that inhibit differentiation and promote pluripotency. ESCs that grow from the inner cell mass are maintained on a layer of mouse embryonic fibroblast feeder cells, which provide nourishment and help to maintain the ESCs in an undifferentiated state. In addition, leukemia inhibitory factor (LIF), one of the most important cytokines required to maintain pluripotency, is added to the culture environment to help preserve the uncommitted state of the ESCs. ESCs grow in rapidly dividing colonies that must be subcultured frequently. For convenience, ESC lines are often frozen

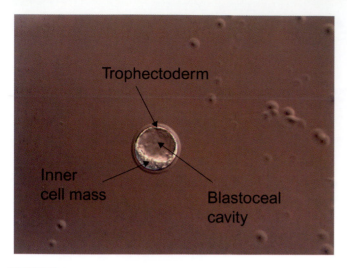

FIGURE 8.12 Mouse blastocyst (3.5 dpc) showing the outer trophectoderm layer, inner cell mass, and blastoceal cavity. Magnification, 200×.

and stored in liquid nitrogen for subsequent use in targeting experiments.

Homologous Recombination

A gene of interest can be introduced into ESCs in culture through a complex process called homologous recombination. Homologous recombination is used for the purpose of knocking out or knocking in a gene at a defined location within the genome. The first step in this process is the design of the engineered targeting construct. Targeting vector design is a very complicated process and a complete description is well beyond the scope of this text. Creating a knockout can be accomplished by introducing a drug selectable marker, such as the neomycin resistance gene (neoR), into a coding exon of a gene that will ultimately disrupt the transcription of the targeted gene. A DNA vector that comprises the neoR gene insertion followed by the addition of homologous arms that are identical in DNA sequence to the genome area where that gene resides provides the template for the targeting vector. Knocking in a gene is more complex and depends on the goal. Typically, knock-ins are used to replace a gene with a reporter gene such as LacZ or GFP, or to replace a gene with a mutated version or the same gene from a different species. This has the effect of knocking out and replacing the targeted gene. In some cases, knock-ins are used to insert genes in locations of the genome that are noncritical, thus allowing precise transgene placement. Regardless of the outcome, knock-in targeting vectors are generated in much the same way as the knockout vectors described above. A drug resistance marker is still needed, along with the reporter or mutant gene of interest, followed by homology arms to the genomic region. Inclusion of a negative selection

marker such as tk (the gene encoding thymidine kinase) at the end of the targeting vector can be employed to determine targeting efficiency. Cells that undergo homologous recombination will lose this marker, as the homologous recombination event has to occur upstream of this DNA sequence.

Vectors are usually transferred to ESCs via electroporation, a process by which an external electrical field is applied to a suspension of ESCs, significantly increasing the permeability of the cells plasma membrane and allowing the targeting vector to enter the cell. Once the targeting vector enters the ESC nucleus, there are three possible outcomes: (1) the DNA is degraded before it has a chance to integrate into the genome and those cells will not survive drug selection; (2) the targeting construct finds the targeted gene of interest and undergoes homologous recombination, replacing the original DNA template with the engineered version and the loss of the negative selection marker; or (3) the targeting construct randomly integrates somewhere else in the genome (nonhomologous), incorporating the negative selection marker (tk) as well.

Following electroporation, the products from homologous and nonhomologous recombination events are subjected to selection in culture medium containing the antibiotic G418 and the drug gancyclovir. Transformed ESCs that contain the targeting construct and neoR are identified by their survival in the presence the G418 (an aminoglycoside related to neomycin). This is referred to as positive selection. As the majority of cells surviving positive selection will have integrated the targeting vector in a random fashion by nonhomologous recombination, a negative selection scheme is also employed. Most of the cells taking up the targeting vector randomly (nonhomologously) into their genomes retain and express the tk gene. Thymidine kinase is an enzyme that phosphorylates the nucleoside analog gancyclovir. DNA polymerase fails to discriminate against the resulting nucleotide and inserts this nonfunctional nucleotide into freshly replicating DNA. The ganciclovir, therefore, kills cells that contain the tk gene. In the much smaller proportion of cells that integrate the vector by homologous recombination, the tk gene is lost in the homologous crossover event, thereby allowing the targeted cells to survive in the presence of ganciclovir. The processes of homologous and nonhomologous recombination are illustrated in Fig. 8.13. Despite this double enrichment for targeted cells, a DNA screening method such as Southern blot assay is also typically utilized to identify and confirm correctly targeted cells.

Collection of Blastocysts

Blastocysts are harvested from the reproductive tract of a superovulated donor. Typically, the blastocyst

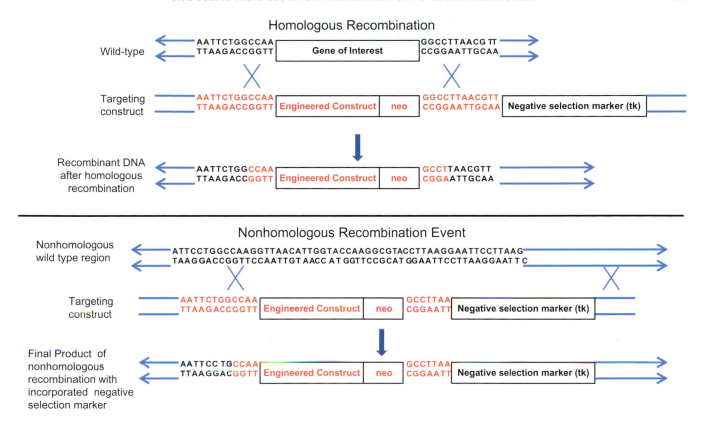

FIGURE 8.13 Simple schematic of gene targeting. (Top panel) Homologous recombination. A targeting construct containing the engineered construct and neo resistance gene cassette are incorporated into wild-type embryonic stem cells (ESCs) via electroporation, and homologous recombination occurs between the engineered construct and the gene of interest. Homologous regions are indicated by the sequences surrounding the gene of interest and the engineered construct. Crosses indicate the areas of recombination. The product of homologous recombination is shown as the engineered construct incorporated into the wild-type genome. (Lower panel) Nonhomologous recombination. In this case, the engineered construct is randomly incorporated into a nonhomologous region within the wild-type genome, along with the negative selection marker. The products of homologous and nonhomologous recombination are subjected to selection in culture medium such that transformed ESCs that contain the engineered construct and the neomycin (neo) resistance gene will survive, while random products of nonhomologous recombination incorporating the negative selection marker will be killed and eliminated from the pool of modified ESCs. See text for details.

donor mouse strain will have a coat color different from that from which the ESC line was derived. For example, if the ESC line is derived from a mouse with a black coat color (e.g. C57BL/6), then the blastocyst donor should be a mouse with a white coat color (e.g. Swiss Webster, BALB/c, or albino B6). This coat color difference can be utilized later to determine the ESC contribution to the genome of the resultant mouse, or chimera, as will be discussed later in this section. At the 3.5-day developmental stage, the blastocysts have migrated to the junction between the oviducts and the uterus (uterotubal junction). To recover the blastocysts, the oviduct and a small portion (1−2 mm) of the uterus is dissected from the donor female and placed in the appropriate medium. The blastocysts are then flushed from the lower portion of the oviduct with a small amount of medium. Approximately 5−10 blastocysts can be obtained from each female.

Injection of Embryonic Stem Cells into Blastocysts

Genetically altered ESCs are injected into blastocysts using a technique similar to the one used for pronuclear microinjection. The injection needle for ESC injection has a much larger inner diameter to accommodate the loading of ESCs into the needle. An area on the trophectoderm is selected for penetration and the injection needle is sharply "jabbed" through the trophectoderm and into the blastocoel cavity. Approximately 10−12 manipulated ESCs are then carefully expelled into and deposited onto inner cell mass area of the blastocyst (Fig. 8.14A−C). The injected blastocysts are then returned to the incubator and cultured at 37°C under 5% CO_2 for several hours. The majority of blastocysts will survive the microinjection process and can be implanted into pseudopregnant recipient females the same day as injection.

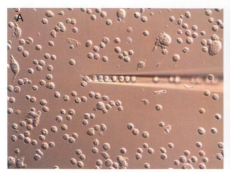
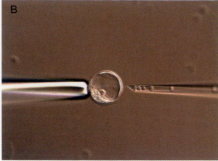

FIGURE 8.14 **Microinjection of genetically modified embryonic stem cells into mouse blastocysts.** (A) Loading of embryonic stem cells (ESCs) into microinjection needle. (B) Blastocyst prior to microinjection, showing holding pipette on the left and microinjection needle with ESC on the right. (C) Successful microinjection of ESCs, showing needle and ESCs being deposited into blastoceal cavity. Magnification, 200×.

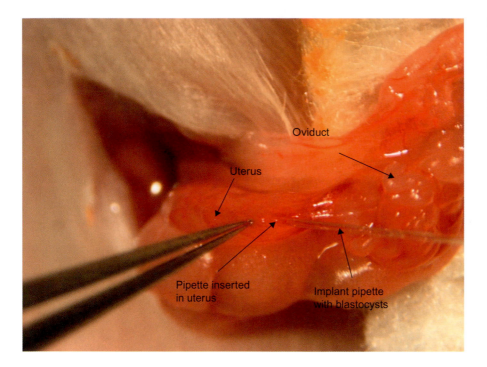

FIGURE 8.15 Photograph of uterine implant surgery showing exposed oviduct and upper portion of the uterus. The implant pipette loaded with injection blastocysts is shown on the right side being inserted into a hole in the uterus. Magnification, 20×.

Transfer of Injected Blastocysts to Recipient Females

As blastocysts are 3.5-day developed embryos, it is imperative that they are placed into the correct location in the reproductive tract of a recipient female at the appropriate age. Blastocysts are typically transferred into the upper uterus of pseudopregnant recipient females that are 2.5 dpc. It is best to use recipient females that are at a slightly earlier stage of pseudopregnancy than the development stage of the embryos. This gives the transferred embryos time to develop to the appropriate stage for implantation in the uterus of the recipient female. The recipient female is prepared for surgery as previously described and the oviduct, associated fat pad, and upper portion of the uterus is exposed. A small hole is made in the upper part to the uterus just below the uterotubal junction and, using a transfer pipette, the blastocysts injected with modified ESCs are expelled into the uterus (Fig. 8.15). Approximately 10–15 blastocysts are transferred into each pseudopregnant recipient female.

Identification of Chimeric Animals

Pups from the implantation surgery are typically born 17–18 days following the surgical procedure. Approximately 60–80% of surgically implanted females will successfully carry the transferred blastocysts to term. A typical surrogate female will have a litter of 3–8 pups, with approximately 10–50% of pups being genetically altered (chimeras) via ESC integration. This rate depends greatly on the quality of ESCs microinjected.

FIGURE 8.16 **Chimeric mice produced by injection of genetically modified embryonic stem cells into recipient blastocysts.** (A) Chimeric mouse produced by the injection of altered 129 Sv/J (agouti) embryonic stem cells (ESCs) into C57BL/6 (black) blastocysts. (B) Chimeric mice produced by the injection of C57BL/6 (black) ESCs into BALB/c (white) blastocysts. Note that the single white mouse is a non-chimeric littermate.

Successful integration of the altered ESC line into the genome of the host blastocyst is determined by the coat color of the resultant mouse. If the ESCs have contributed to the genome of the host, then the coat color will be a combination of the ESC donor strain and the strain of the host blastocyst, or a chimera. The most common strains and resultant outcomes are: 129/Sv (agouti) ESCs injected into C57BL/6 (black) host blastocysts, resulting in a black mouse with agouti markings (Fig. 8.16A), and C57BL/6 (black) ESC injected into Albino B6 (white) host blastocyst will result in a white mouse with black markings (Fig. 8.16B). The greater the contribution of ESC-derived coat color, the higher the contribution of the ESC to the genome. For the chimeric mouse to pass the genetic alteration onto the offspring, the ESC must be incorporated into the germ cells of the chimera. To test for transmission, the chimeric mouse is mated to the same strain of mouse as the blastocyst donor strain. If the gonads from the chimeric mouse were derived from the recombinant ESCs, then the offspring will retain the coat color of the ESC donor strain and every cell in the offspring will be heterozygous for the homologous recombination event. This is referred to as successful germline transmission and is directly correlated to the quality of the ESC used for microinjection.

The average rate of germline transmission for a quality line of ESCs is in the range of 50−80%. If the offspring have the same coat color as the host blastocyst donor, then nothing from the ESC was contributed to the germline of the chimera and it will not pass the genetic alteration to subsequent generations. The heterozygous germline transmitted mice can then be mated to establish a pure homozygous knockout or knock-in mouse line (see Fig. 8.11 for details).

TRANSGENIC ANIMALS AND BONE BIOLOGY

The understanding of bone diseases has been greatly accelerated by transgenic animal models. Many genes implicated in bone metabolism have been overexpressed or mutated, creating novel animal models. One of the first genes targeted for deletion in the study of bone disease was the *Pthlh* gene, encoding parathyroid hormone-related protein (PTHrP). Ablation of this gene in the mouse results in pup lethality shortly after birth associated with a cartilage developmental defect in the differentiation of growth plate chondrocytes, implying a role of PTHrP in the inhibition of chondrocyte differentiation. Subsequent deletion of *Pth1r* [encoding PTH/PTHrP receptor (PTH-1 R)] further supported this finding and also implicated PTH-1 R as the primary receptor in skeletal development.

The function of other important hormones in bone growth and differentiation has been tested in vivo through gene targeting. Thyroid hormone was suggested to play a role in skeletal development and homeostasis; however, evidence to support a direct role for this hormone in skeletal function in vivo was lacking. Mice carrying deletions of the thyroid receptors alpha or beta (TR1α or TRβ) have altered bone growth and development. Similarly, somatotropin/growth hormone receptor (GH) receptor-deficient mice display reduced bone size and bone turnover, as well as reduced chondrocyte proliferation. The insulin-like growth factors, IGF-I (somatomedin-C) and IGF-II (somatomedin-A), are produced primarily in the liver, but are also expressed in bone. These factors predominantly circulate complexed with IGF-binding proteins (IGFBPs) to facilitate their transport to tissues. IGFBPs can either enhance or restrict IGF activity. IGF-I and -II act through the IGF-I receptors (IGF-IR and IGF-IIR) and these ligand-receptor interactions promote cell proliferation and differentiation. The *Igf1*-null mouse has reduced cortical bone and femur length, however, trabecular density is increased. In vitro findings suggest that IGF-I also increases osteoclastogenesis, and *Igf1*-null mice have reduced levels of RANK ligand [tumor necrosis factor ligand superfamily member 11/ receptor activator of the NF-κB ligand (RANKL)] in osteoblasts isolated from bone marrow. Therefore, animal models have revealed that IGF-I may regulate osteoclastogenesis through direct and indirect actions. Overexpression of IGF-I in osteoblasts through transgenic approaches leads to increased BMD and increased trabecular volume, although osteoblast numbers are not increased. These studies suggest that IGF-I acts directly on osteoblasts to enhance their function. Specific deletion of IGF-IR from osteoblasts in mice results in decreased trabecular number and volume, and a dramatic decrease in bone mineralization, further supporting the role of the IGFs on osteoblasts.

The discovery and description of the receptor activator of nuclear factor-κB [tumor necrosis factor receptor superfamily member 11 A (RANK)]-RANKL-OPG pathway that collectively mediate osteoclast activity was very important to the field of bone biology. Genetic alteration of each of these mediators has played a significant role in the elucidation of the mechanisms of osteoclastic bone resorption. The identification of OPG was first reported in 1997. This was followed by the creation of the OPG transgenic mouse that displayed systemic overexpression of rat *Tnfrsf11b/Opg* cDNA under the control of the *APOE* gene [encoding human apolipoprotein E (ApoE)] and its associated liver-specific enhancer. OPG is a decoy receptor that prevents RANKL from binding to RANK and serves as the key physiologic inhibitor of osteoclastic bone resorption. The OPG transgenic mouse exhibits a phenotype of profound yet nonlethal osteopetrosis and increased bone density, associated with a significant decrease in osteoclasts, and suppressed osteoclastic activity. No difference was noted in the relative size and shape of long bone, compared to control animals.

As rats are the preferred species for certain models of bone disease due to their larger size and increased flexibility in study design, an OPG transgenic rat was created. OPG transgenic rats display systemic overexpression of rat OPG, resulting in complete and continuous inhibition or RANKL. This inhibition results in significant suppression of bone turnover associated with increased bone density and shortened long bones, compared to control rats. This phenotype can be clearly seen in X-ray radiographs (Fig. 8.17). OPG knockout mice exhibit a decrease in total bone density and a high incidence of fractures, similar to

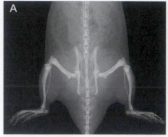 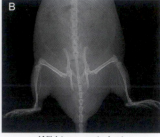

OPG transgenic rat   Wild-type control rat

FIGURE 8.17 The phenotype in OPG-overexpressing animals can be seen in radiographs. X-rays show increased radiodensity in the long bones of 8-week-old OPG transgenic rats (A) compared to wild-type controls (B).

postmenopausal osteoporosis. This phenotype is a result of increased bone resorption associated with increased numbers and activity of osteoclasts. OPG-deficient mice also exhibit arterial calcification of the aorta and renal arteries. The ablation of RANK results in no RANK-RANKL interaction and therefore no differentiation or activation of osteoclasts. Initially, the early phenotype observed in RANK-deficient pups is small body size, shortened limbs, domed skull, and failure of tooth eruption. In adulthood, the RANK knockout (RANK-KO) mouse shows severe osteopetrosis as evidenced by shortened long bones and increased bone density, accompanied by an absence of osteoclasts. These mice also show a complete lack of peripheral lymph nodes, indicating a role of RANK in lymph node organogenesis. In addition, there are a reduced number of B lymphocytes, with normal thymus size and T-cell development. The phenotype of the RANKL-deficient mice is very similar to that of RANK-KO mice, in that they show severe osteopetrosis, lack osteoclasts, an absence of peripheral lymph nodes, and reduced number of B lymphocytes. In addition, RANKL-KO mice have a reduced thymus size and impaired T-cell development. Knock-in technology has been used to create a human RANKL (huRANKL) knock-in mouse. This mouse was created as a model to develop a human RANKL assay and to study the OPG-RANKL interaction. This mouse model endogenously secretes human RANKL that will readily bind to human OPG, and treatment of huRANKL mice with huOPG results in increased bone density. These models have helped to unravel the components of this complex bone resorption pathway.

Mouse models have also been critical in deciphering the functional roles of sets of genes required at the earliest stages of bone development. The process of osteoblastogenesis is characterized by cell proliferation, matrix development and maturation, mineralization, and apoptosis. These stages are associated with activation of specific transcription factors and genes leading to the expression of osteoblast phenotypic markers. Transcription factors such as RUNX2 and Sp7/osterix (OSX) are very early osteoprogenitor factors and are essential for establishing the osteoblast cell phenotype. RUNX2 is expressed prior to Sp7/OSX, but both proteins control the expression of factors that regulate bone formation and remodeling, including osteocalcin and RANKL. RUNX2 regulates osteoblast differentiation and function by several signaling pathways, including those activated by Wnts and bone morphogenic proteins (BMPs), as well as the differentiation and survival of osteoblasts induced by integrins and the PTH-1 R. Genetic deletion of RUNX2 or Sp7/OSX from mice results in no mineralized bone being formed; only a cartilage matrix. These knockout animals provided the important finding that RUNX2 and Sp7/OSX expression is required for the early differentiation of preosteoblasts to osteoblasts, which are necessary for forming the mineralized skeleton.

Bone Cell-Specific Promoters

The expression of genes or detectable reporters in bone can be undertaken in genetically modified mice using promoters from genes known to become activated at specific stages of cellular differentiation or within specific cell lineages in bone. One of the best characterized methods of controlling expression of transgenic genes ectopically in bone uses the promoter of the *Col1a1* gene (which encodes type I collagen). Type I collagen is a major secretory component from osteoblasts that is required for tensile strength and mineralization of bone. A 2.3 kb fragment of the rat *Col1a1* gene promoter potently drives gene expression in mature osteoblasts. Interestingly, a 3.6 kb *Col1a1* promoter fragment will drive cDNAs or reporter genes, such as fluorescent proteins that can be imaged at distinct wavelengths to produce distinct colors, in both early and late osteoblasts (Fig. 8.18). Osteocytes can be targeted for gene expression using the mouse 10 kb promoter of the *Dmp1* gene [encoding dentin matrix acidic phosphoprotein 1 (DMP-1)], and osteoclasts using the *Acp5/Trap* promoter [of the gene encoding tartrate-resistant acid phosphatase type 5 (TRAP)]. Certainly, the use of any transgenic animal must be approached with the forethought that not all gene promoters are absolutely cell and tissue specific, and therefore may have expression outside of bone. This has been shown to be the case with many promoters including an 8 kb fragment of the *Dmp1* promoter, which also displays modest expression in skeletal muscle.

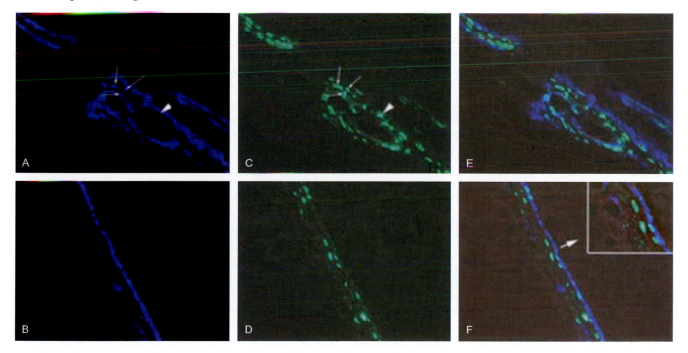

FIGURE 8.18 **Distinct expression of pOBCol2.3-GFPcyan, and DMP1-GFPtopaz in double transgenic mouse tissue.** (A and B) Osteoblasts lining the bone surfaces of calvariae are positive for osteoblastic pOBCol2.3-GFPcyan directed expression. (C and D) DMP1-GFPtopaz expression is restricted to cells within the bone matrix (osteocytes and periosteocytes). (E and F) Overlaid images taken under topaz- and cyanGFP-specific filters. Dual Col2.3-GFPcyan and DMP1-GFPtopaz-expressing osteocytes are indicated by arrows in (A and C), while some of the osteocytes express DMP1-GFPtopaz with no signal of Col2.3-cyanGFP (A and C; indicated by arrowheads). A higher magnification shows that the expression of DMP1-topaz within the bone matrix is localized to osteocyte dendritic extension connecting osteocytes and extending to osteoblasts on the bone surface (F, inset image). Magnification, 20×. *Reproduced with permission from Bone. 2009 Oct;45 (4):682—92.*

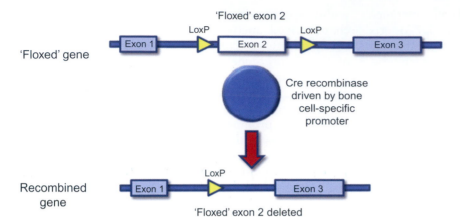

'Floxed' gene

'Floxed' exon 2

LoxP　　LoxP

Exon 1　　Exon 2　　Exon 3

Cre recombinase
driven by bone
cell-specific
promoter

Recombined
gene

LoxP

Exon 1　　Exon 3

'Floxed' exon 2 deleted

FIGURE 8.19 In the Cre-LoxP conditional deletion system, the bacterial Cre recombinase is expressed in bone through the activity of a cell-specific promoter. Cre interacts with the LoxP sites that flank a critical gene exon and delete the exon, leaving behind one LoxP site.

Another gene expression approach that has proven reliable for expressing cDNAs within bone in a cell- and temporal-specific manner has been the use of inducible systems such as the tetracycline (Tet)-On/Off transgenic cassettes. In the Tet-On system, doxycycline (Dox) or tetracycline (Tet) are provided to mice carrying a transgenic cassette that express a *reverse* tetracycline-controlled transactivator (rtTA) under the control of a cell-specific promoter. When Dox is provided to the animal by injection, the rtTA binds to a tetracycline-responsive element (TRE) and gene transcription of a cDNA downstream of the TRE is initiated. In the Tet-Off system, a tetracycline-controlled transactivator (tTA) is prevented from binding to the TRE through a conformational change induced by Dox binding. One disadvantage of this system is that the production of rtTA and tTA can be "leaky" and have inappropriate transcriptional activation; therefore, matching littermate controls are crucial for phenotype interpretation. Characterized inducible systems useful in bone are Tet-OSX-Cre, used to inactivate genes in the early osteoblast lineage, and a Tet-2.3Col1a1 cassette, used to activate genes in osteoblasts.

Conditional Models Using Cre-LoxP Technology

The most common strategy for gene targeting is the elimination of a specific gene to create mutations resulting in animals with corresponding phenotypes. However, many of these mutations result in embryonic lethality at some point during development or soon after birth. To overcome this problem, the Cre-LoxP system has been applied to gene targeting to create conditional (or inducible) models. The Cre-LoxP system forms a natural part of the P1 bacteriophage viral life cycle and is necessary for the circularization, replication, and development of new viruses. The use of Cre-LoxP recombination allows the gene to be knocked out in a specific tissue or at a specific time. This not only eliminates the problem of embryonic lethality but

also enables the creation of models that more precisely mimic human disease states and those that greatly enhance the study of developmental genes. The process of conditional model production begins by the use of gene targeting to create a mouse that has the gene of interest flanked on both sides by LoxP sequences, referred to as a *floxed* gene. The Cre protein is a site-specific DNA recombinase that catalyzes the recombination event between the two LoxP sites, thus "cutting out" the gene of interest (Fig. 8.19). The Cre protein is introduced by mating the mouse with the floxed gene to a mouse expressing Cre driven by a tissue-specific (e.g. bone, heart, lung, liver, or T cells) or ubiquitous (i.e. whole body) promoter. When cells that have LoxP sites flanking the gene of interest in their genome are exposed to Cre, by mating to mice expressing Cre in a tissue-specific fashion, the recombination event can occur between the LoxP sites.

Using regions of gene promoters that are active within specific populations of bone cells, genes can be deleted during almost any stage of cellular maturation. Therefore, bone-specific Cre-expressing mouse lines utilize promoters from the transgenic mice described previously. To delete genes from mesenchymal stem cell lineage, Prx1-Cre and Sox9-Cre mice can be utilized. To target the chondrocyte cell lineage, the *Col2a1* and Col10a1 promoters are available. Runx2-Cre mice can be used to delete genes in the very earliest stages of osteoblast differentiation. Similarly, Sp7-Cre mice can be used to delete genes within the initial osteoblast lineage. Ocn-Cre and 2.3Col1a1-Cre mice and can be used to delete genes in the mature osteoblast stage, and the Dmp1-Cre mice can be used to delete genes in the osteocyte. Osteoclasts can be targeted through use of the Trap-Cre and lysozyme-Cre mice.

A recent application of the Cre-LoxP system in bone revealed that deletion of the primary ciliary protein, polycystin-1 (PC1; encoded by *Pkd1*) specifically from osteocytes results in the inability of bone to respond to applied strain. The use of a conditional system was

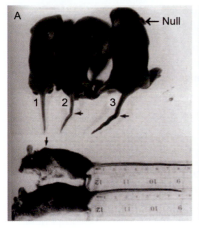

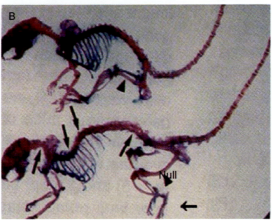

FIGURE 8.20 FGFR-3 knockout mice it was revealed that mice are larger (A) with greater bone length (B; arrows), than wild-type littermates, indicating that FGFR-3 acts as a negative regulator of bone growth. *Reproduced with permission from Deng et al., Cell. 1996 Mar 22;84 (6):911–21.*

particularly important in this example due to the fact that loss of PC1 in kidney in mice and humans results in renal failure from polycystic kidney disease. The profound secondary effects on bone from the renal disease are eliminated through deletion of *Pkd1* only in bone.

Another important aspect in conditional mutations is to precisely control the temporal activity, or timing, of the genetic mutation. Inducible, cell-specific conditional deletion of genes for most cell lineages is possible through the use of combination systems of tissue-specific promoters and inducible transcription proteins. This is accomplished fusing the Cre protein to a mutated ligand-binding domain of the human estrogen receptor (ERT). These Cre-ERT fusion proteins reside in the cytoplasm but become localized to the nucleus following the delivery of a synthetic steroid, such as tamoxifen, which then allows Cre to undergo a conformational change, and then recombine and inactivate genomic DNA sequences containing LoxP sites. This approach temporally and spatially controls gene expression, which allows the animal to develop normally before gene deletion to decrease the chances of unknown developmental effects of a particular gene. Some characterized inducible systems include Col2a-CreER (which targets chondrocytes), 2.3 kb Col1a1-CreER (which targets osteoblast lineage), and Dmp1-CreER (which targets osteocytes).

Bridging Human Disorders and Mouse Models of Bone Diseases

Traditional knockout models have revealed important aspects of bone structure and function and, besides providing biological information regarding the function of genes in vivo, can reveal novel disease mechanisms. One classic example of this value is the global deletion of fibroblast growth factor receptor-3 (FGFR-3) in mice. The human disorder achondroplastic dwarfism is caused by gain-of-function mutations in FGFR-3. Deletion of FGFR-3 from mice results in bones that are longer than those of wild-type mice, demonstrating that indeed FGFR-3 is a key receptor whose activation by local bone FGFs negatively regulated bone length (Fig. 8.20).

DMP-1 is an extracellular bone matrix protein that is highly expressed by osteocytes. DMP-1 is secreted in bone and dentin as 37 kDa N-terminal (residues 17–253), and 57 kDa C-terminal (residues 254–513) fragments derived from a 94 kDa full-length precursor. Recombinant DMP-1 appears to have multiple functions: it binds calcium-phosphate ions and the N-telopeptide region of type 1 collagen with high affinities. However, initially the physiologic role of DMP-1 was far less clear. Potential functions for DMP-1 in bone and teeth were thought to include regulating hydroxyapatite formation and, depending upon proteolytic processing and phosphorylation, regulation of local skeletal mineralization processes. Inactivating mutations in *DMP1* result in the metabolic bone disease, autosomal recessive hypophosphatemic rickets (ARHR1). The global DMP-1 knockout mouse was valuable in that it showed that loss of DMP-1 results in a marked increase in FGF-23 mRNA and protein. FGF-23 is known to cause renal phosphate loss, thus explaining the hypophosphatemic rickets phenotype of ADHR (i.e. autosomal dominant) patients. The DMP-1 knockout mouse was used to show that the primary cellular defect due to loss of DMP-1 may be a disturbance in osteoblast to osteocyte maturation, leading to inappropriate expression of typically *osteoblastic* or *early osteocyte* genes such as those encoding type I collagen, alkaline phosphatase, and FGF-23.

Secondary Challenges and Genetic Background Impacts on Skeletal Phenotypes

Animals that have been genetically recombined to either overexpress or delete genes causing skeletal diseases have been very useful for determining the mechanisms that underlie the way bone and bone cells function. In some cases, it has been revealed that further challenges on top of the genetic changes of either knockout or overexpression of genes in mouse models can reveal important phenotypes. This has been demonstrated in a mouse model of the ADHR metabolic bone disorder. Patients with ADHR have gain-of-function mutations in the *FGF23* gene that disrupt a furin-like protease site ($R_{176}XXR_{179}$) by substituting glutamine (Q) or tryptophan (W) for the arginine residues (R_{176} or R_{179}). FGF-23 inhibits renal phosphate reabsorption; therefore, patients with elevated FGF-23 have hypophosphatemia and develop osteomalacia and rickets. What makes ADHR unique among the disorders associated with FGF-23 is that patients can develop the disease from birth or it can be late onset, usually with disease developing after puberty or pregnancy. These physiologic states are associated with low serum iron concentrations. A knock-in mouse model of ADHR (R176Q) was produced that had a mild phosphate phenotype at birth; therefore, based upon the knowledge that iron deficiency may be a biological "trigger" for ADHR, the mice were provided with iron-deficient diets. This treatment resulted in markedly elevated *Fgf23* mRNA in bone and increased circulating concentrations of active FGF-23 hormone. Further, the mice showed hypophosphatemia and had reductions in the renal sodium-phosphate cotransporter, NaPi-2a (sodium-dependent phosphate transport protein 2A), and inappropriate vitamin D metabolism that has previously been associated with FGF-23 excess. The skeletal features of the ADHR mouse receiving low iron included severe metabolic bone disease, including osteomalacia, reflecting the phenotype of ADHR patients.

Another important consideration in mouse genetics is the background strain. Individual mouse strains have differing BMD and bone mineral content, which may be important for understanding mechanisms associated with the variable penetrance of skeletal diseases. Osteogenesis imperfect (OI) is a disease in which thin bones break easily, and is due to mutations in *COL1A1* or in genes encoding collagen prolyl hydroxylases required for normal collagen synthesis, folding, and assembly. In humans, the OI disease phenotype is highly variable. A mouse model of OI was developed by knocking in a Gly349Cys OI mutation into the *Col1a1* gene. Interestingly, the early founder animals were found to have high lethality when mated to C57BL/6 and CD-1 genetic backgrounds. With further matings of the survivors of the founders onto the CD-1 genetic background, a greater proportion of the mice were viable, supporting the importance of genetic background in the expression of the OI phenotype and the severity of the disease.

STUDY QUESTIONS

1. Discuss the differences between simple and complex traits. How do the two differ in relation to disease risk? Which one (simple or complex) might be more easily addressed from a public health standpoint to mitigate risk of osteoporosis, and why?
2. A GWAS identified a SNP within a gene of interest on chromosome 18 for a phenotype related to bone stiffness. The researchers subsequently confirmed the gene of interest by performing a GWAS in a second unrelated cohort. According to the literature, the gene of interest encodes an extracellular matrix component. What might the authors do next to confirm the function of the gene in relation to the specific phenotype (bone stiffness)? Describe how the authors would perform the subsequent functional experiments.
3. Researchers performing transgenic mouse experiments found that when their gene of interest was knocked out, the mouse pup died within hours of being born. They determined that the knocked out gene is critical for development and that without the gene the mouse would not survive. They thought, however, that the gene was not essential for survival of the adult. How could the researchers get around this problem and still determine the effect of the knockout on the mouse's phenotype? Describe the experimental details required to bypass this issue and how the technique works.
4. Describe why the following are important and how they are used in different aspects of transgenic mouse technology.
   a. Viral vectors
   b. Antibiotics and antibiotic resistance genes
   c. ES cells
   d. Restriction enzymes
   e. Southern blotting
5. List and discuss/describe three examples of how genetic techniques and transgenic mouse technology have furthered our understanding of bone biology.

Suggested Readings

Balemans, W., Van Hul, W., 2007. The genetics of low-density lipoprotein receptor-related protein 5 in bone: a story of extremes. Endocrinology. 148 (6), 2622–2629, Epub 2007 Mar 29. Review. PMID: 17395706.

Elefteriou, F., Yang, X., 2011. Genetic mouse models for bone studies—strengths and limitations. Bone. 49 (6), 1242–1254.

Galli-Taliadodos, L.A., Sedgwick, J.D., Wood, S.A., Korner, H., 1995. Gene knock-out technology: a methodological overview for the interested novice. J. Immunol. Methods. 181, 1–15.

Kuhn, R., Wurst, W. (Eds.), 2009. Gene Knockout Protocols. second ed. Humana Press, New York, NY.

Nagy, A., Gertsenstein, M., Vintersten, K., Behringer, R., 2003. Manipulating the Mouse Embryo: A Laboratory Manual. third ed. Cold Spring Harbor Press, Cold Spring Harbor, New York.

Paic, F., Igwe, J.C., Nori, R., Kronenberg, M.S., Franceschetti, T., Harrington, P., et al., 2009. Identification of differentially expressed genes between osteoblasts and osteocytes. Bone. 45 (4), 682–692.

Pease, S., Sanders, T.L. (Eds.), 2011. Advanced Protocols for Animal Transgenesis: An ISTT Manual. Springer Press, New York.

Ralston, S.H., Uitterlinden, A.G., 2010. Genetics of osteoporosis. Endocr. Rev. 31 (5), 629–662, Epub 2010 Apr 29. Review. PMID: 20431112.

Zheng, H.F., Spector, T.D., Richards, J.B., 2011. Insights into the genetics of osteoporosis from recent genome-wide association studies. Expert Rev. Mol. Med. 13, e28, Review. PMID: 21867596.

SKELETAL ADAPTATION

Mechanical Adaptation

Alexander G. Robling[1], Robyn K. Fuchs[2] and David B. Burr[1]

[1]Department of Anatomy and Cell Biology, Indiana University School of Medicine, Indianapolis, Indiana, USA

[2]Department of Health and Rehabilitation Sciences, Indiana University School of Medicine, Indianapolis, Indiana, USA

WOLFF'S LAW—A HISTORICAL PERSPECTIVE

Although it appears dead and unresponsive, bone is one of the most vital and adaptable tissues and organs in the human body. This was recognized nearly 200 years ago in an abstract sort of way, when pathologists and surgeons noticed that trabecular bone in the femoral head and neck was oriented in particular directions that seemed to make engineering and mathematical sense (Fig. 9.1). Although the fundamentals of this observation were outlined as early as 1834, the observations culminated in a treatise by Julius Wolff, and were codified by Wolff's Law, which we still cite frequently today. In translation, Wolff's Law states:

> Alterations of the internal architecture clearly observed and following mathematical rules, as well as secondary alterations of the external form of the bones following the same mathematical rules, occur as a consequence of primary changes in shape and stressing or in the stress of the bones.

This seemed to suggest that bone was adaptable to mechanical loads that create stresses in the structure. In reality, Wolff's interpretation was based on observation and not on empirical evidence. Moreover, his interpretation was based on developmental processes; it is not at all clear that Wolff understood that bone remodeling, and adaptation to mechanical loads, could occur in the adult skeleton. Wolff's Law became an important conceptual underpinning for further experimental work that attempted to identify the mathematical rules that he proposed. The important concept that he presented so clearly was that bone structure can be predicted using mathematical rules, based on mechanical stresses imposed on the structure. This was not completely new, as a German mathematician named Culmann, who had predicted

stresses in other more homogeneous materials, had proposed 25 years earlier that bone trabeculae had trajectories (orientations) that were mathematically determined.

The paradigm that Wolff made famous is rather simple: a bone, or bone tissue, is subjected to mechanical stress, which is sensed in some manner by bone cells, which subsequently act to add more bone where it is needed, to prevent fragility, or remove bone where it is not needed. Some of the most important questions of this model include:

1. What are the signals that the cells sense?
2. Which cells detect these signals, and how?
3. How do cells know where and when to add or remove the bone?
4. How do the cells know when to stop making or removing bone?

The answer to the first of these questions was suggested in 1917 by D'Arcy Thompson, the eminent mathematician and author on form and function, who stated:

> The origin or causation of the phenomenon would seem to be partly in the tendency of growth to be accelerated under strain... accounting therefore for the rearrangement of... the trabeculae within the bone.

This still frames the adaptive process in the light of development, but raises the important issue that strain may be the signal. Although some have subsequently claimed to have developed the idea that strain is the driving force in the mechanical adaptation of bone, and the signal that bone cells sense, Thompson was the first to state this explicitly. However, again, he did not demonstrate this empirically. That was to come with experimental methods (i.e. strain gauging) to measure actual bone strain during various types of activities, and with computational approaches such as

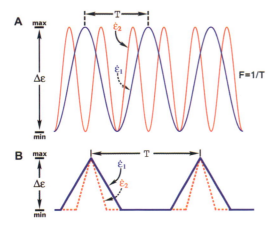

FIGURE 9.3 (A) Two strain-time curves, showing that as frequency increases, strain rate increases. This is because strain rate is proportional to frequency (see Eqn 9.1). Therefore, strain rates are always higher when you run than when you walk, even if the strain magnitude on the bone is the same. Frequency $(f) = 1/$the interval between peak load (T). $\Delta\varepsilon$ = strain magnitude; $\dot{\varepsilon}$ = different strain rates at low $(\dot{\varepsilon}_1)$ and high $(\dot{\varepsilon}_2)$ values, respectively. (B) Experimentally, it is possible to dissociate strain rate from frequency if a triangular waveform is used and the time between individual load cycles is altered.

highlighting the differences between the two concepts. A particular loading bout could last for 1 h (long duration) and be populated with 90 loading cycles. Although the duration is long, the frequency is low (1 cycle every 40 seconds, or 0.025 Hz). Conversely, one could populate a 5-min loading period (short duration) with 900 loading cycles. In this case, the duration is short, but the frequency is high (3 cycles every second, or 3.0 Hz). During our usual gait cycle, strain frequency and strain rate are closely related, and it is impossible to alter one with changing the other (Fig. 9.3). However, under experimental conditions, it is possible to separate these two factors of the strain environment, and this allows us to determine the relative importance of each in driving bone's adaptive mechanism.

Polarity

Polarity refers to the kind of deformation and the nature of the stresses generated within the bone when it is loaded, e.g. compression, tension, or shear. Most bones are loaded in bending, which generates all three kinds of stresses at different locations within the bone's structure. Polarity determines the location of the response, whereas magnitude, rate, and frequency play a critical role in determining whether there will be a response at all. The idea that strain distribution— strain gradients—is important to the adaptive process, and understanding that the polarity of loading will alter both distribution and gradients in characteristic

ways, provides a clue to the signaling mechanisms that define the location of bone adaptation. Early observations that trabeculae were oriented in the principal stress directions (which, unlike strain, could be calculated) also suggested that the polarity of loading and the nature of the stress might play a role in optimizing the orientation of structures within the bone.

Strain Energy

All of the features described above combine to determine the total amount of energy created when a load is applied to a bone. Strain energy is the product of stress and strain. Therefore, some have dispensed with trying to parse out the effects of individual factors, in favor of the idea that bone responds to the amount of energy imparted to it. However, strain energy has no direction and is always positive, so even if it were important in initiating a response, it could have little effect on the location of the adaptive response and no effect on trabecular orientation. Nevertheless, it can be helpful in modeling bone's response to strain because it is relatively simple.

ANIMAL MODELS OF SKELETAL MECHANOTRANSDUCTION

Our discussion thus far has focused on properties of the mechanical signal and/or environment that have important effects on bone tissue. Before continuing deeper into this realm, it would be pertinent to review some of the experimental models that have been (and continue to be) used to determine those and other properties. Bone mechanobiology is often studied by deforming bone tissue in vivo using a variety of techniques, and quantifying the response via a number of histologic and biochemical methods. Experimentally, the force required to induce bone deformation can come from intrinsic sources, such as voluntary muscle contraction during a vigorous exercise session (intrinsic noninvasive models), or from normal activity following the surgical removal of a nearby bone that formerly shared the load (intrinsic invasive models). Conversely, the load can originate from extrinsic sources, such as loads applied to surgically implanted pins (extrinsic invasive models) or pressure applied to skin adjacent to bone (extrinsic noninvasive models).

Intrinsic Loading Models

Intrinsic animal loading models are defined as those in which forces imposed on the skeletal element of interest are generated by the animal's own activity.

Noninvasive intrinsic loading models usually involve conditioning an animal to engage in some type of enhanced physical activity, which can alter a number of components of the typical mechanical loading environment (e.g. number of cycles, peak strain magnitudes, rates, and orientations). Many different species have been trained to run on treadmills, swim in pools, and jump up to or down from elevated platforms. Additional ambulatory models that do not require animal compliance with a specific exercise protocol have been developed to increase mechanical loading. Rats can be forced to adopt a bipedal posture for brief periods by raising the height of the food tray in their cages, or they can be constrained to use three rather than four legs for locomotion, by either casting or bandaging one of the hind limbs to the body, thereby increasing the loads on the functioning hind limb. In addition, centrifugation—rotation of the entire habitat to simulate the effects of increased gravity—can be used to enhance skeletal loading generated from otherwise normal functional activities.

The advantage to using noninvasive intrinsic loading models is that there are no surgical complications; surgical manipulation can cause bone gain or loss that has nothing to do with the mechanical environment that is being studied, thus confounding interpretation. The loads are derived from muscle contraction and substrate reaction forces, thereby providing a reasonable estimate of what humans could expect to gain in bone mass under similar exercise conditions; and, unlike most extrinsic loading models, trabecular responses in the limb bone metaphyses can be studied because the muscle and ground reaction forces are transmitted through the joints and underlying epiphyseal/metaphyseal trabeculae. Limitations of this approach include incomplete control over the mechanical inputs to the bone. The same exercise protocol can produce a wide variation in peak strains and strain distributions in different animals within the same experimental (age-matched and weight-matched) group. Additional limitations of the exercise models include:

1. The lack of an internal control bone (nonloaded contralateral bone)—running, swimming, and jumping require loading of both right and left limb bones; consequently, there is no normal bone within the same animal to which the loading response can be compared; and
2. It is difficult to isolate the effects of mechanical loading per se as the cause of the adaptive response and exclude those influences or factors related to a general physiologic response to exercise.

An alternative intrinsic loading model to an exercise protocol is the osteotomy procedure (Fig. 9.4A). In the forearm of most quadrupedal mammals, both the radius and ulna transmit the weight of the thorax from the distal humerus to the carpus. When one of these elements is removed or resected (typically the ulna), all of the force must be transmitted through the remaining intact bone. In the osteotomized animal, exercise programs are not required (but can be used in conjunction) to elevate strains because normal activities will elicit a greatly enhanced strain environment in the intact bone. Osteotomy experiments have been conducted in a wide range of species, including rats, rabbits, guinea pigs, dogs, sheep, and pigs. The standard site for osteotomy in larger animals is the radius or ulna, although in the rat the central metatarsals have been overloaded by surgical removal of the peripheral metatarsals or by removing the upper limbs, which forces the animal to assume a bipedal posture. Osteotomy models are associated with many of the same limitations described for noninvasive exercise models, particularly the lack of control over mechanical inputs. A particular disadvantage to these models, however, is the potentially inflammatory effects of surgical intervention, which can result in injury-induced bone formation.

Extrinsic Loading Models

Extrinsic animal loading models are defined as those in which forces imposed on the skeletal element of interest are generated by a mechanical actuator. Extrinsic loading models can be classified as invasive, which use the surgical implantation of pins to transduce the force generated in the actuator to the bone, or noninvasive, which avoid surgical intervention and typically transduce the mechanical signal through the skin and soft tissues.

Invasive (Surgical) Models

Three main surgical models have been developed to alter the mechanical environment of axial and appendicular bones. These models all involve the surgical implantation of steel pins or fixtures (e.g. caps) within the bone. After healing, well-controlled mechanical signals can be applied to the rigid pins or fixtures via an actuator. Forces applied to the pins are transmitted directly to the bone, resulting in bending (if unilateral force is applied) or axial compression (if bilateral force is applied). Surgical pin models of bone adaptation include the rabbit tibia, the avian ulna, and the rat caudal spine (Fig. 9.4B–D). A major advantage of the surgical pin model is that the mechanical signal generated in the actuator is preserved with great integrity in the bone diaphysis, because the signal travels through very rigid materials rather than through soft tissue and joints, which

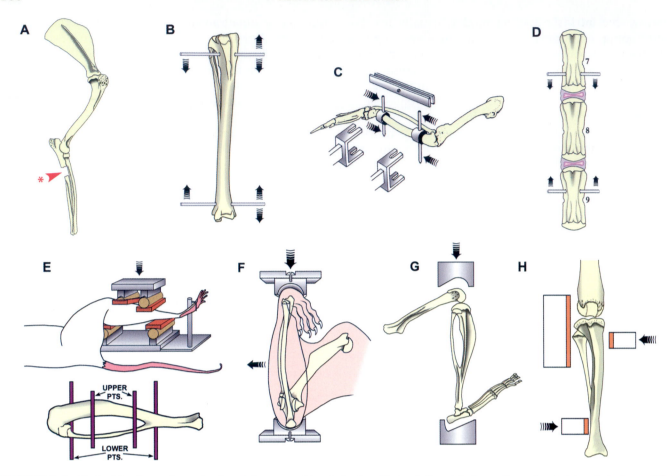

FIGURE 9.4 Skeletal loading using animal models are used to study the effects of altered (enhanced) mechanical inputs to bone. (A) A dog forelimb that has undergone ulnar osteotomy (red arrow) causes the radius to bear increased loading. (B) A rabbit tibia, (C) turkey ulna, or (D) rat caudal spine can undergo enhanced compressive loading via implanted pins by mounting the pins in a material testing machine. (E) The rodent tibia can undergo four-point bending in the mediolateral direction. (F) The rat or mouse forelimb can be loaded in axial compression to generate bending at the midshaft ulna. The mouse tibia can be loaded in (G) axial compression or (H) mediolateral cantilever bending to generate mechanical stimulation. PTS, platens. *Panel C is modified from Rubin CT et al., J Bone Joint Surg [Am] 1984; 66:397–402.*

tend to dampen the signal. This attribute affords the investigator great control over the mechanical environment produced in the tibial diaphysis. Another advantage is the opportunity to use the contralateral skeletal element, or an adjacent one in the case of the spine, as a normally loaded (via habitual cage activity) internal control. This contralateral control is subject to the same systemic factors as the loaded limb, thereby isolating the mechanical input from other nonmechanical factors that may affect bone. The main disadvantage to the surgical pin models is the potentially confounding effects of inflammation from the pin-bone interface during loading, which have not been controlled for experimentally. This limitation has been overcome to some extent in the pinned caudal vertebra model, in which the bone being studied for adaptation (caudal vertebra 8) is not subjected to surgical manipulation; rather, the two adjacent vertebrae

(CV7 and CV9) are pierced and pinned, leaving CV8 undisturbed (Fig. 9.4D).

Noninvasive Models

There is considerable appeal in the development and use of animal loading models that are capable of applying a relatively well-defined mechanical signal to bone, without the potential complications of surgically induced irritation or inflammation. Nonsurgical models are technically simpler, less expensive, and do not rely on healing processes, as compared to the surgical models. Turner described one of the first noninvasive extrinsic loading models, which entailed subjecting the rat tibia to four-point bending in the mediolateral direction (Fig. 9.4E). In this model, which has been scaled down for the mouse, a hind limb of an anesthetized animal is placed between pairs of upper and lower padded load points. The points are offset so that when a downward-directed force is applied to the upper points, the load is transmitted to the tibia through the skin, fascia, muscle, and periosteum intervening between the load points and the bone surface, resulting in the production of a bending moment in the region between the two upper points. To reveal the osteogenic effects of pressure on the force-transducing soft tissues without applying a bending moment, a sham configuration has been implemented in which the upper and lower points directly oppose one another. The rat tibia model, while useful for studying endocortical adaptation, is limited in its utility for studying periosteal bone formation, as a woven bone response is typically elicited on that surface, whether bending or sham loading is applied.

Another rodent tibial bending model was introduced by Gross and colleagues at the University of Washington. In this model, the proximal mouse tibia is fixed to a platform via padded grips (Fig. 9.4H). The distal end of the bone is pushed medially via an actuator, thus producing a cantilever bending configuration that results in mediolateral bending. Unlike Turner's rodent tibia mediolateral bending model, the cantilever model can be used to study periosteal and endocortical adaptation.

Perhaps the most widely used in vivo loading model is the rodent ulna axial loading model developed in the Lanyon laboratory in the UK. In this model, the forearm is secured between two small metal cups—one receiving the elbow and the other receiving the dorsal surface of the volar flexed wrist—which are mounted to the platens of a materials testing machine or other actuator (Fig. 9.4F). Compressive forces applied to the platens are transmitted to the ulnar diaphysis through the skin, fascia, articular cartilage (at the distal end), and ulnar metaphyseal bone. The natural curvature of the ulnar diaphysis translates most (approximately 90%) of the axial compression into a mediolateral bending moment. Between loading sessions, the animals are permitted normal cage activity, and show no signs of gait modification or lameness from loading. The main limitation to the rodent ulna loading model is the lack of a sham-loaded control. It is generally assumed that the osteogenic response observed along the ulnar diaphysis is not influenced by trauma or soft tissue pressure, but this has never been tested experimentally.

The rodent ulna model was more recently adapted to the tibia. In this model, axial compression is applied to the tibia through fixtures that apply force to the distal femur and the calcaneus (Fig. 9.4G). Like the ulna model, the tibial compression model produces a bending moment due to its natural curvature. However, this model has advantages over the ulna model in that

force is never applied directly to the bone under study (the ulna model applies force directly to the olecranon) and also that trabecular bone adaptation in the proximal tibia can be studied. Like the ulna model, there is no currently used sham control configuration.

RULES FOR BONE ADAPTATION

Since the early 1970s, a multiplicity of experiments has allowed us to identify key concepts that define the manner in which bone adapts to its mechanical environment. There is now sufficient experimental verification for each of these such that they can be considered established laws for bone adaptation. Not all of the relevant experiments will be described here, but additional support for these concepts can be found in the suggested readings at the end of this chapter.

Bone Responds only Above (or Below) a Threshold of Strain or Strain Rate

It had been surmised for years that bone does not respond within all ranges of strain. There was an intuitive understanding that bone is in maintenance mode, unless the mechanical stimuli are outside the range of normal activities. This was, and is, the basis for many of the computational models used to predict the bone response to mechanical loads, including those devised by Cowin et al., Carter and Beaupre, and Frost. The idea is that if the mechanical signal is too low, then bone will be lost, but if the mechanical signal is high enough, then bone will respond by adding new bone in locations that effectively reduce the strain to within the usual range. Two notable studies demonstrated that a threshold does indeed exist, and both were consistent in identifying the threshold for skeletal adaptation to increased strain.

Using a rat four-point tibial bending model, Turner loaded a series of animals at strains of $400-2000\ \mu\varepsilon$ and measured bone formation rate (BFR) using fluorochrome labels on the endocortical surface of the tibia. There was no difference in BFR in the loaded tibia when compared to the contralateral tibia that was not loaded, up to about $1050\ \mu\varepsilon$. However, above this threshold, there was a linear and significant increase in BFR (Fig. 9.5). The idea that bone formation is linearly related to strain magnitude was not new, but the experimental demonstration that there is a threshold, and identification of where that threshold lies, was novel. Earlier, Rubin and Lanyon had performed a similar experiment in which they applied loads to avian forelimbs from zero (nonloaded) to $4000\ \mu\varepsilon$ and

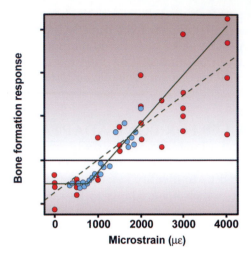

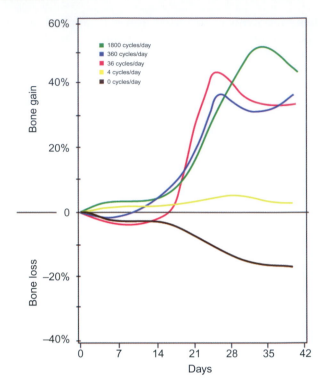

FIGURE 9.5 The results of several experiments suggest that it is necessary to achieve a threshold of strain before bone formation will begin to occur. This threshold appears to be approximately 1000 με (or 0.1% change in length of the bone). Below this, resorption and formation are either in equilibrium or there is bone loss if strains are low enough. There is some disagreement about whether the adaptive bone formation response is linearly related to strain magnitude or represents an actual threshold response. *Red dots and dotted line are based on data in Rubin CT et al., Calcif Tiss Int 1985; 37:411—417. Blue dots and solid line based on data from Turner CH et al. J Bone Miner Res 1994; 9:87—97.*

FIGURE 9.6 As long as the mechanical strain is sufficiently high, very few cycles are necessary to prevent bone loss or to cause bone formation. Moreover, once the threshold for bone formation is achieved, adding additional loading cycles (of 36-1800 cycles/day) has no effect on the amount of bone formed. These data suggest that the bone's response to mechanical loading can saturate. *Graph redrawn from data presented in Rubin CT et al., J Bone Jt. Surg 1984; 66 A:397—402.*

noted a linear increase in cortical bone area related to increasing strain (Fig. 9.5). They did not identify a threshold for response in their earlier experiment, but presented a linear response over the entire range of loads. However, when their data are viewed more closely, it becomes clear that there is a threshold for response slightly above 1000 με. This suggests a threshold, and implies that it may be similar across species.

Bone Responds Only to Dynamic Loads

For reasons that are probably related to the way that bone cells receive mechanical signals, it is now clear that bone does not adapt to loads unless they are applied cyclically. Static loads, even those that are quite high, will not elicit any adaptive response. Early experiments by Hert in 1969 first suggested that dynamic loads are necessary for bone response. Later, this was underscored by Lanyon and Rubin, who applied static loads to the avian ulna for 2 months and observed endocortical and intracortical bone loss, changes identical to those found in immobilized animals. Dynamic loads at equivalent peak strain increased net bone formation on the periosteal surface, and prevented endocortical and intracortical loss. Subsequently, an experiment performed using the rat axial ulnar loading model further supported this concept. Bone formation on the periosteal

surface in rats loaded statically at 8.5 N and 17 N showed the same amount of bone formation, which was not different to formation on the nonloaded contralateral limb. This showed that even with increasing loads, a static load does not generate a response. However, when a dynamic load was applied at 17 N for an equivalent length of time, there was a large increase in periosteal bone formation. This showed that a load of equivalent magnitude can have very different effects, depending on whether it is applied statically or cyclically.

The Loading Period can be Short

The third key concept in mechanical adaptation was first proposed by Rubin and Lanyon, again, using the avian model. By loading the rooster ulna for various cycles/day, ranging from none (disuse) to 1800 cycles/day, they discovered that 4 cycles/day at 2050 με was sufficient to maintain bone (Fig. 9.6). Moreover, they found that 36 cycles/day caused an increase in bone mineral content (BMC) of the ulna, and that this response saturated, such that more cycles, up to 1800/day, did not produce a further increase in BMC.

Subsequent experiments in rats by others have verified this result, indicating that bone is exceedingly sensitive to small amounts of mechanical stimuli of sufficient magnitude.

Rate-Related Phenomena are Critical to Response

The requirement that dynamic loads are required for bone's mechanical adaptation implies that the rate of application of a load may be an important component of the response. Studies show that exercises involving loads applied at higher rates, such as running, jumping, and so forth, may be more osteogenic than other forms of exercise. Even short periods of impact loading can stimulate increased bone apposition, or prevent decreased bone apposition rate in immobilized animals. Moreover, more bone formation is seen in response to 50-ms impact loads than in response to less impulsive 500-ms loads, even though the manner in which the load is applied and the magnitudes of the load and strain are equivalent. Rough estimates have suggested that 68—81% of the variance in bone surface adaptation may be accounted for by strain rate. Experiments using the rat ulnar axial loading model have verified the importance of strain rate to bone response. When loaded to a high strain, at strain rates corresponding to walking, (0.01/s), running (0.03/s), or jumping, the response is higher at the jumping strain rate than at the strain rate for running or walking. Thus, given an equivalent strain, bone will respond more markedly at higher rates of load application.

Strain Rate is a Function of Strain Magnitude and Loading Frequency

The characteristics of loading discussed above were synthesized by Turner based on several of his own experiments to demonstrate how bone responds to strain rate, and the nature of the relationship between strain magnitude, rate, and frequency. These three variables are linearly proportional to each other:

$$\text{Strain rate} = 2\pi f \varepsilon \qquad (9.1)$$

where f = frequency and ε = strain. This proportionality shows that if either the strain magnitude or frequency of application is increased, then strain rate will increase; therefore, both of these factors are reflected in the single parameter of strain rate. Now, during our normal gait cycle, strain is applied in a format similar to a sine wave, and any increase in frequency, e.g. when we are running, will be translated into an increased strain rate (Fig. 9.3A). Experimentally, however, rate

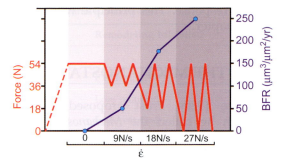

FIGURE 9.7 **Strain rate significantly affects bone formation response to loading.** Using the rat four-point tibial loading model, strain rate ($\dot{\varepsilon}$) was varied independently of strain magnitude by either static loading ($\dot{\varepsilon} = 0$) or cycling at 9, 18 or 27 N/s to a maximum strain of 54 N (left axis). The results showed that there was a fivefold increase in bone formation rate (BFR; right axis) on the endocortical surface of the loaded tibia, with a threefold increase in strain rate. Static loading to the same maximum strain was not associated with new bone formation. This demonstrates the profound effect of strain rate on stimulating bone formation. *Based on data in Turner CH, et al., Am J Physiol 269 (Endocrinol Metab 32):E438—E442; 1995.*

and frequency can be dissociated by utilizing a triangular waveform for loading, and varying the period between cycles to alter one variable (strain rate or frequency) while keeping the other constant (Fig. 9.3B). Using this concept, Turner showed that at equivalent strains and strain rates, frequencies greater than 0.2 Hz (i.e. 1 load cycle every 5 s) will generate an adaptive response, whereas those lower than that will not. In a second experiment, Turner loaded the ulna to the same strain magnitude (54 N), but then cycled the ulnae to that magnitude over a different strain range (0, 18, 36, or 54 N), effectively changing the strain rate. In this experiment, cycling over a strain range of 0 is equivalent to static loading (loading to 54 N and holding it there), and one would not expect any response. This is what occurred. However, over the range of 18—54 N, there was a dose-response effect, with increasing BFR as strain rate increased (Fig. 9.7).

Taking these two experiments, one in which frequency was varied, and one in which strain rate was varied, Turner calculated the total strain stimulus (E) as:

$$E = k_i \sum \varepsilon_i f_i \qquad (9.2)$$

This equation looks very much like Eqn 9.1 for strain rate, indicating that over a series of cycles, the adaptive response (E) will be proportional to strain and frequency. When the strain stimulus is calculated for the two experiments in which either frequency or cyclic strain magnitude are varied, the relationship between strain stimulus and BFR is almost identical. This demonstrates that it is the combination of strain magnitude and frequency—or strain rate—that drives

may be different because of their location. For instance, it is much easier to cause removal of bone from the endocortical surface than it is to cause addition of bone to this surface; the reverse is true of the periosteal surface—bone apposition is much easier periosteally than is bone removal, although both can occur. The reason for this is that the endocortical surface is exposed to the marrow, which has many cytokines, as well as precursors for osteoclasts and osteoblasts. The periosteal surface, on the other hand, has quiescent osteoblast precursors in its deep cambium layer; osteoclasts have to be recruited from the vascular system and are more difficult to activate on this surface. Of course, some of the difference in response on these two surfaces could also be related to strain—for mechanical reasons, the periosteal surface of a long bone will generally be under greater strain than the endocortical surface (see Chapter 6). Nevertheless, this difference in strain is unlikely to entirely explain the differences observed on these two surfaces in response to a common load, and it has been shown that, even under equivalent strain, they do not respond in the same way.

CELL SENSITIVITY AND REFRACTORY PERIODS

The Mechanostat does a good job of explaining when and where bone formation will occur, but implies that the only feedback mechanism that stops formation is re-equilibration of the strain stimulus. This suggests that cells will be active until strain returns below (modeling), or above (remodeling), a given threshold. However, there is good evidence that cells lose their sensitivity to mechanical loading signals after a period, and that this period can be quite short, as suggested by one of the four rules for bone adaptation. How quickly cells saturate their response, and how long the refractory period is before they become resensitized, has only recently been discovered. The early Rubin and Lanyon experiments suggested saturation could occur but because those experiments were focused on the initiation of bone adaptation, the saturation point was never recognized. Saturation limits

were also discussed by Parfitt at about the same time those experiments were performed, but in this case saturation referred to the maximum speed at which cells could adapt, generally assumed to be about 3–4 μm/day of lamellar bone (but more quickly in younger individuals). However, saturation in the present case refers to the limits of loading beyond which cells will not respond at all.

Several studies since the early Rubin and Lanyon experiments, using different experimental designs and different animal models, suggested that when bone is loaded to about 1000 με, the adaptive response (measured by BFR, or gauged by bone mass gained over time) will begin to saturate within about 200–400 loading cycles. These experiments led to others that used the four-point bending model in rats, in which 360 load cycles were delivered either all at one time or divided into 2, 4 or 6 bouts of loading equaling 360 load cycles/day. One would expect that, without cell saturation, it wouldn't matter how the 360 cycles are delivered, since the amount of energy imparted to the bone in each case is identical. However, these experiments showed that time between loading cycles is important, and that BFRs are significantly higher when the load is delivered in 4 or 6 bouts, with 90 or 60 cycles within each bout, compared to groups in which the 360 cycles are delivered at one time (Fig. 9.9).

This convincingly demonstrated that cells saturate to repeated loads and that, at these levels of strain, saturation occurs within the first 200 load cycles (because the loading cycles were delivered at 2 Hz, this represented <1 min of cyclic loading). However, this experiment did not show how long it took for cells to become resensitized to loads once they had become saturated. So, a second set of experiments was performed in which all groups were given four bouts of 90 cycles per bout, with the time between the bouts varying from 30 min to 8 h. These experiments demonstrated that bone formation is significantly greater when the bouts are separated by 4 h compared to a group that received 360 cycles at one time, and that when they are separated by 8 h, bone formation is significantly greater than in those that received their loads only 30 min apart (Fig. 9.9). These experiments

CELL SATURATION AND REFRACTORY PERIODS

1. Mechanical loading is more osteogenic when divided into discrete loading bouts, allowing a recovery period between bouts.
2. A recovery period of about 4–8 h appears sufficient to recover most of the cell's sensitivity to loading,

although full sensitivity may not return until 12 h following loading.
3. Several shorter intervals of daily exercise, rather than a single sustained session, may be more effective for optimizing net bone formation.

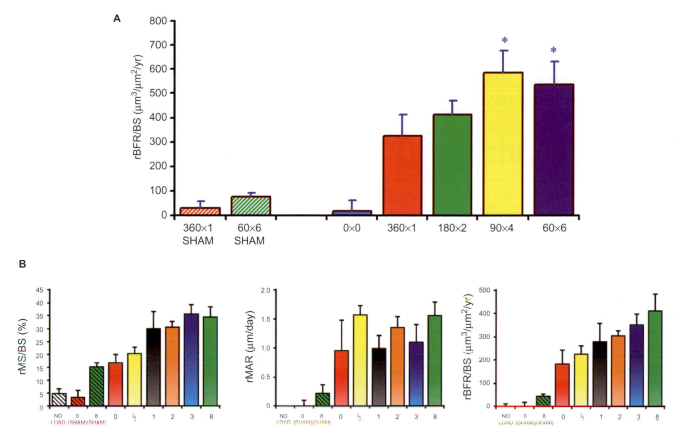

FIGURE 9.9 **Bone formation saturates quickly in response to a load.** Significantly greater bone formation can be achieved when a small number of cycles are delivered in separate bouts of loading, allowing time between the bouts for the bone cells to recover. (A) This graph shows significantly greater response when 360 cycles of loading are delivered in groups of 60 cycles or 90 cycles at four different times during the day. (B) The refractory period for bone cells lasts about 4−8 h. Therefore, if periods of loading are separated by about 8 h, the bone formation response will be significantly greater, even if the total number of load cycles is identical. MAR, mineral apposition rate; rBFR/BS, relative rate of bone formation per unit bone surface; rMS/BS, relative mineralizing surface per unit bone surface. *Redrawn from data presented in (A) Robling AG et al., J Bone Miner Res 2000; 15:1596−1602 and (B) Robling AG et al. J Exper Biol 2001; 204:3389−3399.*

suggested that most of the recovery of cell sensitivity occurs within 4−8 h, but that full recovery may require as long as 12 h (see box below).

Cells also demonstrate refractory periods on shorter time scales. Several sets of animal experiments have demonstrated a more prolific bone response and greater formation rates (measured histologically) when the frequency of the applied loading cycles is reduced. When individual cycles within a loading bout are applied at intervals of 10−14s, bone formation is significantly more rapid than when loads are applied at 1 or 2 Hz. The reason for this is not entirely clear, but cell saturation on this shorter time scale may be related to voltage-gated or calcium-sensitive channels that regulate the transport of minerals and proteins in and out of the cell, and have an effect on the cell response. There has been shown to be a marked decline in the rate of fluid flow within the canalicular system when the frequency of loading is increased from 1/sec to 20/s. If the strain-induced flow of fluid through bone is the local mechanical stimulus for cell response (see below), then insufficient time for the fluid in bone to "relax" and return to its resting state could impair the mechanical signals, so that the cells perceive a more static loading situation and fail to respond to their full potential. Thus, cell saturation occurs both on longer and shorter time scales, probably driven by different causes and perhaps with somewhat different long-term effects.

A TEST OF WOLFF'S LAW

Let's now revisit the fundamental tenet of bone adaptation, Wolff's Law, and determine whether Wolff's Law predicts bone response based on experiments that have defined some of the rules for bone adaptation. Recall that Wolff's Law stated that stresses imposed on bones change architecture, and that the change can be predicted using mathematical rules.

This implies that there is a unique adaptive solution to any given set of mechanical inputs. If Wolff's Law is correct, then even though adaptation may occur more quickly in some cases than in others, over a period of time bone mass and architecture should achieve identical structural changes. However, it appears that this may not be correct. When the rat ulna was cyclically loaded 3 days/week using either 360 cycles delivered at one time, or 90 cycles delivered four times throughout the day, periosteal bone formation was still threefold higher after 4 months (four remodeling cycles in a rat) in the 90×4 group than in the 360×1 group. Moreover, BMC, whole bone areal bone mineral density (aBMD), cross-sectional area, and rigidity in the direction of bending at the midshaft were all significantly greater—approximately 40–50% greater—in animals loaded four times throughout the day. This was true, in spite of the fact that an identical number of cycles was applied in each case; the only variation was the time over which the load was delivered, allowing the cells to recover sensitivity before applying the next set of loads. This suggests that Wolff's Law is incomplete (or that we don't fully understand the mathematical laws yet). The practical implication is that short bouts of exercise, with rest time between them, are more osteogenic than are single, longer periods of exercise.

MECHANOTRANSDUCTION ON A SMALLER SCALE: CELL TYPES AND THEIR ENVIRONMENT

The discussion thus far concerning mechanical adaptation of bone tissue has been largely phenomenological—that is, if biologically appropriate loads are applied to living bone, we can expect to see some type of change in the size or shape. Loading has clear effects on the resident cells per se, since the extracellular matrix is incapable of metabolism. How then do the cells in and around the bone tissue sense the applied load, or some physical consequence of the applied load? Moreover, how do the cells translate that stimulus into a series of intracellular biochemical events that ultimately leads to bone gain or loss? These are two very involved questions, and can perhaps be more readily explored as a series of smaller, interrelated questions.

Before embarking on an exploration of how resident cells in and around the bone might sense applied loads, it would be prudent to focus on the "sensor" cell type in bone. Bone tissue is embedded with, and surrounded by, a multitude of cell types, particularly in the marrow cavity. An understanding of the cell type involved in mechanical signal reception is not

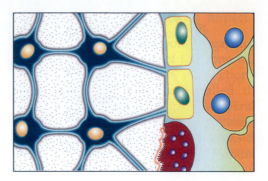

FIGURE 9.10 **The different cell types that inhabit bone tissue are situated in a wide range of physical environments.** Marrow cells (orange cells to the right) are housed in a wall-to-wall fashion inside the medullary cavity. These cells do not physically attach to a mineralized matrix and are consequently accessible from all sides by fluid pressures that develop inside the medullary cavity. Osteoblasts (yellow cells), osteoclasts (maroon cell), and bone lining cells (not shown) attach to the mineralized bone matrix at its surfaces. These cells are exposed to medullary fluid pressures and also to bone tissue strains where they attach to the matrix. Osteocytes (blue cells to the left) are completely surrounded by mineralized bone. They attach to the matrix intermittently and are subject to tissue strains that develop, but also to high velocity fluid movement in the spaces between the plasma membrane and the mineralized matrix walls of the canaliculae and lacunae. This tight space serves to increase fluid velocity during loading.

simply an academic exercise; it dictates how we proceed experimentally for two main reasons. First, different cell types have different transcriptional signatures. Osteoblasts have a different "toolbox" of transcribable genes at their disposal than do osteocytes, bone lining cells, or stromal cells in the marrow. Knowing which cell type is the sensor cell then allows us to narrow our exploration of the relevant genes, proteins, and lipids for delineating molecular mechanisms of mechanotransduction. Second, the physical environments of the different types of bone cells are vastly different, so the relevant mechanical stimulus is necessarily cell-type dependent. For example, stromal cells in the marrow cavity of long bones are housed in a wall-to-wall fashion. While fluid movement certainly exists in the marrow, it is most likely to be at lower velocities, since there are few restrictions on fluid movement in that locale (Fig. 9.10). Hydrostatic pressure of the marrow cavity (e.g. from end-loading of long bones), however, might be a more meaningful and potent stimulus for these cells. Alternatively, osteoblasts, osteoclasts, and bone lining cells adhere to the bone surfaces and would therefore experience bone tissue surface strains and some fluid movement where pores in the bone tissue open to the surface. The physical environment of the osteocyte is completely different than that of bone cells localized on the bone surface or of marrow cells. Osteocytes are entombed in a small, form-fitting cavity within the

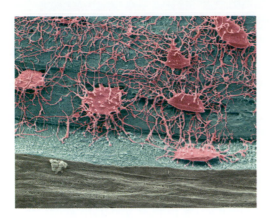

FIGURE 9.11 **Pseudocolored scanning electron micrograph of a plastic-infiltrated, acid-etched cortical bone sample from a mouse long bone.** The red structures represent a plastic cast of the lacuna-canalicular network that exists within the cortical bone. The green background represents mineralized bone matrix, and the gold region toward the bottom of the micrograph represents the periosteum (bone edge). In life, the red network would be mostly occupied by osteocytes and their long cell processes. The micrograph highlights the preponderance of connections among neighboring osteocytes via the cell processes, and the large number of processes emanating from each osteocyte. A recently buried osteocyte can be seen toward the bottom of the image, sending cell processes into the periosteum to communicate with surface cells (osteoclasts, osteoblasts, bone lining cells).

bone matrix (the lacuna). As the osteocyte cell processes emerge and course away from the cell body and lacuna, they travel in microcanals within the bone matrix (the canaliculi) (Fig. 9.11). The canaliculi are approximately 260 nm in diameter. The tight space between the osteocyte cell membrane and the lacuno-canalicular wall (approximately 80 nm) restricts extracellular fluid flow and serves to enhance its velocity when driven by load-induced pressures or matrix strains.

Decades ago, it was postulated that the osteocyte was the best candidate for a sensor cell type, for several reasons. First, the osteocytes are regularly distributed throughout cortical and trabecular bone, even in areas of mineralized matrix devoid of vasculature. Consequently, the network of osteocytes provides a widespread load-monitoring "net" that infiltrates every cubic millimeter of bone tissue. Second, osteocytes are connected to one another through long cellular processes that course through the bone and project to the bone surfaces. Osteocytes have a large number of these cell processes (approximately 50/cell) emanating from the cell body and coursing in all directions. The cell processes join similar cell processes from neighboring osteocytes and transmit information intercellularly via gap junctions, which facilitate rapid cell-cell communication. Third, it is clear that osteocytes are not effector cells, as they are entombed in a bony

matrix and are therefore incapable of adding or removing very much matrix, and can only remove it in the small area surrounding their lacunae. This very localized activity of osteocytes, while potentially meaningful for regulating serum calcium levels, has little or no effect on bone size, shape, and structural properties. Because their role as an effector cell is precluded, osteocytes have been thought of historically (somewhat by default) as a sensor cell.

Beyond teleological arguments, experiments have supported the role of the osteocyte as the primary mechanosensory cell type in bone. Compared to osteoblasts, osteocytes are much more sensitive to shear stress induced by fluid flow in vitro. Further, a very clever in vivo experiment has highlighted the importance of functioning osteocytes in sensing changes in the mechanical loading environment. Kyoji Ikeda's group in Japan engineered a transgenic mouse model that expresses an inducible suicide gene in the osteocyte population. They induced the suicide gene (which ablated the osteocytes) in a set of mice and conducted a mechanical disuse experiment, which normally results in bone loss. Despite having intact and functional osteoblasts and osteoclasts, mice in which the osteocyte network was wiped out failed to lose bone as a result of disuse—suggesting that an intact osteocyte syncytium is required to sense the change in the mechanical environment. While an accumulating body of evidence points to the osteocyte as the primary mechanosensory cell type in bone, other cell types have not been definitively excluded.

FLUID FLOW IN BONE TISSUE HAS MANIFOLD CELLULAR EFFECTS

Assuming that current and future research continues to support the assertion that the osteocyte is the primary mechanosensory cell type in bone, we may now advance our inquiry into looking at how the osteocyte might become exposed to mechanical stimulation. We have already established that the osteocyte encounters a unique physical environment among cells—indeed even among bone cells. Osteocytes are attached to the lacunocanalicular walls via transmembrane integrin dimers and a meshwork of extracellular glycoproteins collectively known as the glycocalyx. This tethering apparatus—particularly the integrin receptors—should be capable of sensing changes in tissue strains of the surrounding bone matrix as the bone deforms during loading. But is tissue strain the mechanical stimulus to which osteocytes respond? Numerous in vitro experiments have been conducted to address this question, but before proceeding we must first address some limitations of cell culture models in the field of osteocyte mechanobiology. Although a strong case can be made for the osteocyte as the primary mechanosensory cell type in bone, a survey of the published literature indicates that the vast majority of in vitro experiments on mechanotransduction in bone have been conducted using osteoblast or osteoblast-like cells. More recently, osteocyte cell lines have been cloned and distributed among investigators, but it is unclear how well they mimic the in vivo osteocyte gene profile and morphology. Moreover, it is difficult to culture these cells in an engineered physical environment that promotes the full development of the presumed in vivo mechanosensory apparatus (e.g. circumferential tethering filaments).

Mindful of these model-specific caveats, we can now cautiously turn back to the question of tissue strain as a driving stimulus for mechanotransduction in bone. Tissue strain measurements collected on the periosteal surface of long bones from a wide range of animals indicate that the peak strains that vertebrates can generate voluntarily during vigorous activity is around 3000 με. However, when cultured bone cells are exposed to 3000 με (via bending a rigid culture substrate or stretching a flexible culture substrate), no measureable response is elicited. It turns out that one must apply over 10,000 με in order to elicit a response in most culture models. In vivo, application of 10,000 με is beyond bone's yield point and would eventually result in fracture. So, it appears that the strains required to elicit a response in bone cells would not exist in nature, except perhaps during a catastrophic event. At the outset, this observation casts severe doubt on the hypothesis that tissue strain per se is a driving stimulus for mechanotransduction. But a challenge to this conclusion has arisen in recent years, based on a series of tediously collected local strain measurements around the osteocyte lacunae. The Nicolella group, using a digital micrograph correlation strain measurement technique, reported that loads generating approximately 2000 με on a bone's surface (measured grossly with traditional strain gauge techniques) can actually generate up to 30,000 με on the osteocyte lacunar wall. This disparity in gross versus local tissue strains is probably due to the stress-concentrating effects produced by the tissue voids created by osteocyte lacunae, which are undetectable at the macro-level.

While the role of tissue strain in bone mechanotransduction remains controversial, a much less equivocal view of mechanical stimulation involves cell stimulation by drag forces resulting from fluid flow. Fluid flow arises in bone tissue when extracellular fluid moves from areas of high pressure to areas of low pressure. These pressure differentials are created in long bones by bending. Bending simultaneously generates areas of high pressure (e.g. in the portion of the cortex experiencing compression) and low pressure (e.g. cortical areas experiencing tension). As discussed earlier, bending is the predominant mode of loading for skeletal elements of the limbs: across a wide range of body sizes (from turkeys to elephants), the percentage of total strain that is due to bending ranges from 75% to 90% in the typical long bone.

Bending and consequent fluid flow are a necessary outcome of skeletal loading, but are fluid flow-derived drag forces (e.g. shear stress) the mechanical stimulus to which osteocytes respond? Shear stress is only one outcome of fluid flow. When fluid moves through bone pores, it does indeed generate shear stresses on cell membranes that lie in its path, but it also enhances chemotransport to the osteocytes (e.g. more rapid delivery of growth factors and nutrients, and removal of cellular metabolic waste) and generates streaming potentials (i.e. flow of charged particles in the fluid) which alter the electrical environment around the cells. These two latter possibilities could be responsible for the skeletal changes induced by fluid flow, without having to invoke any mechanical stimulation whatsoever on the bone cells. Fortunately, several key experiments—described next—have been conducted to sort out the relative contribution of shear stress, chemotransport, and electrical potentials in the response of bone cells to fluid flow.

The first comes from the Brighton group, where the investigators exposed cultured osteoblasts to convective currents with or without superimposed fluid flow. They found that when the streaming current (i.e. electrical potential) that accompanies fluid flow was neutralized via running identical current (using an external DC power supply) in the opposite direction of fluid flow, the cellular response to fluid flow was maintained (Fig. 9.12A,B). Moreover, when they doubled the streaming current by running the same current in the same direction as the fluid flow, the cellular response was the same. Those data suggest that fluid flow has significant effects on the cellular response, but that streaming potentials are not the underlying cause of the effects observed. Next, both electrical potentials and chemotransport effects were tested by the Chambers group. They exposed osteoblasts to shear stress at a substimulatory flow rate, and observed no effect in the cells (as expected). Using the same substimulatory flow rate, they added methylcellulose (MC) to the medium to increase viscosity. Changes in viscosity affect the shear stress on cell membranes, independently of flow rate, which can be appreciated from the mathematical relation describing shear stress in the parallel plate flow chamber:

$$\tau = \frac{6\,Q_u}{w\,h^2} \tag{9.3}$$

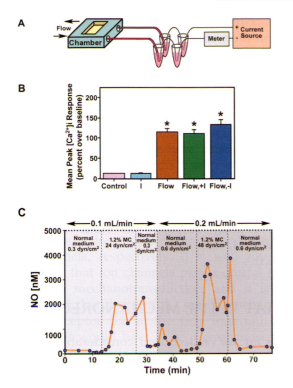

FIGURE 9.12 To address whether convective electrical currents affect cultured bone cells' response to fluid movement, Hung et al. designed a clever fluid flow apparatus (A) that allowed them to alter the electrical currents produced during a typical fluid flow experiment. Culture medium was routed through a parallel plate flow chamber (shown to the left of the blue chamber) containing osteoblasts. An electrical current, generated via an external DC power supply, was applied to the chamber via a series of agar/medium bridges that ultimately reached the flow chamber and applied current along the flow path within the chamber. The power supply was modulated such that the convective current density produced by flow was either doubled by running the current in the same direction as flow (Flow, + I), or neutralized by running current in the opposite direction as flow (i.e. *upstream*; Flow, -I). (B) The intracellular calcium response, an early mechanotransduction readout, was monitored in real time via imaging. Whereas flow alone produced the expected increase in calcium responsiveness, neither neutralization nor doubling of the convective current density had any additional effect on the calcium response to flow, suggesting that shear and not streaming potentials were responsible for biological signaling. (C) Another parallel plate flow chamber experiment that addressed a different confounding factor—chemotransport—was conducted by Smalt et al. They looked at the release of a nitric oxide (NO) metabolite into the medium of flowed cells, as a mechanotransduction readout. In the first 31 min of the experiment, the flow rate was kept at a constant 0.1 mL/min. At 11 min, methylcellulose (MC) was added to the incoming culture medium. MC increases the viscosity of the medium and consequently increases the cellular shear stress in this model. At 21 min, the medium was changed back to MC-free medium. One can observe a spike in NO production when the MC is added, and a return to normal NO production after MC washout. In the remaining portion of the experiment, they repeated the MC-infusion and washout at a higher flow rate (0.2 mL/min), which yielded a similar (but heightened) outcome. *Panels A and B redrawn after Hung CT, et al. J Biomech 1996; 29:1403-9. Panel C redrawn after Smalt R, et al., Am J Physiol (Endocrinol & Metab) 1997; 273:E751—8.*

where τ is the resulting shear stress, Q is the flow rate, μ is the viscosity of the fluid, and w and h are the width and height (depth) of the chamber bed. Upon addition of 1.2% MC, fluid shear was increased two orders of magnitude under the same flow rate, and a dramatic increase in cellular response to flow was observed (Fig. 9.12C). Upon washout of the MC, the response returned to baseline. Because the electrical properties of the medium were not altered by MC, and because the flow rate was unchanged throughout the experiment, both chemotransport *and* electrical effects could be ruled out as causes for the increase in responsiveness after "gumming up" the medium with MC. A similar conclusion was reached in a later experiment by the same investigators, in which they kept the flow rate constant, but reduced the level of shear on the cells by mounting them in a deeper flow chamber (consider the effect of increasing the h term in the above equation, on resulting shear stress, with all other parameters held constant; Fig. 9.12A). In that experiment, the cells lost their responsiveness despite experiencing an identical flow rate. The obvious conclusion from the combined experiments is that shear stress per se is driving the responsiveness of the cells to fluid flow. More broadly, these experiments illustrate why strain rate in bone has such a large impact on bone adaptation: higher strain rates drive fluid through the bone lacuna-canalicular system at higher velocities, resulting in higher shear stress and greater osteocyte stimulation.

CELLULAR DEFORMATION AND STIMULATION

While fluid flow-induced mechanical stimulation of osteocytes appears to be a potent stimulus for bone cells, it is unclear whether the nature of the stimulation can be fully attributable to shear stress on the osteocyte cell membrane. Membrane shear is by definition the force per unit area in a direction tangential to the membrane, i.e. along the direction of flow. This is an intuitive interpretation of the effects of fluid movement around osteocytes, but since the early 1990s a new model developed by the Weinbaum group has offered a persuasive challenge to whether shear stress is the main stimulus to osteocytes during mechanotransduction in vivo. The Weinbaum model incorporates recently discovered micromorphological properties of the osteocyte in situ. These cells, particularly the cell processes, are suspended from the lacuna-canalicular bony wall via integrin complexes

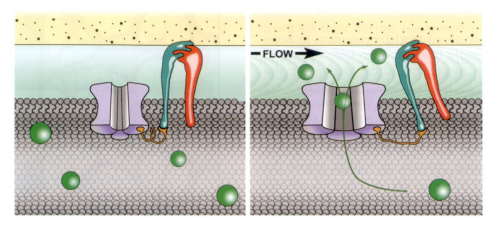

FIGURE 9.15   In bone cells, gap junction alpha-1 protein/connexin 43 (Cx43) is highly expressed compared to the other known connexins. Because efficient communication among osteocytes is a prerequisite for mechanotransduction to work, a number of investigators have looked into the role that Cx43 might play in mechanotransduction. Connexons (assembled protein structures comprising several individual connexins) can exist unopposed from a neighboring cell, in which case the connexon is considered to be a hemichannel (shown in purple). Hemichannels have been proposed as a main mechanism by which bone cells release PGE2 (green spheres) upon mechanical stimulation. Mechanically stimulated osteocytes release PGE2, presumably via the opening of Cx43 hemichannels. When these channels are blocked, PGE2 release is inhibited. Moreover, mechanically induced opening of the Cx43 hemichannel appears to be controlled by physical perturbation of the $\alpha 5\beta 1$ integrin (blue and red structures) through a direct mechanical linkage.

the cell (e.g. actin filaments) through a series of adaptor proteins (Fig. 9.14). Many years ago, in a set of experiments that were technologically far ahead of their time, the Ingber group studied the effects of mechanically perturbing one of the integrins—the β_1 integrin, by treating cultured cells with ligand-coated magnetic nanobeads. Once bound to the β_1 integrin, the beads were twisted in situ using magnetic tweezers, and large changes in cellular mechanical properties were measured. Remarkably, not only were whole cell properties altered, but specific changes in the nucleus (including nucleolar realignment and chromatin remodeling) occurred immediately upon stimulation. These studies provided evidence that a direct mechanical linkage might exist between the extracellular domain of the integrins and gene expression, requiring no "shuttling" proteins to relay mechanical information from the matrix to the nucleus. Further, direct linkage from mechanically activated integrins extends not only to the nucleus but also more tangentially to pull on other proteins on the plasma membrane, including nearby channels and hemichannels (Fig. 9.15).

While a direct mechanical linkage effect of integrin stimulation might explain some of the mechanotransduction effects in bone, there is also significant and growing evidence for a signal transduction role for integrins in mechanobiology. Mechanical stimulation of cells results in conformational changes of integrins at the transmembrane and cytoplasmic domains of the receptor. These conformational changes expose binding sites in the short (<50 amino acids) cytoplasmic tails for active kinases or inactive adaptor proteins to

bind, ultimately activating several signaling cascades inside the cell. The integrin-generated signaling cascades begin in discrete foci populated by numerous integrins and associated adaptor and signaling proteins, collectively known as focal adhesions. An interesting observation concerning many focal adhesion-associated proteins [e.g. focal adhesion kinase (FAK), zinc finger protein 384/nuclear matrix protein 4 (Nmp4), zyxin, and breast cancer antiestrogen resistance protein 1 (p130cas), among others] is that they are also found in the nucleus under certain circumstances. Their ability to localize in both focal adhesions and the nucleus highlights a novel and alternative relay mechanism from the direct mechanical linkage model described by Ingber. Thus, mechanical information might be relayed from the integrins to the nucleus in the form of mobile nuclear-cytoplasmic shuttling proteins. A great deal more is known about the key mechanically activated focal adhesion-associated signaling proteins in other cell types (e.g. endothelium) than is currently known for bone, but a number of key discoveries are now being made in bone.

G Protein Signaling

A third model for mechanoreception in osteocytes involves the activation of G protein machinery inside the cell. G protein-coupled receptors (GPCRs) represent the largest family of cell surface receptors, and are activated by a variety of ligands including neurotransmitters, hormones, small peptides, local cytokines, amino acids, and fatty acids, among others. Activity of

GPCRs can be monitored by measuring hydrolysis of the trimeric guanine nucleotide-binding proteins (G proteins) that they activate. In the 1990s, it was demonstrated that fluid flow activates G proteins in osteoblasts, and that pharmacologically preventing G protein activation prevents the normal response to fluid shear. Could one or more GPCRs be the initial mechanoreceptor, kicking off a cascade that begins with G protein activation? The Frangos group has generated some interesting data regarding the role of G protein signaling in mechanoreception. Using GPCR conformation-sensitive fluorescence resonance energy transfer (FRET) in MC3T3 osteoblastic cells and bovine aortic endothelial cells, they found that fluid shear stress leads to conformational changes in the parathyroid hormone/parathyroid hormone-related peptide receptor (PTH1-R) and the B2 bradykinin receptor. These conformational changes occurred within milliseconds and were independent of the presence of either receptor's ligand. It was also reported that the responsiveness of the energy transfer signal could be modulated by membrane fluidity (e.g. modulation of membrane stiffness), indicating that these GPCRs might be direct sensors of mechanical perturbation of the membrane. While mechanotransduction may or may not involve either of these two particular receptors, the data make the larger point that other GPCRs more crucial to the mechanotransduction response might undergo similar conformational changes when the cell is mechanically stimulated. The premise that a ligand is not necessary for shear to activate intracellular G proteins has been taken one step further, and it might be true that even the receptor itself is not necessary for shear-induced G protein activation. This system might have even fewer necessary components. Beyond G protein activation through a ligand independent receptor, G protein activation has been reported to occur through a receptor-independent mechanism. The same group showed that when purified G proteins were reconstituted into otherwise empty phospholipid vesicles, they were activated [by guanosine diphosphate (i.e. GDP) hydrolysis] almost immediately upon fluid shear. This flow-induced activation was independent of a GPCR presence but, rather, was modulated by membrane stiffness. Thus, the role of G protein activation in mechanoreception is clearly intriguing, but it is far from settled.

COORDINATING THE BIOCHEMICAL RESPONSE TO MECHANICAL STIMULATION

Once the mechanical signal is received by the local bone cell population and translated into an initial biological signal, a series of secondary biochemical signaling events must occur to propagate the signal within the cell and to other sensor and effector cells. Efforts to understand the signaling pathways involved in mechanical signal propagation have uncovered a multitude of changes in the mechanically stimulated osteocyte/osteoblast, including gene expression changes, protein and lipid modifications (e.g. phosphorylation events), protein degradation, intracellular translocation events, release of secreted factors, and alterations in cell shape and size, among others. The challenge presented by these observations is to determine which among them are critical for mechanotransduction to occur, and which are simply auxiliary events that have few functional consequences for the mechanotransduction process. This distinction is useful for our discussion of the pathways involved in mechanotransduction, as we will limit our scope to those pathways that have been shown through in vivo functional studies to be important in mechanotransduction, rather than pathways that are simply altered by mechanical stimulation (thus having more uncertain consequences).

One of the earliest pathways identified to be involved in bone cell mechanotransduction is the prostaglandin G/H synthase [or cyclooxygenase (COX)]-prostaglandin E2 (PGE2) pathway. PGE2 is a hormone-like lipid that is generated from arachidonic acid by the COX enzymes and secreted in response to a number of stimuli. Vigorous jumping exercises in humans induce an immediate release of PGE2 from lower limb (loaded) bone tissue. In rodents, loading upregulates the mRNA and protein levels of COX-2, the inducible isoform of COX, whereas the constitutive isoform (COX-1) remains unchanged. The importance of PGE2 signaling has been demonstrated in vivo by depleting the intracellular PGE2 pool prior to mechanical loading. Pharmacologic inhibition of both COX-1 and COX-2 via indomethacin treatment, or selective inhibition of COX-2 alone via NS-398 treatment, was found to reduce the osteogenic response to loading conducted several hours after administration of the inhibitors (Fig. 9.16A). This result has been confirmed in vitro using fluid shear and stretch, where PGE2 levels can more easily be measured from cell culture medium. The mechanism of PGE2 release from mechanically stimulated cells is controversial, and might involve the opening of large, pore-forming gap junction alpha-1 protein/connexin-43 (Cx43) hemichannels or the purinergic P2×7 protein complex. Once released, PGE2 binds in an autocrine or paracrine fashion to the heptahelical EP receptors (EP1−4) that mediate its effects.

Another pathway activated by mechanical stimulation is the nitric oxide (NO) pathway. NO is a free radical and as such can diffuse through the plasma

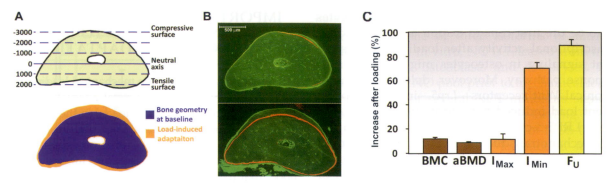

FIGURE 9.18 **The adaptive response of bone to mechanical loading is site specific such that only those regions within a bone that experience sufficient microstrain will adapt.** (A) Schematic diagram of the rodent midshaft ulnar cross section, showing the strain distributions generated (upper panel) and resulting bone formation (lower panel) after ulnar loading (using the model shown in Fig. 9.4E). (B) Mechanical loading of the right forearm (lower panel) 3 days/week for 16 weeks resulted in apposition of new bone predominantly on the periosteal (outer) surface. There was minimal new bone apposition on the endosteal surface or in the contralateral, nonloaded left arms (upper panel). The new bone formed as a result of the experiment can be visualized between the red alizarin labels and the outer bone surfaces. (C) Load-induced increases in bone mineral content (BMC) and areal bone mineral density (aBMD) assessed by dual-energy X-ray absorptiometry are mild in rat ulna loaded 3 days/week for 16 weeks; however, mechanical testing of the same bones revealed much more robust increases in the maximum force that the loaded ulnas could resist (ultimate force; Fu) and the energy that they could absorb before breaking (energy to failure; U).

bone mass, whereas soccer players have greater lower limb bone mass, when compared to controls. An eloquent example of specificity is evident in limb-dominant sports where unilateral loading demands are present (Fig. 9.17). The dramatic bone hypertrophy in the playing arm of tennis players may be dependent on the age at which training is initiated. The optimal time to maximize bone structure with exercise will be discussed later in this section.

The site-specific nature of bone adaption can be broken down further into focal regions within a targeted bone. Bone is laid down only where mechanical strains are greatest. For example, when a bone undergoes bending, different regions within the cross section of the bone will experience different levels of strain. Bone will only be added where it is needed most, which is typically in the plane of bending. The concept of localized bone formation in response to changes in strain is illustrated in Fig. 9.18, based on the rodent ulna axial compression model. Recall from page 181 that the rodent ulna bends laterally under axial compression. This generates compressive forces on the medial surface, tensile forces on the lateral surface, and no strain through the neutral bending axis (Fig. 9.18A). New bone is preferentially added to those surfaces that experience high strain, namely the lateral and medial surfaces (Fig. 9.18A,B). A negligible amount of bone is formed on the neutral axis, where the bone experiences the least amount of strain. Interestingly, modest increases in BMC and aBMD were identified by dual-energy X-ray absorptiometry (DXA) in the loaded ulna compared to the nonloaded ulna. However, mechanical testing in compression (the direction of loading) of the loaded ulna revealed substantial increases in rigidity (I_{min}, I_{max}) and the maximum force

(F_u) among loaded, compared to the nonloaded, limb (Fig. 9.18C). These data in animals can be translated to exercise trials in adults, which have been found to improve bone geometry of the tibia without a corresponding significant increase in bone mass.

Overload

The load magnitude or strain applied to the skeleton is important when considering how a bone will respond. As noted earlier, bone cells respond to dynamic, but not to static, loading and will disregard a stimulus that is below a certain strain threshold. Mechanosensing cells interpret static loads as inconsequential and will quickly ignore a prolonged mechanical signal. An exercise program that includes only static load-bearing activities or one that includes too many loading cycles over a short period of time will diminish the osteogenic potential of bone to adapt, because of cellular accommodation. The combination of load magnitude and rest are key elements in determining how the skeleton will, or will not, respond. When introducing a skeletal region to a specific load, the key is to overload the skeleton to a strain threshold that will elicit a meaningful change without causing pathologic damage, such as a stress fracture. In humans, the exact strain magnitude required to induce a change in either mass or structure is unknown, and in fact may differ throughout the skeleton. However, there is clear evidence that activities imposing higher loads or rates of loading (i.e. jumping or plyometrics) will result in a greater strain stimulus and more robust skeletal changes than lower load activities (i.e. walking; see p. 176). Activities that place limited loads on

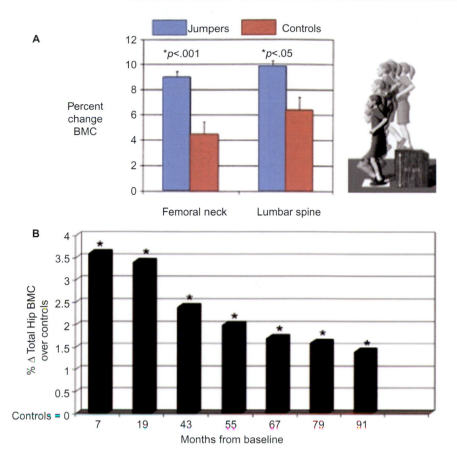

FIGURE 9.19 (A) Exercise-induced bone gains from a high-impact jumping program (100 jumps/day, three times per week for 7 months) initiated during childhood has been found to significantly increase hip and spine bone mass when compared to controls. The landing impact was $8.5 \times$ body weight, with an average load rate of $472 \times$ body weight/s. (B) However, the benefits may not persist long term when exercise is removed. After 7 years, the skeletal benefits declined by 60% in the exercise group following the cessation of the intervention. At 7 years, the bone mass differences between the exercise group and controls remained statistically significant. A limitation of this study was that only dual-energy X-ray absorptiometry of the hip was used; no structural measures were used. It is possible that structural gains were maintained to a higher level. *A and B reproduced with permission from Fuchs RK, et al., J Bone Miner Res 2001;16:148–156.*

the skeleton such as swimming and cycling fail to stimulate significant changes in bone, and in fact can cause bone loss if a large amount of time is spent participating in these activities.

Clinically, it is important to consider initial bone density and structure prior to initiating an exercise program. Individuals with smaller bones or lower baseline bone density have a greater bone response to exercise and greater reductions in fall risk. For example, a sedentary individual would have a greater initial bone response to a specific exercise program than an athlete exposed to the same program. Further, regardless of an individual's initial density or geometry, the skeleton needs to be continually challenged with new, novel loading activities of greater strain magnitude to create further gains in either mass or structure.

It is important to recognize the potential to "overload" the aging skeleton as a means to preserve bone mass. Important work by Snow and colleagues found older postmenopausal women failed to respond to a 9-month resistance training program combined with weighted-vest jumping activities. However, after continuing the exercise program for 5 years, those women who continued to participate showed attenuated age-related declines in hip bone mass associated with menopause. Thus, as we age, overloading the skeleton

provides a nonpharmacological "antiresorptive" option for minimizing bone loss.

Reversibility

Reversibility pertains to how bone responds when an endogenous agent such as exercise is removed. Generally, skeletal adaptations that occur during childhood and adolescence are less reversible than adaptations that occur in adulthood. Structural adaptations made during the early growing years persist into adulthood, and are largely maintained in the adult, even in the face of decreasing activity. This highlights the importance of engaging in regular exercise when the skeleton is capable of undergoing significant modeling to maximize the accumulation of bone mineral and allow appropriate adaptive changes in skeletal morphology. Fuchs, Gunter, and Snow have showed this in a landmark randomized controlled trial of high-impact jumping exercise in prepubertal boys and girls. A 7-month jumping intervention resulted in significant gains in bone mass at the hip and spine, which are clinically relevant fracture sites (Fig. 9.19A). These children were found to still have higher bone mass than the controls after reassessment 8 years later, although the differences between the groups were smaller than at the end of the experiment (Fig. 9.19B). It is unclear whether this is because the exercise cohort lost some of the bone mass that they had gained or that the nonexercized cohort had simply "caught-up." Unfortunately, this study did not look at true bone structure via quantitative computed tomography. Thus, despite a reduction in exercise-induced bone mass as assessed by DXA, structural adaptations may have been maintained.

In contrast, discontinuing exercise in adulthood can result in a loss of skeletal benefits. Published exercise trials investigating the impact of exercise in pre- and postmenopausal women have failed to find persistent skeletal effects when exercise training stops. Unfortunately, exercise-induced gains in bone mass quickly revert toward baseline values. But how fast can we lose the exercise benefits? Winters and colleagues evaluated the skeletal effects of exercise and detraining in premenopausal women who engaged in a program of weighted vest plus impact exercise. After 12 months, the exercise group had a significant gain of 2.5% in hip bone mass; however, after only 6 months of detraining, the exercise-induced gains were lost. In this particular study, the women did not continue to engage in the exercise program. It is currently unclear whether or not structural changes (e.g. periosteal apposition), which may be more important than density for maintaining bone strength, persist after the cessation of exercise. For the aging skeleton, the important take home message is to continue exercise training to maintain the musculoskeletal benefits of exercise.

The amount of bone mass lost after exercise is discontinued and may depend on the age at which an individual initiates an exercise program. Athletes who commence training earlier in life have a higher bone mass in adulthood, but also lose some of it when exercise is stopped. Those who begin exercising as adults, and then stop, are more likely to lose all of the gains. Still, despite some loss of benefit with the discontinuation of exercise, those who engage in regular exercise do have the added benefit of reducing fall risk, and thus fracture risk. The complete removal of weight-bearing activity, which occurs during prolonged bed rest or in space flight, results in dramatic losses of both cortical and trabecular bone that are more severe than the consequences of simply being inactive. These losses constitute a type of osteoporosis (disuse osteoporosis), and would not be considered an aspect of the concept of reversibility.

SKELETAL BENEFIT OF EXERCISE DURING INFANCY AND CHILDHOOD

The younger skeleton is more responsive than the older skeleton to mechanical loading. Hence, it is crucial to capitalize on the growing years to maximize both mineral accumulation and structural adaptations. During growth, approximately 95% of the adult skeleton is formed, with 25–30% of our adult bone mineral being accrued during the 2–3 years surrounding puberty. Importantly, the amount of bone gained during the 2 years surrounding peak bone accrual approximates the amount of bone lost in older adulthood. This underscores the importance of maximizing peak bone mass during growth. Although 60–80% of our adult skeletal mass is determined by our genetic make-up, the inclusion of physical activity during the growing years presents a unique opportunity to achieve the maximum potential size and skeletal density allowable by our genetic background. Maximizing peak bone mass reduces the risk of fracture during aging and delays the onset of osteoporosis. Retrospective studies report fewer fractures in those individuals who engaged in greater physical activity during their youth.

Infancy

Optimal bone health in adulthood begins in utero. A multitude of intrinsic and extrinsic factors impact skeletal growth in utero, which can leave a lasting impression on the skeleton. The importance of movement in utero for achieving optimal skeletal maturation can be observed by evaluating the skeletal impact of prenatal muscular and neurological pathologies, which limit movement and muscular activity. Reduced fetal movement has been associated with skeletal abnormalities, hypomineralized bone tissue, reduced muscular strength, and a higher risk for fracture of the fetal skeleton. Optimal fetal movement is particularly important during the last trimester, when bone mineral accretion is greatest. Traumatic fractures can occur in fetuses with intrauterine akinesia (lack of movement).

During the first few weeks to a year of life, the administration of assisted physical activity is an effective way to increase skeletal mineralization and bone strength. A total of 5–11 min of daily passive resistance range of motion exercise and massage has been found to increase bone mass in the exercised limb of preterm infants and in infants exposed to limited intrauterine movement in as little as 4–8 weeks. Providing infants with daily passive physical activity and adequate nutrition may result in greater increases in bone mass and size than those obtained with adequate nutrition alone.

Childhood

This period of the lifespan is commonly referred to as the "window of opportunity." For reasons not entirely understood, bone cells are more mechanosensitive during growth than in the adult skeleton. During this period, the skeleton is more amenable to altering its structure by accelerating the process of modeling. It has been demonstrated that children who engage in weight-bearing exercises have a higher bone mass than do less active children. Those who engage in high-impact activities (i.e. gymnastics, ballet, and dancing) have higher bone mass and more positive structural changes compared to those who participate in low- or nonimpact activities (i.e. swimming and cycling).

Observational and longitudinal studies provide strong evidence supporting school-based physical education programs as an important means of promoting bone health. Finding specific exercises capable of inducing changes in bone in a short period is very important for many school programs, where time is limited. The addition of physical education classes offered to children in elementary school programs around the world have found consistent improvements in bone health with as little as 30—41 min/day of general exercise. In addition, numerous targeted exercise programs involving high-impact jumping exercise (i.e. jump rope, jumping off boxes, and jumping on the ground) for as little as 11 min/day is even more effective in increasing bone mass. Thus, it is feasible to stimulate a large increase in bone mass in a short period. In fact, a larger bone response is found in those children who perform shorter duration high-impact activities such as jumping, compared to those who perform low-impact exercises that are more similar to the loading demands of normal daily activities. The amount of time the skeleton is given to rest between loading bouts is also crucial for maximizing the osteogenic potential of the exercise program. Thus, exercise programs devised for children demonstrate many of those principles of mechanical adaptation first identified by research in animal models: the benefits of a high strain rate for maximal skeletal response; the smaller importance of exercise duration; and the roles played by cellular habituation and refractory periods.

The age at which a child incorporates regular weight-bearing exercise into a daily routine is instrumental in the attainment of peak bone mass, the ability to meet or exceed one's true genetic potential, and reducing future fracture risk. We can see this most robustly in racquet sport players who started their playing careers prior to puberty when compared to those who commenced training after puberty. Individuals who initiated training more than 5 years prior to puberty have the largest changes in bone mass, when compared to those who commenced training after puberty (Fig. 9.20). This same phenomenon has been found in exercise trials comparing bone response to impact exercise across all stages of

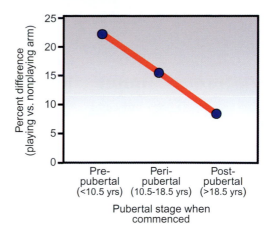

FIGURE 9.20 **The growing skeleton is more responsive to mechanical loading than the adult skeleton.** This figure illustrates the importance of when exercise is initiated during the growing years. A study of competitive female tennis and squash players showed that those who started playing at an earlier age (several years before menarche [Pre]) had more than twice as much differential (playing versus nonplaying arm) in mineral accrual than those who started playing during their adult years [Post]. *Modified from Kannus et al. Ann Intern Med 123:27—31; 1995.*

pubertal development in girls. Longitudinal studies of female tennis players have found the period when training is initiated to be critical in determining how the dominant playing arm adapts. Initiating training prior to puberty enhances periosteal apposition, with a corresponding expansion of the endocortical surface. In contrast, initiating exercise after puberty results in a smaller increase in periosteal expansion, with a corresponding contraction of the endocortical surface. From an engineering perspective, the dominant playing arm of the girl who initiated training prior to puberty will be stronger, despite being thinner, because the mass is distributed further from the bending axis.

Physiologically, estrogen also plays an important role in how the skeleton will adapt. Estrogen functions to blunt periosteal expansion and stimulate endocortical bone formation. In contrast, the absence of estrogen promotes bone resorption on the endocortical surface. Differing responses to exercise across the developmental period highlights the importance of the "window-of opportunity" for optimizing bone health. The introduction of estrogen causes a blunting of the exercise response, when compared to prepubertal girls. Periosteal expansion that occurs in response to exercise is tied to a significant reduction in fracture risk in the aged skeleton, as a larger skeleton is harder to break (Fig. 9.21). During aging, bone is preferentially lost from the endocortical surface, while the periosteal surface is preserved. Thus, if periosteal bone apposition is augmented with exercise during the critical growing years, bone strength will be maintained in the aging skeleton despite accelerated loss on the endocortical surface. Data

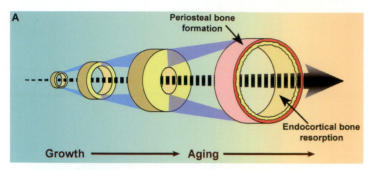

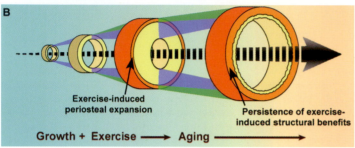

FIGURE 9.21   (A) Bone loss during aging occurs primarily via bone resorption on the endocortical surface. There is concomitant bone formation on the periosteal surface, which helps to maintain bone structure, but this is insufficient to maintain bone mass. (B) Exercise during growth facilitates periosteal bone formation, which optimizes bone structure. As bone loss during aging occurs from the inside out, the enhance structure induced by exercise during growth has the potential to remain intact irrespective of age-related changes in bone mass. *Redrawn after Warden JS and Fuchs RK Br J Sports Med 43:885—7; 2009, and Seeman E, "Periosteal bone formation—A neglected determinant of bone strength." NEJM 349: 320—2; 2003.*

in boys on the optimal time to commence training is more limited, but the same general principles apply, namely the response to exercise is maximized when starting earlier in life rather than after the peak of the growth spurt. During childhood and adolescence, there is not just one specific exercise program capable of improving bone health. The goal is to encourage participation in weight-bearing exercises that place unique mechanical loads on the skeleton to induce adaptation. General guidelines for the mode and intensity are to perform dynamic, high-impact activities that maximize bone response, keeping in mind that only the skeletal sites that are challenged above a certain strain threshold will adapt. The optimal frequency and duration will depend on the intensity of the activity. Low to moderate intensities requires 40—61 min a minimum of three times per week. High intensity jumping exercises that generate loads between 4—8 × body weight result in a similar effect with only about 11 min of exercise three times per week.

SKELETAL RESPONSE TO EXERCISE IN ADULTHOOD

All is not lost as we age. Although the osteogenic potential of the adult skeleton is reduced compared to the pediatric skeleton, exercise-induced gains in bone mass and structure can still occur. Perhaps more importantly, however, physical activity can prevent some bone loss with aging. According to the mechanostat, remodeling suppression should be achievable with moderate exercise because the "set point" for altering remodeling is lower than that required for osteogenesis. Moderate levels of physical activity (2—4 h/week) during aging have been found to reduce the risk of hip fractures by approximately 25%.

To date, there is not just one specific exercise program directed at improving bone mass and reducing fall risk in adults. A multifactorial program of lower impact exercises combined with resistance training may be the most effective strategy not only to preserve muscle strength and bone mass but also to improve balance and flexibility. Exercise programs for adults should include exercises that not only prevent bone loss but also improve balance and muscle strength to reduce the risk for falls. It is important to carefully evaluate what exercises an individual can safely tolerate. General guidelines include moderate intensity exercises such as jumping from floor height (with or without a weighted vest), tennis, and resistance training (both upper and lower body, with the use of Thera-Bands). Walking is often prescribed as an osteogenic exercise; however, walking may not be sufficiently intense to maintain bone mass because it does not provide a significant bone stimulus.

Most exercise studies in older adults are associated with small improvements in bone mass (1—2%), but few studies have evaluated structural changes (e.g. periosteal expansion) associated with various exercise regimens. Such structural changes can improve bone strength significantly, even with only small additions of new bone. Small changes in bone mass are clinically meaningful and can translate to large increases in bone strength

Although bone density is important, improving bone mass and structure in patients with diagnosed

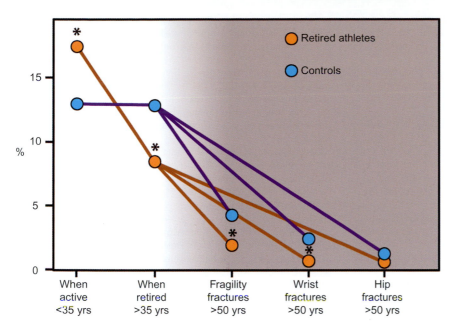

FIGURE 9.22 Former retired soccer and ice hockey players were found to sustain a greater number of fractures prior to 35 years of age. However, upon retirement the prevalence of fractures was lower in the retired players compared to controls both when the players were over 35 years and over 50 years. This study provides a good example of how exercise during youth may protect the skeleton and result in few fragility fractures in the aging skeleton. *Adapted from Nordstrom et al., J Bone Mine Res 20:202–207; 2005.*

osteoporosis may not be the most important goal to prevent fracture. Individuals with osteoporosis may have significant functional impairments that create significant declines in quality of life. Evidence from systematic reviews and meta-analyses of randomized trials shows that at least 15% of falls in older people can be prevented. Hence it is important to develop an exercise program that improves overall function and the ability to perform activities of daily living. With aging, our focus shifts toward improving muscle strength and coordination, which reduces fall risk and the risk of hip and forearm fracture. The components of a successful fall prevention program should include exercises that not only build bone but also enhance strength, endurance, balance, and flexibility.

Overall, exercise during the growing years promotes structural optimization, with a shift toward maintaining bone health, minimizing fall risk, and preventing fractures in adulthood. The ultimate goal is to build a solid skeletal foundation during growth, which will persist long term to have antifracture benefits later in life. Long-term participation in exercise can attenuate bone loss in both males and females. More importantly, engaging in exercise throughout life may result in fewer fractures when compared to having a sedentary lifestyle (Fig. 9.22).

STUDY QUESTIONS

1. Define strain rate and its relationship to strain magnitude and loading frequency. How can strain rate and frequency be dissociated experimentally?

2. What are the four different mechanical usage windows and the minimum effective strains needed to stimulate each window? Describe the cellular activity and tissue outcomes of these windows.

3. What elements unique to osteocytes support the hypothesis that they act as mechanical sensors?

4. Describe how ion channels, adhesion/cytoskeletal molecules, and G protein-related molecules are involved in mechanical sensing.

5. What components (e.g. strain magnitude, distribution, loading rate, and refractory periods) would you include in an in vivo or clinical experimental designed to optimize the amount of bone formation that occurs in response to mechanical stimulation?

Suggested Readings

Batra, N., Burra, S., Siller-Jackson, A.J., Gu, S., Xia, X., Weber, G.F., et al., 2012. Mechanical stress-activated integrin α5β1 induces opening of connexin 43 hemichannels. Proc. Natl. Acad. Sci. U.S.A. 109, 3359–3364.

Burr, D.B., Robling, A.G., Turner, C.H., 2002. Effects of biomechanical stress on bones in animals. Bone. 30, 781–786.

Burr, D.B., 1992. Orthopaedic principles of skeletal growth, modeling and remodeling. In: Carlson, D.S., Goldstein, S.A. (Eds.), Bone Biodynamics in Orthodontic and Orthopedic Treatment, vol. 27. Craniofacial Growth Series. Ann Arbor: Center for Human Growth and Development, The University of Michigan, pp. 15–50.

Fuchs, R.K., Bauer, J.J., Snow, C.M., 2001. Jumping improves hip and lumbar spine bone mass in prebuescent children: a randomized controlled trial. J. Bone Miner. Res. 16, 148–156.

Han, Y., Cowin, S.C., Schaffler, M.B., Weinbaum, S., 2004. Mechanotransduction and strain amplification in osteocyte cell processes. Proc. Natl. Acad. Sci. U.S.A. 101, 16689–16694.

Hung, C.T., Allen, F.D., Pollack, S.R., Brighton, C.T., 1996. What is the role of the convective current density in the real-time calcium response of cultured bone cells to fluid flow? J. Biomech. 29, 1403–1409.

Koch, J.C., 1917. The laws of bone architecture. Am. J. Anat. 21, 179–293.

Martin, R.B., Burr, D.B., Sharkey, N.A., 1998. Skeletal Tissue Mechanics. Springer-Verlag, New York.

Parfitt, A.M., 1983. The physiologic and clinical significance of bone histomorphometric data. In: Recker, R.R. (Ed.), Bone Histomorphometry: Techniques and Interpretation. CRC Press, Boca Raton, pp. 143–223.

Robling, A.G., Castillo, A.B., Turner, C.H., 2006. Biomechanical and molecular regulation of bone remodeling. Ann. Rev. Biomed. Eng. 8, 455–498.

Rubin, C.T., Lanyon, L.E., 1982. Limb mechanics as a function of speed and gait: A study of functional strains in the radius and tibia of horse and dog. J. Exp. Biol. 101, 187–211.

Rubin, C.T., 1984. Lanyon. Regulation of bone formation by applied dynamic loads. J. Bone Jt. Surg. 66 A, 397–402.

Smalt, R., Mitchell, F.T., Howard, R.L., Chambers, T.J., 1997. Induction of NO and prostaglandin E2 in osteoblasts by wall-shear stress but not mechanical strain. Am. J. Physiol. 273 (4 Pt. 1), E751–E758.

Snow, C.M., Shaw, J.M., Winters, K.M., Witzke, K.A., 2000. Long-term exercise using weighted vests prevents hip bone loss in postmenopausal women. J. Gerontol. A Biol. Sci. Med. Sci. 55, M489–M491.

Turner, C.H., 1998. Three rules for bone adaptation to mechanical stimuli. Bone. 23, 399–407.

Tveit, M., Rosengren, B.E., Nilsson, J.A., Ahlborg, H.G., Karlsson, M.K., 2012. Bone mass following physical activity in young years: a mean 39-year prospective controlled study in men. Osteoporos. Int. in press.

US Department of Health and Human Services, 2004. Bone Health and Osteoporosis: A Report of the Surgeon General. US Department of Health and Human Services, Office of the Surgeon General, Rockville, MD.

Warden, S.J., Fuchs, R.K., Castillo, A.B., Nelson, I.R., Turner, C.H., 2007. Exercise when young provides lifelong benefits to bone structure and strength. J. Bone Miner. Res. 22, 251–259.

Winters, K.M., Snow, C.M., 2000. Detraining reverses positive effects of exercise on the musculoskeletal system in premenopausal women. J. Bone Miner. Res. 15, 2495–2503.

Wolff, J., 1986. The Law of Bone Remodeling. Trans. by P.G.J. Maquet, R. Furlong. Berlin, Springer-Verlag.

Fracture Healing

Jiliang Li and David L. Stocum

Department of Biology and Center for Developmental and Regenerative Biology,
Indiana University-Purdue University, Indianapolis, Indiana, USA

INTRODUCTION

A bone fracture or an osteotomy causes a break in the bone, which leads to the loss of anatomic continuity and/or to mechanical instability of the bone. Fractures commonly happen because of falls, car accidents, or sports injuries. Fractures are often associated with penetration injuries on the battlefield. Other factors such as lower bone density and osteoporosis increase the incidence of fracture.

Bone fractures are common and costly to the public due to high health care expenditures. According to the 2004 Surgeon General's Report, about 1.5 million Americans suffer a fracture because of osteoporosis each year. Hip fractures are the most devastating type of bone fracture and account for almost 300,000 hospitalizations per year. Among patients with osteoporotic fractures, 20% die and another 20% end up in a nursing home within a year of the fracture. Many become isolated, depressed, or afraid to leave home because they fear falling. Care for bone fractures from osteoporosis costs nearly US$18 billion each year. In order to find ways to treat bone fractures, it is essential to understand how fractures normally heal and what factors interfere with fracture healing. This chapter provides an overview of the fracture healing process at the cellular and molecular levels and a discussion of several key situations that complicate healing. Current therapeutic strategies that are aimed at accelerating fracture repair are also discussed.

TYPES OF BONE FRACTURE

All fractures can be broadly described as closed (no skin break) or open (skin break). Open fractures are always associated with more damage to the surrounding soft tissue, including the periosteum, have a higher risk of infection, and often have a higher incidence of nonunion than closed fractures. Fractures of long bones, such as the femur, humerus, tibia, and other long bones can be classified according to the characteristics of the force that causes them. Simple and comminuted fractures in which the bones are broken into two or several pieces, respectively, are caused by a single injury. Stress fracture, which is an overuse injury, results from repetitive loading.

Simple fracture occurs when a bending force or twisting force is applied to a bone, resulting in two fragments with transverse, oblique or long curved (spiral) edges of the broken bones. This type of fracture heals through the spontaneous repair processes we will discuss later.

Comminuted fracture is characterized by the breaking of a bone into several small pieces and is the result of high velocity injuries, like car accidents, or falls from a height. Repair of comminuted fractures follows a healing pattern similar to that of simple fractures, but on a larger scale. Such fractures generally are very difficult to treat, and may result in a deformity of the injured part even after treatment.

Stress fractures result when low magnitude cyclically repeated force is applied over a long period of time, causing progressive accumulation of microdamage. Unlike simple and comminuted fractures, stress fractures and their associated fatigue damage heal via normal bone remodeling. This process involves the sequential and coordinated activity of osteoclasts and osteoblasts that remove and replace the damaged bone, respectively. If the repetitive loading is prolonged and/or microdamage cannot be repaired, the bone may eventually fail through propagation of the microdamage.

Basic and Applied Bone Biology.
DOI: http://dx.doi.org/10.1016/B978-0-12-416015-6.00010-1

PRIMARY AND SECONDARY REPAIR MECHANISMS

Repair of long bones after fracture is a unique process that results in the restoration of normal bone anatomy and function after injuries. This repair can be divided into primary healing and secondary healing based on differences in the local motion between the fracture fragments.

Primary healing involves a direct attempt by the cortex to reestablish continuity between the fracture fragments (Fig. 10.1). This process seems to occur only when the alignment stability and decrease in interfragmentary motion of the fracture fragments are established by rigid internal fixation. Osteoblasts derived from mesenchymal stem cells (MSCs) lay down osteoid on the exposed bone surfaces. New haversian systems will be reestablished across the original fracture line through intracortical remodeling.

Secondary (spontaneous) healing involves a response of the periosteum and surrounding soft tissues at the fracture site (Fig. 10.2). The response from the periosteum is a fundamental reaction to bone injury; it is enhanced by limited fragment motion and inhibited by rigid fixation. Mesenchymal cells and osteoprogenitor cells contribute to the process of repair

by a recapitulation of embryonic intramembranous ossification and endochondral bone formation. The new bone formed by intramembranous ossification is found peripheral to the site of the fracture. Osteoblast progenitor cells in the inner layer of the periosteum differentiate into osteoblasts in response to molecular signals produced during fracture and directly synthesize new bone matrix on the bone surface without first forming cartilage. This process does not contribute to directly bridging the fracture. Callus that forms by endochondral ossification is formed within the fracture site and involves the development of cartilage in response to hypoxia caused by the lack of blood supply. The chondrocytes are derived from MSCs in the periosteum and endosteum. They proceed through a state of hypertrophy and the cartilage matrix becomes calcified. The hypertrophic chondrocytes undergo apoptosis and the calcified matrix is removed by invasion of osteoclasts and blood vessels, followed by osteoblast-induced bone formation.

Fracture repair is clearly related to external factors, including the mechanical environment at the fracture site. Motion at the fracture site results in healing primarily through cartilage formation (endochondral ossification), and stability favors the direct formation of bone (intramembranous ossification). Most long bone fractures heal by a combination of intramembranous and endochondral ossification.

Both endochondral and intramembranous ossification produce *woven bone* with poorly organized hydroxyapatite matrix. This is extremely important during fracture healing, since rapid new bone formation is required in order to quickly consolidate fracture fragments to restore the mechanical stability of bone. The mineral appositional rate of woven bone formation can be 2−4 times greater than the lamellar bone formation. The woven bone will be later remodeled by osteoclasts to achieve lamellar bone.

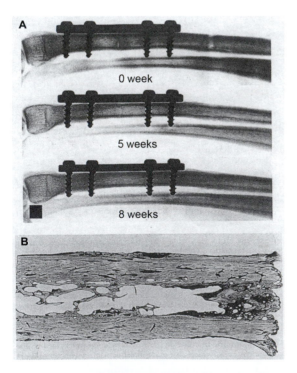

FIGURE 10.1 Directing healing of a transverse osteotomy in dog radius. (A) In the radiographs, there is no external callus formation. The fracture line disappears 5−6 weeks after production of osteotomy. (B) The longitudinal section at 10 weeks shows minimal callus formation around the fracture site. *Reproduced with permission from Schenk R et al., 1963 Experientia 19:593−595.*

STAGES OF FRACTURE REPAIR

Secondary fracture healing of long bones can be considered as a series of four discrete stages occurring in sequence and overlapping to a certain extent (Fig. 10.3).

Inflammatory Response

When trauma occurs, the continuity and vascular supply of bone is disrupted and a hematoma forms at the site of the injury. This leads to a loss of mechanical stability, a lack of local oxygen and nutrients, and a release of various factors from platelets. Macrophages, leukocytes, and other inflammatory cells then invade the area. The damage also sensitizes the surviving local cells and enables them to respond better to local and systemic messages. This inflammatory response peaks within the first 24 h and lasts for about 7 days.

Soft Callus Formation (Cartilage Formation)

The cells that are stimulated and sensitized during the inflammatory stage begin producing new vessels, fibroblasts, intracellular material, and supporting cells. The hematoma is replaced with fibrovascular tissue, a fibrin-rich granulation tissue. Fibrocartilage then develops and stabilizes the bone ends (Fig. 10.4). In mouse and rat models, the peak of soft callus formation occurs 7−10 days postfracture with a peak in both type II collagen and proteoglycan production.

Hard Callus Formation (Endochondral Ossification)

After the soft callus forms, replacement of the cartilage and fibrovascular tissue occurs via vessel invasion

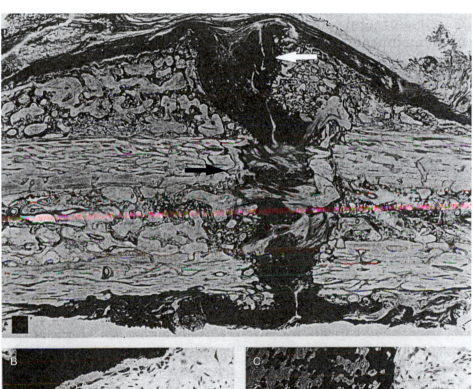
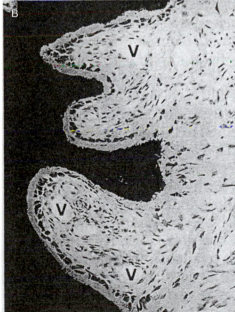
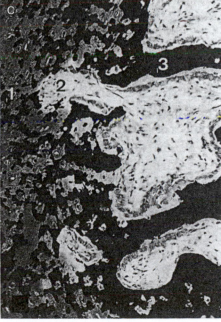

FIGURE 10.2 Secondary (spontaneous) fracture healing in dog radius after transverse defect-osteotomy. (A) Longitudinal section 14-weeks postfracture shows extensive bony callus formation along the periosteal surface and within the bone marrow cavity. (B) The area indicated by the black arrow in Image A contains fibrous tissue that is sufficiently vascularized to permit intramembranous ossification. (C) The area indicated by the white arrow in Image A shows that fibrocartilage has to mineralize (1) before undergoing resorption and vascular invasion (2). At other sites of the endochondral ossification, bone is deposited on persistent calcified cartilage or bone. (3). *Reproduced with permission from Schenk R et al., 1963 Experientia 19:593−595.*

and endochondral ossification (Fig. 10.4). Periosteal bone apposition also occurs, contributing to formation of the hard callus. The calcified cartilage is replaced with woven bone. The peak of hard callus formation usually occurs around 14 days postfracture in animal models, as indicated by callus volume as well as osteoblast markers, such as type I collagen, alkaline phosphatase (ALP), and osteocalcin. Intramembranous

ossification peripheral to the site of the fracture also contributes to the hard callus.

Bone Remodeling

In this last stage of fracture repair, woven bone is gradually replaced by lamellar bone via bone remodeling

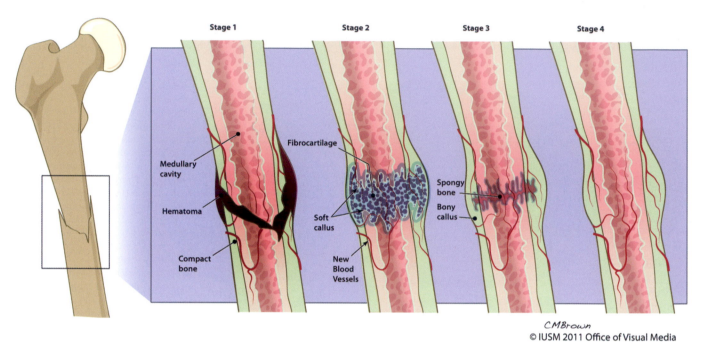

CMBrown
© IUSM 2011 Office of Visual Media

FIGURE 10.3   **Stages of fracture repair.** Stage 1: Following fracture or osteotomy, blood supply is disrupted and a blood clot (hematoma) forms. Stage 2: Progenitor cells in the periosteum and marrow differentiate into osteoblasts to facilitate intramembranous bone formation where an intact blood supply is preserved. In the fracture space where the tissue is hypoxic the progenitor cells undergo chondrogenesis. The chondrocytes hypertrophy and their matrix becomes calcified, leading to chondrocyte apoptosis and invasion of the matrix by periosteal and marrow blood vessels. These vessels are accompanied by perivascular mesenchymal stem cells that differentiate into osteoblasts. Stage 3: The osteoblasts form woven bone on the calcified matrix. Stage 4: The remodeling process proceeds with osteoclasts and osteoblasts facilitating the conversion of woven bone into lamellar bone and eventually recreating the appropriate anatomical shape.

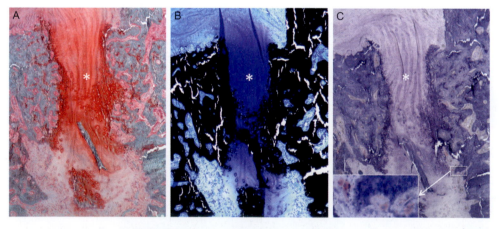

FIGURE 10.4   **Fracture repair of the closed femoral fracture in female Sprague-Dawley rats at 10 weeks postfracture.** (A) Cartilage (star) is detected by the red stain with safranin O staining. (B) Endochondral ossification demonstrated by von Kossa staining shows cartilage (blue) and bone (black). (C) Tartrate-resistant acid phosphate staining reveals osteoclasts stained red with multiple nuclei, as seen in the enlarged insert.

(see Chapter 4). Hard callus resorption by osteoclasts is followed by lamellar bone formation by osteoblasts to restore the anatomical structure of the preinjured bone and support mechanical loads. This process, also sometimes referred to as *secondary bone formation*, starts after 3–4 weeks and may take years to be completed before the original anatomic structure is restored. In adults, the original shape and cavity of new bone may never be fully restored.

The new outer cortical shell develops over the cartilage core. The ischemic bone tissue (the original cortex) at the fracture site is resorbed as the outer cortical shell remodels inward to become the new diaphyseal bone (Fig. 10.5A). Osteoclasts identified by tartrate-resistant acid phosphatase (TRAP) staining carry out two tasks during fracture healing: removal of endochondral matrix (Fig. 10.4) and remodeling of woven bone (Fig. 10.6). Inhibition of bone resorption, for example with bisphosphonate treatment, prevents the removal of the ischemic bone tissue and restoration of diaphyseal bone (Fig. 10.5B).

Fractures are usually considered healed when the bone stability has been restored by the formation of new bone that bridges the area of fracture even before the final shaping of the bone is achieved. An adequate blood supply and a gradual increase in mechanical stability are crucial for successful fracture healing. However, the interruption of normal healing processes results in fracture nonunion. Nonunion fractures, which are defined as the cessation of all reparative processes of healing without bone union, are traditionally classified as *atrophic* or *hypertrophic*. Atrophic nonunion, which typically shows little callus formation, results from poor vascularization at the fracture site. Hypertrophic nonunion is linked to inadequate immobilization (unstable fixation) and appears to have adequate blood supply and cartilage formation that leads to pseudarthrosis, a false joint associated with abnormal movement at the unhealed site of bone.

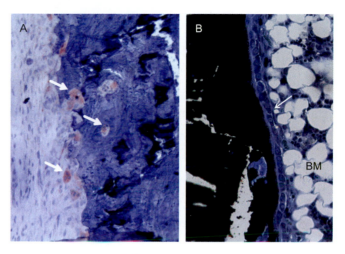

FIGURE 10.6 **Remodeling of woven bone in the outer cortical shell after fracture.** (A) Tartrate-resistant acid phosphatase staining shows osteoclasts (arrows) with multiple nuclei on the periosteal surface of the bone, indicating ongoing bone resorption. (B) On the endosteal surface of the outer cortical shell, a group of active osteoblasts (arrows) are observed, indicating ongoing new bone formation. The outer cortical shell remodels inward to become the new diaphyseal bone, while the original fracture cortex is resorbed.

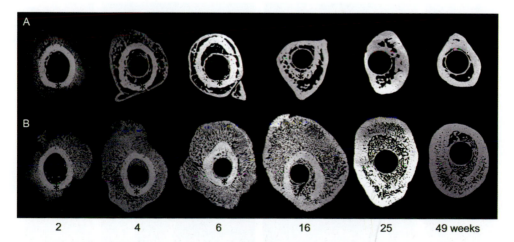

FIGURE 10.5 (A) Contact microradiographs of cross sections at the fracture sites in a rat open femoral osteotomy model show formation of an outer cortical shell and resorption of the original fracture cortex (star) from 2 to 49 weeks following surgery in Sprague-Dawley rats. (B) Cross sections from rats continuously treated with bisphosphonate (incadronate, 100 μg/kg daily) show a larger callus area compared with the control animals. There is a highly porous shell in which the endocortical border is unclear at 25 weeks. The remnant of the original cortex is still observed at 49 weeks. *Adapted from Li J et al., 1999 J Bone Miner Res. 14: 969–979, Li J et al., 2000 J Bone Miner Res 15:2042–2051 and Li C et al., 2001 J Bone Miner Res 16:429–436 with permission of the American Society for Bone and Mineral Research.*

ASSESSMENT OF FRACTURE HEALING

Methods of Evaluation

Fracture healing involves a dynamic interplay of biological processes to restore the original anatomic structure and mechanical function of bone. Therefore, both structural and biomechanical evaluations are used to assess fracture repair. The extent and quality of structural repair can be evaluated using radiographic and histologic methods. The mechanical properties of a healing bone are assessed by mechanical tests. There is a consensus that mechanical tests provide the gold standard measures of healing in a laboratory setting. Clinical assessment of healing requires noninvasive methods, including clinical symptoms (pain or tenderness when bearing weight) and radiographic indicators.

Plain radiography is a ubiquitous method used to evaluate fracture healing in both laboratory and clinical settings, due to its noninvasive nature. The most common radiographic definitions of fracture healing involve the bridging of fracture site by callus, obliteration of the fracture line, and continuity. Recently, computed tomography has been used to define union as bridging of >25% of the cross-sectional area at the fracture site. Callus volume can also be calculated using three-dimensional (3D) reconstructed images.

Bone histology is an invasive method used to study the bone structure during fracture healing in the laboratory. Conventional bone histomorphometry can be applied in fracture studies. Compared to single longitudinal sections of a healing fracture, which do not capture all of the tissue heterogeneity within callus, transverse sections at the fracture line level can provide the more accurate measurement of cross-sectional area, as well as an estimation of tissue heterogeneity. On cross sections of callus, fibrous cartilage and bone tissues can be measured and calculated as the percentage of the total cross-sectional callus area. Osteoblasts and osteoclasts can be specifically stained and their activities can be quantified, as shown in Fig. 10.6. In addition, fluorochromes, used widely for the measurement of bone formation-related parameters to estimate bone metabolism, can also be used to examine callus remodeling, in particular, during the middle and late stages of fracture healing (Fig. 10.7).

To test the mechanical function of a healing bone, torsion and four-point bending tests are logical choices when studying fracture healing in long bones. The choice of the type of test is mainly determined by technical considerations. In general, torsion is a better choice than four-point bending because torsion tests subject every cross section of the callus to the same torque, whereas four-point bending might create a nonuniform bending moment throughout the callus. As a result, failure of the callus during a four-point bending test does not necessarily occur at the weakest cross section of the callus. It is important to remember that three-point bending should not be recommended to estimate the mechanical properties of a healing bone, especially during early stages of the healing process, because the site to which the force applied is located at

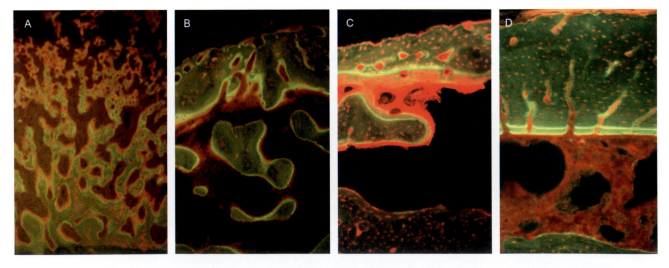

FIGURE 10.7 **Fluorochromes can be used to examine callus remodeling during fracture repair.** Calcein is injected before rats are sacrificed 2, 4, 6, and 16 weeks after femoral osteotomy. (A) Diffuse calcein labeling (green) suggests woven bone formation at 2 weeks. (B–D) Linear calcein labeling indicates lamellar bone formation during callus remodeling at 4, 6, and 16 weeks, respectively. These data suggest that measurement of fluorochrome labeling is a useful tool to estimate the rate of bone formation during normal fracture repair and repair occurring after various therapies.

the original fracture line, which is composed primarily of cartilage, calcified cartilage, or less mature bone tissue depending on the healing stages.

The outcome measures that can be obtained from mechanical tests, such as ultimate strength, stiffness, energy to failure, and torque in the torsion test are structural rather than material properties. These structural properties of a fracture callus depend collectively on the individual tissues, including cartilage, calcified cartilage, and woven bone, and the spatial distribution of these tissues, as well as the overall geometry of the callus. Since the callus geometry can now be easily measured using computed tomography, it is possible to roughly estimate the overall callus tissue material properties by normalizing the structural properties with the callus geometric parameters. However, the true measurement of the material properties of the callus requires direct testing of the individual tissues in the callus. For instance, nanoindentation can be used to measure the elastic properties of the individual tissues of callus.

Biomechanical Stages of Fracture Healing

It is apparent that mechanical properties improve with the progress of fracture healing. During the ossification process of external callus, a fourfold increase in the total amount of calcium per unit volume and a twofold increase in hydroxyproline (an indicator of total collagen content) lead to a threefold increase in the breaking strength of the callus in a tensile test. Studies also suggest a high correlation between the hardness of fracture callus and its mineral content per tissue volume. In 1977, White et al. divided the whole period of secondary fracture healing into four biomechanical stages, based on the results of torsion tests performed on healing rabbit tibiae at multiple time points (Table 10.1).

In stage 1, the bone fails through the original fracture with a low stiffness.
In stage 2, the bone fails through the original fracture site with a high stiffness.

In stage 3, the bone fails partially through the original fracture site and partially through the previously intact bone with a high stiffness.
In stage 4, the bone fails though the previously intact bone with a high stiffness.

These stages correlate with progressive increases in average torque and energy absorption to failure as healing progresses, but do not map onto the four biological stages in a one-to-one manner.

It is important to note that assessment of fracture healing must be based on both bone structure and mechanical properties. In some cases, especially with the treatment of antiresorptive agents, the recovery of mechanical properties of a healing fracture alone does not represent the restoration of original bone structure. Bisphosphonate treatment has been shown to enhance callus strength though inhibition of callus remodeling, resulting in a larger callus and larger proportion of mineralizing cartilage and woven bone (Fig. 10.5). In terms of material properties, callus tissues are not as strong as the well-organized lamellar bone tissues. However, the larger cross-sectional area and moments of inertia of the callus resulting from bisphosphonate treatment compared to intact bone can compensate for the inferior material properties. When being tested, the healing bone may fail at least partially through the previously intact bone. Although the histologic progress is delayed due to the suppression of callus remodeling by bisphosphonates, the recovery of biomechanical properties may not be affected, suggesting an inconsistency between restoration of anatomic structure and recovery of strength of healing bone caused by bisphosphonate treatment.

CELLULAR EVENTS OF FRACTURE REPAIR

Bone formation during fracture healing is accomplished primarily by MSCs in the periosteum, with lesser contributions from the endosteum and marrow stroma. The first event in fracture repair is hemostasis to stop bleeding from damaged blood vessels in the bone and periosteum, resulting in the formation of a hematoma (clot) within the break. Platelets trapped in the hematoma degranulate, releasing platelet-derived growth factor (PDGF) and transforming growth factor beta (TGF-β), which chemoattract neutrophils and macrophages into the fracture zone to initiate an inflammatory phase. Osteocytes die back for a short distance on both sides of the fracture, leaving dead bone matrix that is degraded by osteoclasts derived from the macrophages.

TABLE 10.1   Biomechanical Stages of Fracture Repair

Stages	Site of Failure	Stiffness
1	Original fracture	Low
2	Original fracture	High
3	Original fracture and previously intact bone	High
4	Previously intact bone	High

Adapted from Bone remodeling in fracture repair by T. A. Einhorn (1992).

The fracture is then repaired by both intramembranous and endochondral ossification. Periosteal MSCs proliferate and differentiate directly into osteoblasts to form collars of intramembranous bone on either side of the fracture space (hard callus). The osteoblasts secrete a bone matrix rich in type I collagen and containing osteocalcin, the mineralization-associated glycoproteins osteonectin, osteopontin, and bone sialoprotein 2 (BSP II), and numerous proteoglycans. Within the fracture space, a cartilage template is formed by the proliferation of periosteal MSCs and their differentiation into chondrocytes (soft callus). The chondrocytes secrete a cartilage-specific matrix composed of aggrecan, types II and XI collagens, fibronectin, and hyaluronic acid, and then undergo hypertrophy characterized by the upregulation of type X collagen and downregulation of other types of collagen.

Next, the cartilage template is replaced by bone, a process that requires neoangiogenesis. The cartilage matrix calcifies and the hypertrophied chondrocytes release angiogenic signals that trigger the sprouting of capillaries in the periosteum before undergoing apoptosis. As osteoclasts degrade the calcified matrix, periosteal capillaries accompanied by perivascular MSCs invade the template. Some of these MSCs differentiate into osteoblasts, while others become residents of the reforming bone marrow. The osteoblasts differentiate into osteocytes of the cortical and trabecular bone. In this way, new bone is formed in the fracture space, while at the same time intramembranous bone forms directly from MSCs on either side of the fracture space. At first, the new bone is higher than the preexisting bone, but is subsequently remodeled by osteoclasts to restore the normal shape of the bone.

Besides the MSCs within the local environment adjacent to a bone fracture, systemic recruitment of skeletal progenitors has also been proposed. The presence and increase in circulating osteoblast precursors in response to bone injury suggests a recruitment of these progenitor cells from nonfracture sites to the fracture area. These circulating progenitor cells express the osteoblast marker ALP and are able to home to the bone-lining cells at the fracture sites. However, these cells might not integrate within new bone as osteocytes. It is unclear whether these circulating cells are directly involved in fracture repair by producing new bone matrix or indirectly by secreting osteoinductive factors.

In order to discover the origin of the skeletal progenitors during fracture repair, the green fluorescent protein (GFP)-reporter mouse model has become a useful tool to track cell differentiation during fracture repair. For instance, two GFP reporters, Col3.6GFPcyan and OcGFPtpz, driven by *Col a1* (which encodes type I collagen) and *BGLAP* (which encodes osteocalcin; Oc)

promoter fragments, respectively, are bred into the same mice to visualize both early and late stages of osteoblast differentiation. Following fracture, a histologic method that can preserve fluorescent signals in undecalcified bone sections can be used to observe osteoblasts. Osteoprogenitor cells arise from the flanking periosteum proliferate and migrate to fill the fracture zone by day 6. These cells differentiate to osteoblasts and chondrocytes, to form a new outer cortical shell. The hypertrophic chondrocytes are dispersed and the cartilage matrix is mineralized by young osteoblasts between days 7 and 14. The original fracture cortex is resorbed as the outer cortical shell remodels inward to become the new diaphyseal bone after 35 days. In addition, a variety of different GFP-reporter mouse models can be used to track the origin and fate of stem cells that contribute to fracture repair as well as track chondrocyte differentiation.

MOLECULAR REGULATION OF CHONDROGENESIS AND OSTEOGENESIS DURING FRACTURE REPAIR

Formation of a hematoma after fracture is regulated as in other wounds by the production of tissue factor (TF) by the nonendothelial cells of damaged blood vessels. TF is the first element in the clotting cascade that ends in the production of thrombin, the molecule that induces platelet degranulation. Degranulation involves the release of α-granules and dense bodies. Molecules such as serotonin and thromboxane A2 that contribute to the vasoconstriction of hemostasis are released by the α-granules. The dense bodies contain fibrinogen, which along with plasma fibrinogen is converted to fibrin of the clot by thrombin. In addition, the dense bodies release the PDGF and TGF-β that initiates the inflammatory phase of fracture repair.

During the inflammatory phase, a number of growth factors are released from sequestration in the bone matrix as it is degraded by osteoclasts, and also from macrophages. These factors act as signals to activate the transcription factors that commit MSCs to differentiate along chondrogenic and osteogenic pathways. RUNX2 is a key transcription factor that commits MSCs to skeletogenic differentiation. SOX-9 is the key transcription factor that commits skeletogenic precursors to chondrogenesis and transcription factor Sp7/osterix is the key transcription factor that determines differentiation into osteoblasts.

Table 10.2 lists the molecular factors that regulate formation of the cartilage template during fracture repair. These factors are quite similar to those involved in skeletogenesis of endochondral bones during embryonic development. The expression of RUNX2 is activated by bone morphogenetic proteins (BMPs) 2, 4, 5, and 7. BMP

TABLE 10.2 Signaling Molecules, Transcription Factors, and Differentiation Markers Expressed in the Periosteum and Regenerating Tissues of a Fractured Endochondral Bone

Tissues	Signals	Transcription Factors	Differentiation Markers
Periosteum	BMPs 2, 4,and 7	RUNX2	—
Soft callus	BMPs 2, 4, 5, and 7 PDGF FGF-2 IGF-I	SOX-9	—
Chondro-callus	IHH BMPs 2, 4, 6, and 7 TGF-β FGFs 1 and 2	SOX-9	Aggrecan, types II, IX, X, and XI collagen
Ossification		Osterix	Osteocalcin, type I collagen

Adapted from Regenerative Biology and Medicine by D. L. Stocum (2012).
BMP, bone morphogenic protein; FGF, fibroblast growth factor; IGF-I, insulin-like growth factor I; IHH, Indian hedgehog protein; PDGF, platelet-derived growth factor; TGF-β, transforming growth factor beta.

receptor types IA and IB are expressed in the periosteal cells of uninjured bone and both they and BMPs 2, 4 and 7 are strongly expressed in periosteal mesenchymal cells in both the hard callus and soft callus. The expression of SOX-9 and soft callus formation is regulated primarily by fibroblast growth factor 2 (FGF-2) in conjunction with the BMPs, PDGF, and insulin-like growth factor I (IGF-I). Differentiation of the chondroblasts of the soft callus is regulated by BMPs, FGF-1 and -2, and TGF-β. SOX-9 induces the expression of genes for cartilage markers such as types II, X, IX, and XI collagens and aggrecan. *IHH* transcripts [(which encode Indian hedgehog protein (IHH)] are detected in chondrocytes and *GLI1* transcripts (which encode Zinc finger protein GLI1) are expressed in a population of cells on the periphery of the callus that will reform the periosteum. In the embryonic development of endochondral bone, IHH, parathyroid hormone (PTH), and parathyroid hormone-related peptide (PTHrP) form part of a feedback loop that controls the rate at which chondrocytes mature. During hard callus formation and replacement of the cartilage template with bone, osteoblast differentiation by perivascular MSCs is regulated by RUNX2 and osterix in response to BMPs.

Whether the same upregulation of hyaluronidase and adhesion proteins [neural cell adhesion molecule (NCAM), fibronectin, and CYR61/IGF-binding protein 10 (IGFBP-10)] observed during the condensation of skeletal cells is required for MSC condensation within the soft callus is unknown. It is unlikely that molecules such as the Hox-A, Hox-D, T-box transcription factors (TBXs), sonic hedgehog (SHH), FGF-4, FGF-8, and LIM homeobox transcription factor 1 (LMX-1), which are involved in axial patterning of the skeletal condensations of the limb bud, play a similar role in repair of an endochondral fracture. This is because condensation and chondrocyte differentiation of the soft callus is taking place within a small gap in an already established pattern.

Transcriptional profiling of intact versus fractured rat femur by subtractive hybridization and microarray analysis reveals that gene expression patterns change dramatically during fracture repair. Sixty-six percent of the total number of genes are homologous to multiple families of genes known to be involved in the cell cycle, cell adhesion, extracellular matrix, cytoskeleton, inflammation, general metabolism, molecular processing, transcriptional activation, and cell signaling, including components of the Wnt pathway. Thirty-four percent represent genes with unknown functions. The majority of these are grouped in two clusters marked by a sharp increase in activity at 3 days postfracture that peak at day 14 and then decrease. This pattern suggests that these genes are involved in the proliferation and differentiation of chondrocytes.

Wnt-β-catenin signaling plays a very important role in fracture healing. Treatment with lithium, an agonist of the Wnt-β-catenin signaling pathway, improves fracture healing; whereas treatment with Dickkopf-1, an antagonist of this pathway, suppresses fracture repair. Sclerostin, another antagonist of Wnt-β-catenin signaling, is also involved in fracture repair. Fractures in mice with a null mutation of *Sost* (the gene that encodes the sclerostin protein) show accelerated callus bridging, greater callus maturation, and significantly improved recovery of mechanical strength of repair bone. Similarly, systemic inhibition of sclerostin by administration of a sclerostin-neutralizing antibody greatly enhances new bone formation and the strength of the fracture callus in rats and nonhuman primates.

The role of sclerostin in fracture repair suggests the involvement of osteocytes in fracture healing because sclerostin is primarily made by osteocytes. Furthermore, attention has also been paid to other osteocyte-specific proteins, like dentin matrix acidic phosphoprotein 1 (DMP-1; encoded by the *DMP1* gene) and FGF-23. DMP-1 is one of the acidic phosphorylated extracellular matrix proteins of the small integrin-binding ligand N-linked glycoprotein (SIBLING) family. DMP-1 is expressed in the mineralized tissues and is involved in the mineralization. In situ hybridization demonstrates that *DMP1* mRNA is strongly expressed in preosteocytes and osteocytes in the bony callus during intramembranous and endochondral ossification until 14 days postfracture. FGF-23 is expressed mainly by osteocytes. However, transcript and immunohistochemical analysis have shown a marked increase in FGF-23 production in osteoblasts in the fracture callus during fracture repair. In addition, elevated FGF-23 (C-terminal fragment) in serum is detected in patients following hip arthroplasty surgery. These data suggest that FGF-23 may serve as an indicator of osteoblast differentiation in the early phase of bone healing.

LOCAL REGULATION OF FRACTURE REPAIR

Prostaglandins

The critical role of prostaglandins, in particular the prostaglandin E2 (PGE2) signaling pathway in normal bone repair was well documented in the first decade of the twenty-first century using genetically modified animal models. Prostaglandins are a family of lipid mediators that coordinate cell-cell communication through interaction with specific cell membrane receptors. These have effects on both bone formation and resorption that are mediated through the proliferation and differentiation of osteoblasts and the regulation of differentiation of osteoclasts. The prostaglandin G/H synthase 2/cyclooxygenase (COX) isozymes catalyze the rate-limiting step in the formation of prostaglandins from arachidonic acid. Two distinct *PTGS/COX* genes (encoding isozymes COX-1 and COX-2) have been cloned and characterized. The production of prostaglandins under physiologic conditions from the constitutive expression of COX-1 plays important roles in the cytoprotection of the gastric mucosa and kidney function. In comparison, prostaglandins derived from induced COX-2 expression are present in pathologic conditions such as cancer and during inflammation following acute injury. COX-2 is not normally detectable but is rapidly induced through multiple signaling pathways in various cell types that participate in the inflammatory response. PGE2 is a major COX-2 product at inflammatory sites, where it causes vasodilation and increases local vascular permeability. The critical role of PGE2 induced by COX-2 during fracture healing has been demonstrated using a mouse model

containing a null mutation of the *Ptgs2/Cox2* gene (Cox2$^{-/-}$ mice). In *Ptgs2*-null mice, skeletal healing is significantly delayed compared with *Ptgs1/Cox1*-null mice and wild-type controls. Thus, COX-2 has an essential role during normal fracture healing. This was found to be mediated through effects on osteoblastogenesis in both intramembranous and endochondral ossification. At fracture sites, COX-2 is expressed primarily in early stem cell precursors of cartilage that also express the *COL2A1* gene that encodes a key part of type II collagen in humans and mice. COX-2 has been shown to regulate RUNX2 and osterix expression, which are key factors in chondrogenesis and osteoblastogenesis.

Nonsteroidal anti-inflammatory drugs (NSAIDs) have been widely used to suppress inflammation and reduce pain following bone injuries. NSAIDs exert their predominant effects by inhibiting COX enzyme activity. The newer COX-2-selective agents act to specifically inhibit COX-2. During the repair of complete fractures, inhibition of COX-2-dependent prostaglandin production results in significant healing defects. In animal studies, indomethacin, a commonly used prescription NSAID, delays endochondral ossification. COX-2-selective NSAID treatment can stop normal fracture healing and induces the formation of delayed- and nonunions. Clinical studies show that NSAIDs decrease bone repair, resulting in a higher nonunion incidence of long bone fractures and delayed spinal fusion.

PGE2 is known to bind four different G protein-coupled receptors (GPCRs): EP1, EP2, EP3, and EP4. Interestingly, manipulation of different EP receptor subtypes has shown different effects on fracture healing. The EP1 receptor is involved in regulating intracellular calcium levels. EP2 and EP4 activation stimulates cyclic AMP (cAMP) through the Gs subunit of GPCRs. In contrast, EP3 activation results in a decrease in cAMP levels through the Gi, Gq, or Gs subunits of GPCRs, depending on the EP3 isoform. EP1$^{-/-}$ mice exhibit accelerated fracture healing. Selective agonists for both EP2 and EP4 have positive effects on bone healing, and EP2 and EP4 receptor knockout mice have impaired fracture healing and bone resorption. Treatment of COX-2$^{-/-}$ mice with an EP4 agonist rescues the impaired fracture healing. EP4-selective agonists also accelerated the delayed fracture healing of aged mice. The EP3 receptor, which negatively regulates cAMP levels, is suggested to have negative effects on bone formation. Therefore, the different EP receptors appear to mediate unique effects on the cells and tissues involved in fracture repair.

Bone Morphogenetic Proteins

The activity of BMPs was first discovered by Marshall R. Urist in 1965 because of their capacity to induce ectopic bone formation at extraskeletal sites. BMPs belong to the TGF-β superfamily, which includes TGF-βs, activins/inhibins, nodal, myostatin, and Muellerian-inhibiting factor [or anti-Muellerian hormone (AMH)]. TGF-β superfamily proteins bind to serine/threonine kinase receptors, and transduce signals predominantly through Smad-dependent mechanisms. To date, over 20 BMPs have been identified and characterized. Members of BMP family bind to dimers of types I and II serine/threonine kinase receptors. There are three distinct BMP type I receptors, called activin receptor-like kinase (ALK)-2, ALK-3 (BMPR-IA), and ALK-6 (BMPR-IB), and three distinct BMP type II receptors, BMP type II receptor (BMPR-II), activin type II receptor (ACTR-II), and activin type IIB receptors (ACTR-IIB). Type II receptors possess constitutively active kinase activity that phosphorylates type I receptors upon ligand-receptor complex formation. Phosphorylated type I receptors transduce the signal to downstream target proteins. A major family of downstream targets of BMPs is the Smad proteins. Of the eight Smad proteins identified in mammals, Smad1, Smad5, and Smad8 are receptor-regulated Smads that are phosphorylated by the BMP type I receptor. Smad2 and Smad3 are activated by activin and TGF-β type I receptors. Smad4 is the only common-partner Smad (co-Smad) in mammals, which is shared by both BMP and TGF-β/activin signaling pathways. Smad6 and Smad7 negatively regulate signaling by the other six Smads. However, the detailed downstream signaling molecules of BMPs have not been elucidated.

BMPs and several Smads have been detected during fracture healing. In a rodent fracture model, overexpression of BMPs 2, 4, and 7, common-mediator Smad (Smad4), and receptor-regulated Smads (Smads 1 and 5) versus lower levels of inhibitory Smad (Smad6), can be detected at day 3 in osteogenic cells in the thickened periosteum and bone marrow at the fracture sites. At day 10, Smad6 increases dramatically and Smad4 remains elevated, while Smad1 and Smad5 decrease in the fracture callus. Smad7 is expressed only in vascular endothelial cells. By day 28, when new bone has replaced the fracture callus, the expression of all BMPs and Smads decreases, approaching control levels. During fracture healing, the expression patterns of Smads 1 and 5 are similar to that of BMPs 2 and 7. These data suggest that BMPs and downstream Smad family members play an important role in the early stage of fracture healing.

Among BMPs, BMP-2 particularly plays a critical role in fracture healing. MSCs, osteoblasts, and chondrocytes all produce BMP-2 at the inflammatory stage of fracture healing. BMP-2 may initiate bone formation, including initiating the production of other BMPs involved in bone formation. Mice lacking the ability to produce BMP-2 in limb bones also lack the ability to heal fractures. Furthermore, BMP-2 is extremely important for intramembranous bone formation because BMP-2 is highly expressed in the periosteal MSCs. Animals lacking BMP-2 in a limb-specific manner show an almost complete lack of initial periosteal activation and an apparent failure of the bone to heal (Fig. 10.8).

BMPs induce cartilage and bone formation during fracture repair by stimulating the proliferation and differentiation of stem cells into chondrocytes and osteoblasts, a process called *osteoinduction*. Osteoinduction is one of three requirements for bone regeneration, in addition to osteoconduction and osteogenesis. *Osteoconduction* is a process that supports the ingrowth of capillaries, perivascular tissues, and osteoprogenitor cells into the 3D structure of an implant or bone graft. Osteoconductive properties are determined by the material architecture, chemical structure, and surface charge. Autologous bone grafts have substantial osteoconductive properties. BMPs and bone grafts are usually needed in the treatment of delayed union or nonunion fracture.

Because of the cartilage- and/or bone-inductive activities of BMPs, extensive research on BMPs have been conducted, with the aim of developing therapeutic strategies for the restoration and treatment of skeletal conditions resulting from trauma and degenerative bone diseases. Thus far, two BMPs have been used clinically: recombinant human (rh)BMP-2 and rhBMP-7 [also known as osteogenic protein-1 (OP-1)]. Currently, both rhBMP-2 and rhBMP-7 are delivered on an absorbable collagen sponge during surgery. However, clinical outcomes are not as impressive as those seen in animal studies, in which more robust bone formation has been observed. The reasons for this are unclear. One possibility is a lack of sufficient numbers of responding cells at the site of implantation in the host. Another possibility is that BMPs may need to be delivered in combination with other growth factors as "cocktails." A recent study showed that BMPs may be introduced in a single percutaneous injectable manner to accelerate fracture repair without direct exposure of the fracture site. An injectable strategy may be used in conjunction with use of other growth factors, injectable grafting materials, and/or stem cells to make the use of BMPs more effective and affordable.

Vascular Endothelial Growth Factor

Vascularization is essential to bone formation. The importance of blood vessel formation in bone repair can be observed during endochondral ossification, where the cartilage template is replaced by primary bone formation. Chondrocytes at the growth plate undergo hypertrophy and then apoptosis, leaving mice show a dramatic decrease in the expression of other genes known to contribute to bone formation as well (e.g. those encoding osteocalcin and type X collagen). These data indicate that in aging animals gene expression is altered early in fracture repair, with consequences for the entire healing cascade. Local injection of an EP4 agonist, which directly activates the EP4 receptor in place of the missing COX-2, to the fracture site of aged mice compensated for the reduced fracture repair observed with aging, leading to a significant reduction in immature cartilage and more efficient formation of mature bone.

Alcohol Abuse

A large number of studies have directly linked alcohol abuse to low bone mass, as well as increased fracture risk. The magnitude of these effects is greatest at young ages when bone tissue is still developing. Alcohol alters the normal balance between bone formation and resorption, thus compromising the mechanical properties (elasticity, stiffness, load carrying capacity, and toughness) of bone tissue and leading to a higher number of fractures. Fractures in chronic alcohol abusers are associated with extended hospital stays and increased mortality due to the inability of the bone to heal properly. Alcohol's disruption of the bone remodeling process causes a delay in onset and a significant decrease in endochondral ossification, leading to a high incidence of both delayed unions and nonunions in fracture repair.

The detrimental effects of alcohol abuse on fracture repair result from its negative influence on preosteoblast recruitment and osteoblast differentiation and proliferation. Alcohol abuse affects the differentiation of bone MSCs into osteoprogenitor cells, which is important to bone formation. Acetaldehyde, the main metabolite of alcohol, significantly decreases the number of fibroblast colony forming units in bone marrow cultures of alcoholic patients. These data suggest inhibition of the recruitment of cells destined to become osteoblasts. Furthermore, an increased amount of bone marrow fat is found in alcoholics compared to nonalcoholics. Adipocytes come from the same lineage as osteoblasts and it appears alcohol can decrease osteoblast differentiation in favor of producing more adipocytes.

Alcohol also directly inhibits the proliferation of osteoblasts. In osteoblasts exposed to alcohol, the proliferation rate is significantly decreased, along with decreased ALP, osteocalcin, and TGF-β. Similarly, crucial mitogenic growth factors involved in osteoblast proliferation are inhibited by ethanol. Sodium fluoride and IGF-II are important proliferation initiators in

osteoblasts and the expression of both is significantly reduced in response to alcohol.

Diabetes Mellitus

Diabetes mellitus (DM), or simply, diabetes, is defined as a group of metabolic diseases characterized by high glucose levels that result from defects in the body's ability to produce and/or use insulin. There are two types of diabetes: type 1 DM (T1DM; also called insulin-dependent DM) and type II DM (T2DM; also called noninsulin-dependent DM). Increased fracture risk and low bone mineral density (BMD) have both been linked to T1DM, but in T2DM, BMD is usually normal or elevated, even though the bone is fragile and at risk of fracture, suggesting that factors other than BMD affect bone fragility. The incidence of both T1DM and T2DM is rising worldwide. Estimates predict a steady increase in both types of DM to about 4.4% of the total population by 2030.

One particular aspect of DM of which relatively little is understood is its effect on bone metabolism. The difference in the bone balance profile between T1DM and T2DM suggests that insulin (or the lack of it) has a significant effect on bone cellular processes. Recent studies are beginning to clear up the interplay between bone metabolism and the issues of insulin deficiency and hyperglycemia associated with the disease.

Insulin has been suggested to be an anabolic agent in bone, although the exact mechanisms by which it affects the bone are not yet clearly understood. One aspect of insulin's impact on bone formation is its effect on osteoblast proliferation and differentiation. Insulin has been demonstrated to upregulate osterix expression through the mitogen-activated protein kinase (MAPK) pathway and to downregulate RUNX2 expression through the phosphoinositide 3-kinase (PI3-K) pathway, with the net effect of driving osteoblast precursors to differentiate. Insulin can cause upregulation of the insulin receptor, which subsequently leads to an increase in cell sensitivity to other signaling molecules including IGF-I, which is known to be an important autocrine/paracrine factor for osteoblast proliferation and differentiation. In addition, insulin increases expression of ALP and type I collagen synthesis.

Hyperglycemia, on the other hand, has been linked to the inhibition of osteoblast differentiation. Through osmotic and nonosmotic pathways, hyperglycemia decreases osteocalcin, VEGF, and collagenase 3/matrix metalloprotease 13 (MMP-13) expression, but increases peroxisome proliferator-activated receptor gamma (PPAR-γ) expression in osteoblasts. Experiments using embryonic stem cells have shown that hyperglycemia leads to a reduced formation of mineralized matrix associated with decreased expression of mRNAs for ALP, osteocalcin, osteonectin, and osteopontin. Cells cultured in conditions of high sugar concentration tend to have a higher amount of advanced glycation end product (AGE)-modified proteins, which could in turn negatively influence cell functions. AGEs are the products of collagen bonds, which tend to make the tissue stiffer but also more brittle and less able to resist fracture.

Fracture healing is impaired in patients with diabetes. In the well-established diabetic animal models, fracture healing is also impaired, as evidenced by the reduced external callus area, reduced collagen matrix secretion, and reduced recovery of biomechanical properties. Impaired chondrogenesis and osteogenesis during the healing process results from the combined effects of hypoinsulinemia and hyperglycemia. It is not a surprise that systemic insulin treatment, used to normalize blood glucose, in a diabetic rat has been shown to reverse the deficit in bone healing. Local intramedullary delivery of insulin to the fracture site, which does not provide systemic management of glucose, can also reverse the healing deficit in mesenchymal cell proliferation and chondrogenesis at the early healing stage, as well as in bone mineralization and biomechanical properties at the late healing stage in a diabetic rat fracture model. Several popular growth factors, including rhBMP-2, rhBMP-7, and recombinant human PDGF BB (rhPDGF-BB) have been shown to positively affect fracture healing in diabetic animal models.

Glucocorticoids

Glucocorticoid therapy has negative effects on the skeleton and is the most common cause of secondary osteoporosis. Glucocorticoids are often used to treat diseases like asthma and arthritis, and can also be used to suppress the immune system after organ transplantation. The most prevalent side effects are the increased rate of bone resorption and reduction of BMD, which leads to secondary osteoporosis and ultimately a greater risk of fracture. Glucocorticoid-induced osteoporosis occurs in two phases. First, an initial rapid reduction of BMD occurs due to increased bone resorption within the first year. Second, a slower progressive phase occurs when BMD declines due to impaired bone formation. Glucocorticoid treatment is frequently associated with an increase in the risk of bone fracture, especially in trabecular-rich sites, such as the spinal vertebrae and the femoral head. The risk of fracture escalates by as much as 75% within the first 3 months after the initiation of the therapy, suggesting that glucocorticoids have adverse effects on bone quality (material properties of bone tissues), as well as on bone mass.

To make the situation even worse, fractures resulting from long-term use of glucocorticoid usually do not heal normally under the glucocorticoid treatment. The inhibitory effects of glucocorticoid on fracture healing have been known since the 1950s. Formation and calcification of the external fracture callus are significantly reduced, as evidenced by the smaller callus area and lower bone mineral and BMD in glucocorticoid-treated animals. Glucocorticoid treatment results in a significant delay in endochondral ossification and bone remodeling in callus (Fig. 10.9). Glucocorticoid-treated animals exhibit less extensive bridging of the fracture and osteotomy and a less mature callus. When evaluated biomechanically, both torsional and four-point bending tests show a similar 50–70% reduction in ultimate strength, stiffness, and energy to failure of glucocorticoid-treated animals compared with vehicle-treated controls.

The detrimental effects of glucocorticoid on fracture healing may result from its direct actions on

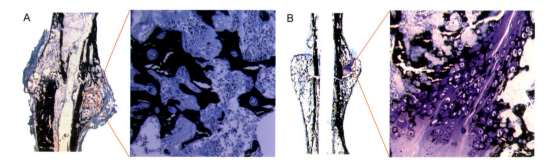

FIGURE 10.9 **Glucocorticoid inhibits callus formation and delays endochondral ossification during fracture repair.** Images show representative histologic images stained with von Kossa and enlarged histologic images from the corresponding squares. (A) Mice implanted with placebo pellets. (B) Mice implanted with slow release prednisolone pellets (1.5 mg/kg/day). No cartilage remains in the large callus of mice implanted with placebo pellets (A), whereas the callus of the prednisolone-treated mice is smaller and contains residual cartilage (B). *Reproduced with permission from Doyon AR et al., 2010 Calcif Tissue Int. 87:68–76.*

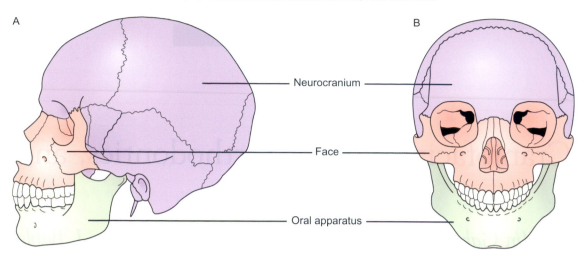

FIGURE 11.1 The craniofacial complex can be divided into three regions according to their functions and structures: the neurocranium, face, and oral apparatus. (A) Lateral view of the skull. Note the sutures connect the bones. (B) Front view of the skull. The midsagittal suture is fused (craniosynostosis).

into the cranial bones. The inner layers of the periosteum and dura mater integrate with the sutures and provide structural stability as well as progenitor cells for osteogenesis. As the brain rapidly grows after birth, the cranial vault expands in response to mechanical forces generated by increasing intracranial pressure. Tension forces stretch suture connective fibers, inducing osteogenic responses. This results in the deposition of osteoid at the suture margins. The osteoid mineralizes, but the suture remains patent for years (until middle or old age). Replication of this process results in the gradual enlargement of the cranial vault. The cranial vault achieves approximately 80% of its adult size by age 4 years and 95% by age 10.

Most sutures remain patent in humans after birth, except for the metopic suture. The metopic suture starts to fuse at age 2 years to form a single frontal bone. Cranial sutures cease growing by the end of the third decade. With age, sutures become more complex and interdigitated, which interlocks the bones and prevents separation. Suture closure (synostosis) starts with the formation of bony bridges across a suture and eventually the suture fuses with bone obliteration (Fig. 11.3). If a suture fuses prematurely due to genetic syndromes (for example, Apert Syndrome, Crouzon syndrome, or Jackson-Weiss syndrome), then craniosynostosis leads to compensatory alterations in the growth of the head or face.

Cranial Base

The cranial base develops and grows through the mechanism of endochondral ossification. In utero, the cranial base forms by merging the cartilage elements from the embryonic chondrocranium (Fig. 11.4). This process begins in the late fourth week of development within the mesenchyme that is situated between the developing anterior neural tube and the foregut. Within this mesenchyme, cells will increase in number and condense into precartilaginous centers on each side of the midline. The paired condensations themselves are named according to their relative position the anterior termination of the notochord (prechordal and parachordal).

Paired parachordal condensations begin to form cartilage and fuse with their contralateral partner to form the basal plate. Prechordal condensations will merge to form the anterior portion of the basal plate. By the twelfth week, all of these precartilaginous condensations have merged to form a single cartilaginous chondrocranium. Ossification of the chondrocranium involves endochondral bone formation. It initiates in numerous ossification centers within the cartilaginous cranial base. These centers form in a caudal-to-rostral direction, starting in the ninth week of prenatal development. The cartilaginous cranial base is ultimately replaced by the major portion of four bones: the occipital, sphenoid, temporal, and ethmoid. However, the fusion of the multiple ossification centers occurs over a long time span, starting before birth and continuing into postnatal life. Areas where cartilage remains form the synchondroses of the skull (e.g. the sphenooccipital synchondrosis between these two bones on the base of the skull; Fig. 11.5).

Synchondroses are temporary cartilaginous joints between cranial bones. A synchondrosis serves as a growth center, similar to the epiphyseal growth plate in the long bone. Unlike the epiphyseal growth plate, each synchondrosis consists of two adjacent back-to-back growth plates (Fig. 11.5). These two plates grow against each other in the same plane. When these two

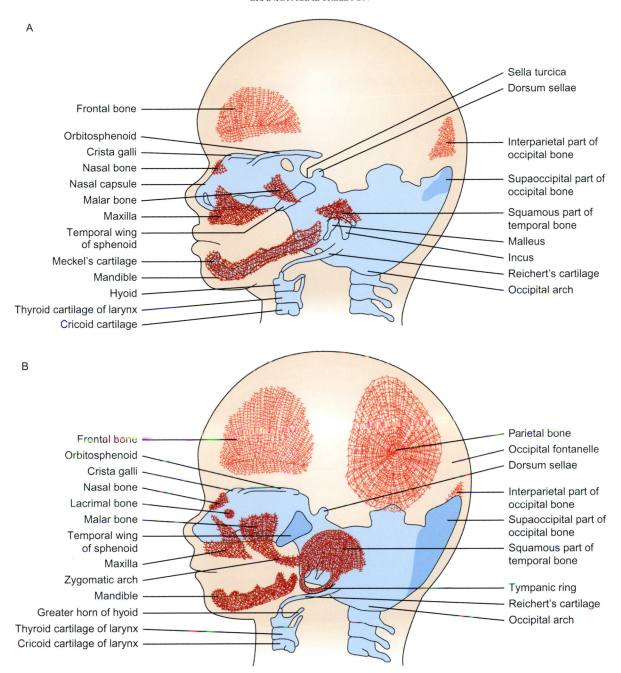

FIGURE 11.2 This illustration demonstrates the developing skull from about 9 weeks to 12 weeks. (A) At 9 weeks, note the small areas of ossification of the frontal bone and lack of a center for the parietal bone at this stage. (B) By 12 weeks, the size of the frontal bone has increased and the parietal bone has started to ossify and cover the brain. As the forming bones approach each other, future sites for sutures and fontanelles become apparent. *Image adapted from Human Embryology (Bradley Patton). Out of print.*

plates grow, the cranial base increases in size and length. Similar to the epiphyseal growth plates, the synchondroses cease growing and fuse when they reach their mature size. The anterior cranial base grows and matures faster than the posterior cranial base. The anterior cranial base growth completes at approximately 7 years and the posterior cranial base completes later in the puberty stage.

Face

Upper Face

The upper face is commonly referred to the region superior to the orbits on the forehead. When the anterior cranial base and cranial vault develop, the frontal bones of the upper face increase in size and height and reach their mature size early. Although the upper face

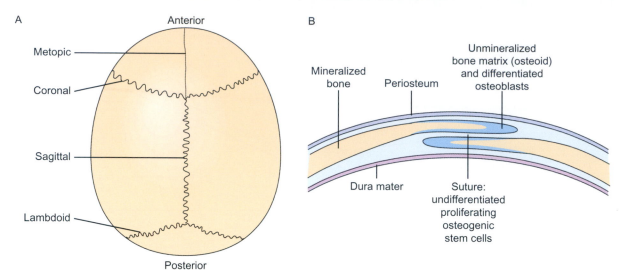

FIGURE 11.3 **Suture closure.** (A) Superior view of the cranial vault with primary sutures identified. (B) Diagram of a cross section through a cranial suture, with ongoing bone apposition. Note the relationships of the dura mater and periosteum to the sutural tissues and the active bone formation at the meeting ends of the articulating bones.

is considered very stable, the frontal bone is slightly enlarged in adulthood because of enlargement of the frontal sinus and bone modeling on the junction between the frontal bone and nasal bones.

Midface

The midface (nasomaxillary complex) connects the neurocranium with circumferential sutures. The nasomaxillary bones develop and grow through intramembranous ossification. At birth, the nasomaxillary complex is small but well developed. The intermaxillary suture system connects the paired maxillae, premaxillary bones, palatal bones, nasal bones, and zygomatic bones. When the anterior cranial base rapidly grows with brain expansion, the nasal septum also grows actively with expansion of the nasal cavity and oral pharynx, leading to anterior and inferior growth of the midface (Fig. 11.6). In response to the midface displacement, bone deposits at the margins of the circumferential maxillary and intermaxillary sutures. This displacement and suture bone growth process repeats to increase bone volume of the nasomaxillary complex until suture closure is completed. Whereas the premaxillary sutures fuse around age 3—5 years, the major intermaxillary sutures do not fuse until the early twenties, when the midface is displaced forward and downward. Bone modeling also occurs; bone deposits on the posterior surface of the maxilla, increasing the length of the maxillary and upper dental arch. At the same time, bone deposits on the roof of the oral cavity, but resorbs on the floor of the nasal cavity. Therefore, the palate relocates downward and the midface grows vertically with increasing height.

While the midface undergoes most of its growth before age 7, when the anterior cranial base growth completes, the midface keeps growing in length and height throughout childhood and adolescence. The rate of midfacial height growth peaks at the adolescent growth spurt. Similar to sex differences in body growth, the midface grows more in males than in females during adolescence. Males generally have a wider, longer, and taller midface than females.

Mandible

The mandible originates from an ossification center in the perichondrial membrane lateral to Meckel's cartilage in the mandibular process. Meckel's cartilage is the structural framework for the developing mandible; it does not form the mandible but serves as a structure adjacent to which the bone will form. Meckel's cartilage develops concurrently with the formation of the chondrocranium, and is associated with the formation of the first branchial (pharyngeal) arch. Each arch has a central core of mesenchyme. This mesenchyme, originating from lateral plate mesoderm, is added to by migration of neural crest cells to form a precartilaginous condensation of cells in the core of the arch. This condensation will differentiate into cartilage within the arch. In the first branchial arch, this structure is termed *Meckel's cartilage*.

Meckel's cartilage forms a single bar of cartilage, bilaterally, that extends from the region of the future mental symphysis (chin) to the middle ear cavity. The two cartilage bars remain separate at the mental symphysis. Within the middle ear cavity, Meckel's

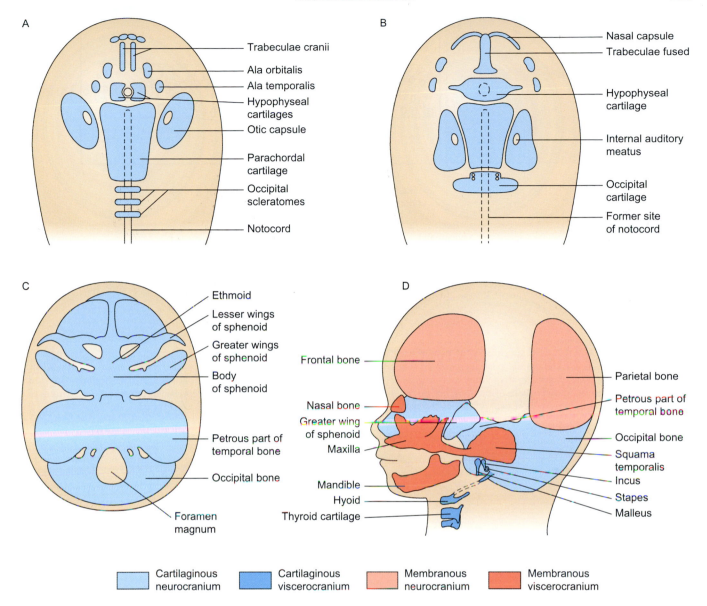

FIGURE 11.4   **Development of cranial base.** (A to C) View of the developing chondrocranium from above; (D) A lateral view. (A) At 6 weeks. Note that the various paired cartilaginous centers that will give rise to the chondrocranium the parachordal centers have already fused. (B) At 7 weeks. Fusion of the paired cartilaginous centers continues. (C) At 12 weeks. The various cartilaginous centers have now fused into a single chondrocranium. (D) At 20 weeks. Note the relationship of the chondrocranium to the developing membranous bones of the vault and face. *Image taken from The Developing Human. Clinically Oriented Embryology (Saunders owned).*

cartilage undergoes endochondral bone formation to form the incus and malleus ear ossicles (the third bone of the middle ear, the stapes, is formed from the second pharyngeal arch). The remainder of the cartilage, however, gradually disappears. During the sixth week of intrauterine development, a condensation of mesenchyme forms near the future site of the mental foramen. Shortly thereafter, intramembranous bone formation is initiated in this condensation and formation of the bony mandible begins. Bone formation spreads both anteriorly and posteriorly from this site along the lateral aspect of Meckel's cartilage. Once the

differentiation of mesenchyme into bony mandible has progressed posterior to the region of the mandibular foramen, Meckel's cartilage begins to disappear.

Each mandibular half consists of the condyle, the ramus, and the mandibular body (corpus). At birth, the two mandibular halves are connected with each other by a fibrous articulation, the mental symphysis, which rapidly fuses within a few months of birth. The mandibular condyle origins from a separate blastema that develops in close relationship to the periosteum associated with the membranous bone that extends from the mandibular foramen to

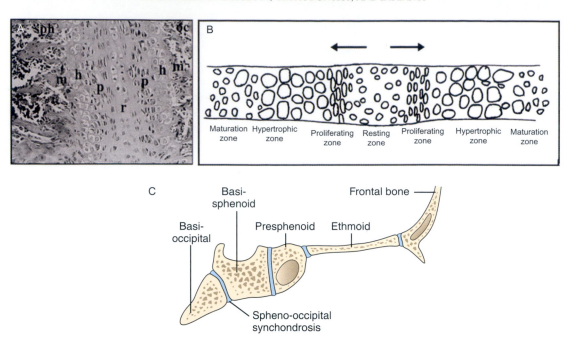

FIGURE 11.5 **Each synchondrosis consists of two adjacent back-to-back growth plates.** These two plates grow against each other in the same plane. As they grow, the cranial base increases in size and length. (A) Histological section through a synchondrosis. h, hypertrophic zone; m. maturation zone; p, proliferative zone; r, resting zone. (B) Diagram of the histology of a synchondrosis. Arrows denote direction of growth, which unlike the growth plate of the long bones, occurs in both directions. (C) Diagrammatic representation of the synchondroses of the cranial base. *(A) and (B) reproduced with permission from Lynne A et al., Seminars in Orthodontics, 11, 4, 2005, 199–208.*

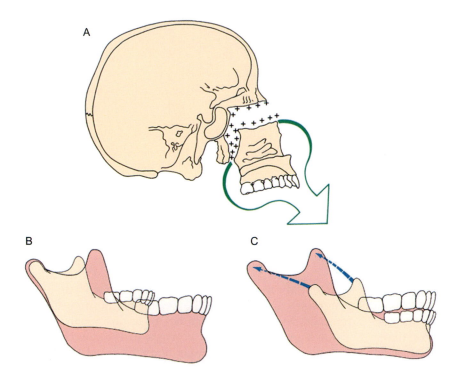

FIGURE 11.6 **Growth of the midface.** (A) Growth of the surrounding soft tissues translates the maxilla downward and forward; new bone is added onto both sides of the sutures. (B) Growth of the mandible, as viewed from the perspective of a stable cranial base; the chin moves downward and forward. (C) Mandibular growth as viewed from the perspective of vital staining studies reveals minimal changes in the body and chin area, while there is exceptional growth and remodeling of the ramus, moving it posteriorly. The correct concept is that the mandible is translated downward and forward and grows upward and backward in response to this translation. *Reproduced with permission from Contemporary Orthodontics fifth Edition.*

the articulation with the developing squamous portion of the temporal bone. The condyle articulates with the temporal bone, and together they make up the temporal mandibular joint. The ramus extends vertically from the condyle to the mandibular body, which provides muscle insertions for the mastication. The mandibular body extends horizontally and anteriorly, carries the lower dentition, and houses the inferior alveolar neurovascular bundles.

Growth of the mandible is greater than for any other component of the craniofacial complex, following its original V-shape through surface periosteal bone modeling (Fig. 11.6). Bone deposition mainly occurs on the posterior and lateral surfaces of the ramus and the posterior and superior surfaces of the condyle and coronoid process. In the first few years of life, the condyle and ramus grow rapidly in the posterior-superior direction. While bone deposits on the posterior surface, bone also resorbs on the anterior surface of the ramus. This relocates the ramus posteriorly, thus increasing alveolar bone length for accommodating the developing lower dentition. Afterwards, the condyle grows more prominently in the superior direction. The greatest growth rates of the condyle occur in early childhood and during the adolescent growth spurt. Similar to growth of stature, mandibular growth is more accelerated in males than in females during adolescence. Mandibular width also increases during childhood and adolescence; however, lateral growth in the mandible is relatively smaller than growth in height and length. No obvious growth spurt is noted in mandibular width.

Oral Apparatus

Dentition

From infancy to early adulthood, humans undergo three dentition stages: primary dentition, mixed dentition, and permanent dentition. Mixed dentition is the transition stage between primary and permanent dentition when the first molars erupt and the primary teeth are replaced by permanent teeth; it usually occurs between ages 6 and 12 years. The upper and lower dentitions keep erupting until they make contact with each other to establish chewing function. Even after functional occlusion is established, the teeth continue to erupt and migrate to compensate for the rapid facial growth during childhood and adolescence. For example, to compensate for the developing vertical nasomaxillary complex and primary mandibular growth, the maxillary molars and incisors erupt 1.2−1.4 mm/year and 0.9 mm/year, respectively, and the mandibular molars and incisors erupt 0.5−0.9 mm/year, primarily in adolescence. As the maxilla grows bilaterally, the maxillary intermolar width increases 4−5 mm between ages 6 and 16 years. Similarly, as the mandible grows bilaterally, the mandibular intermolar width increases 2−3 mm. In the anteroposterior direction, incisors change their inclinations to adapt to differential jaw positions. For example, for individuals with a smaller mandible (relative to the maxilla), the upper incisors tend to tip backward and lower incisors tend to flare forward so the upper and lower incisors can make contact with each other to cut food. In contrast, the upper incisors tend to flare

forward and the lower incisors tend to incline backward in individuals who have a larger mandible.

Alveolar Bone

Alveolar bone is one of three tissues that support the tooth; the other two are the periodontal ligament and the cementum. Alveolar bone is formed by intramembranous bone formation during the formation of the mandible and maxilla. Alveolar bone actually consists of two components (Fig. 11.7). The first is the alveolar process of the two jaws, the maxilla, and mandible. This bony structure forms to house the developing tooth buds and, once erupted, the roots of the teeth. It provides structural support for the dentition. If the teeth are lost, then the need for this process is lost and through time the process is resorbed. The second type of bone is alveolar bone proper, which is the portion of bone that lines the tooth socket. It provides an attachment site for the periodontal ligament and its associated tooth.

The development of alveolar bone is associated with the formation of the membranous bones of the upper and lower jaws in relationship to the concurrent development of the primary dentition. During the developmental process of the mandible, bone forms around the inferior alveolar nerve and its terminal branch, the incisive nerve. This results in the formation of a bony trough within which the nerve will lie. This trough consists of lateral and medial alveolar plates that extend superiorly from the forming body of the mandible. The trough will not only house the nerve as it forms but will also ultimately house the developing tooth buds. The mandibular symphysis anteriorly remains a fibrous union until shortly after birth, when the two membranous bones will unite.

The development of the maxilla and its alveolar process is more complex than the development of the mandible, since the maxillary bone will ultimately give rise to the maxillary sinus and form adjacent to important structures associated with the nasal capsule, including the orbit and nasal region. However, formation of an alveolar process that will enclose the developing primary tooth buds is similar for both jaws. Over time, individual tooth buds become separated from each other by bony partitions, thereby creating the tooth sockets.

Alveolar Bone Proper

The formation of alveolar bone proper is initiated with the eruption of the developing tooth. Once the crown of a tooth has been formed, root formation begins. Formation of the root involves a complex interaction between the mesenchyme of the dental follicle and the Hertwig root sheath. The dental follicle gives rise to cementoblasts that begin to deposit the

FIGURE 11.7 **Anatomical components** of the tooth, the supporting alveolar bone, and their interface. See text for details on structures.

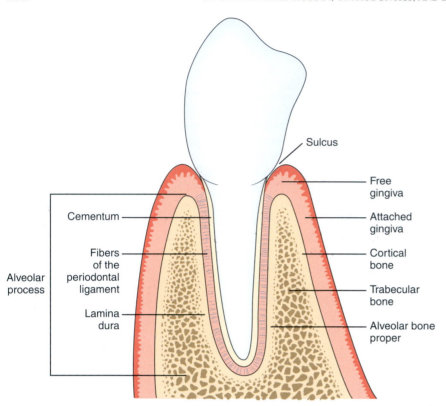

cementum that lines the external surface of the root. Concurrently, other mesenchymal cells in the dental follicle differentiate into fibroblasts, forming the periodontal ligament (PDL), and still other mesenchymal cells differentiate into osteoblasts adjacent to the bone, forming the sockets within the alveolar process. This relationship between cementum deposition, connective tissue fiber formation, and bone deposition facilitates the embedding of PDL fibers into the cementum of the tooth and the alveolar bone proper.

As the root continues to form, the PDL continues to increase in length as the new root portion provides attachment to new fibers of the PDL. Similarly, the alveolar bone lining the socket continues to be remodeled. Bone deposition occurs vertically, thus increasing the depth of the socket. The alveolar bone continues to remodel, filling in around the root as it erupts and lengthens. It is during this process that the true alveolar bone is created to provide support for the tooth. Ultimately, the crown of the tooth emerges from the bony jaw, pierces the overlying gingiva, and moves toward occlusion. As the tooth comes into functional occlusion, the PDL absorbs and then distributes the force placed on the tooth during mastication or other events and distributes it to the surrounding alveolar process via the alveolar bone proper. Alveolar bone proper appears on a radiograph as a thick radiopaque line adjacent to the alveolar socket, termed the *lamina dura*.

The alveolar bone proper provides the attachment site for Sharpey fibers from the PDL. These collagen fibers are organized into bundles and calcified within the bone to provide a strong attachment between tooth and bone. This portion of alveolar bone is sometimes referred to as *bundle bone* due to the presence of the fiber bundles. Bundle bone, in turn, merges with adjacent lamellar bone that comprises the alveolar process. Bundle bone is the most important to tooth movement and disease processes involving the periodontium. The remaining portion of the alveolar bone proper is lamellar bone. It is perforated by numerous small foramina that allow the nerves and vessels within the alveolar process to reach the PDL tissues. This perforated bone is often referred to as the *cribriform plate*. The bone lining the socket is closely contoured with the tooth, and its coronal margin becomes the alveolar crest. The composition of alveolar bone proper is similar to that of other bone. There is some evidence, however, that the alveolar crest is more mineralized than the bone adjacent to the apex of the tooth. Under functional occlusion, the thickness of the alveolar bone also increases. This is unsurprising, since the tension of the PDL is increased with functional occlusion and this in turn stimulates bone deposition.

The Alveolar Process

Bone of the alveolar process is composed of both an outer layer of cortical bone and an inner region of cancellous bone (Fig. 11.7). The cortical bone is similar to that seen in other regions of the skeleton and is composed of lamellar bone; it contains Haversian systems for bone maintenance and remodeling. The alveolar process contains the nerves and blood vessels that support the bone and teeth. Bone marrow is present and contains significant amounts of adipose cells, as well as osteogenic cells and hemopoietic tissue. Cortical bone of the alveolar process tends to be thinner in the maxilla than in the mandible. It is thickest in the mandible adjacent to the premolars and molars. This is an important consideration when planning dental implants.

The alveolar bone proper merges with the cortical bone of the process to form the alveolar crest at the coronal border of the socket. In a healthy individual, the alveolar crest is generally 1−2 mm below the *cement-enamel junction* (CEJ) of the tooth. It becomes a thin margin of bone adjacent to the tooth. The bone between adjacent sockets is termed the *interdental septum*. The shape of the alveolar crest, as well as the interdental septum, is affected by the position of the CEJ in adjacent teeth. The interdental septum is composed primarily of cortical bone. However, toward the apex of the tooth, increasing amounts of cancellous bone occupy the region between cortical plates. The amount of cancellous bone is dependent on the location along the arch. The anterior region of the arch containing the incisors contains very little cancellous bone. In this situation, the cortical plates are in close proximity and merge with the alveolar bone proper. However, the posterior arch, which contains multiple root teeth, contains a significant amount of cancellous bone.

Diseases Affecting Alveolar Bone

Periodontal disease can produce significant changes in alveolar bone proper. Periodontal disease starts as inflammation in the gingival tissues. Left untreated, it can spread to the PDL and alveolar process. The periodontal tissues function in unison to support the tooth. Therefore, damage from disease has significant effects on both its function and the ability to repair lost function. Advanced periodontal disease results in significant alveolar bone loss. This can involve a single tooth but will typically include adjacent teeth. Alveolar bone loss can extend vertically toward the apex of the tooth. In this situation, a periapical abscess may form in the alveolar bone adjacent to the apex of the tooth. Individuals with advanced periodontal disease may have multiple teeth that are quite mobile (sometimes referred to as *floating teeth* due to a loss of the bony support for the teeth). The teeth, however, may remain vital.

Pulpitis is inflammation of the pulp of the tooth and may also result in alveolar bone loss. It can result from trauma to the tooth, but most often reflects invasion of dental caries into the pulp chamber. If inflammatory products exit the pulp canal, they can produce a local inflammatory reaction in the alveolar bone adjacent to the apex of the tooth, called *osteomyelitis*, which, in turn, can result in a periapical abscess. The inflammation may also involve the PDL adjacent to the apex, producing periodontal disease secondary to the bony inflammation. The result of this cascade of events is that the alveolar bone adjacent to the apex of the tooth degenerates. As the inflammation progresses, there is increased ischemia in the area involved; bone tissue becomes nonviable and necrotic bone will be lysed from viable bone. Significant bone resorption will result in the formation of an area devoid of bony tissue. If sufficiently large, the abscessed bone can cause a tooth to become movable.

ORTHODONTIC TOOTH MOVEMENT

Tooth movement requires an interaction between the PDL and both the tooth and bone. When a force is applied to a tooth, the entire bony socket is affected. Continuous forces can reposition the tooth within the alveolar process. These forces can be produced by a variety of factors. Normal biological factors such as tooth eruption, lip pressure, tongue pressure, or even thumb sucking can produce a physiologic response that results in tooth repositioning. Tooth movement can also be produced by the use of external force on a tooth. Orthodontics involves the use of hardware (brackets and wires) to move teeth within the alveolar bone using external forces.

Tooth movement requires changes to both the PDL and the bone (Fig. 11.8). The PDL has a fairly consistent thickness, ranging in humans from 0.15 to 0.38 mm. Force on a tooth first affects the PDL by imparting either compressive or tensile strain that lead to narrowing or lengthening of the PDL space. The PDL responds in order to maintain the optimal space between bone and tooth. The portion of the PDL adjacent to the front edge of the moving tooth is compressed, thus reducing the tensile forces acting on the adjacent alveolar bone. In addition, blood flow to this region of the PDL diminishes. The combined effects of compression and reduced blood flow lead to collagen degradation and cell death in the PDL—a process referred to as *hyalinization*. The tissues respond by stimulating increased osteoclast activity in an attempt to restore a balance in space between the tooth and the bone. It should be noted that if cell death in the tissues is severe, then remodeling will be delayed until sufficient healthy cells are present. If cell death is too severe, then no remodeling will occur. This is a concern for heavy orthodontic force application—trying to move teeth too fast can result in significant damage and delayed healing. At the same time that these effects are

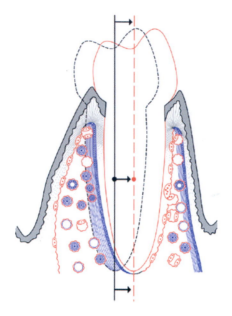

FIGURE 11.8 **Tooth movement requires both changes to the periodontal ligament and bone remodeling.** This schematic diagram shows the bone physiology associated with translation of a tooth. Note that there is a coordinated bone modeling and remodeling response leading and trailing the moving tooth. Arrows indicate the direction of tooth movement. This mechanism allows a tooth to move relative to basilar bone, while maintaining a normal functional relationship with its periodontium. Osteoclastic and osteoblastic activities are shown in red and blue, respectively. *Reproduced with permission from Roberts, W. Eugene, et al. Seminars in Orthodontics. 12. 4. 2006.*

occurring on one PDL-bone interface, the PDL on the opposite side of the moving tooth (with increased strain) is being stretched and experiencing an increased space between tooth and bone. This situation stimulates bone formation. The increased tensile force stimulates bone deposition, and the space between bone and tooth diminishes toward its typical dimension. It is the balance between bone resorption along the front edge and bone deposition along the trailing edge of the tooth that allows the tooth to migrate slowly through the mineralized matrix, while allowing it to maintain its function.

Tooth Eruption

The process of tooth eruption requires continued remodeling of bone around the tooth. As the root of the tooth elongates, the alveolar process increases in height to accommodate the root. The erupting tooth requires continuous bony remodeling to permit the tooth to move out of the crypt, while the facial skeleton is also remodeled to reposition the teeth within the jaws. While the mechanism for generating the eruptive forces is unclear, the PDL is known to be involved. Within the socket, the PDL provides an interface between erupting tooth and bone. The erupting tooth has some mobility within its socket. Therefore, tooth eruption produces

forces that act on the bone and PDL to stimulate bone remodeling. This results in the tooth being repositioned within the jaw. As seen in other mechanisms of tooth movement, it is the tooth that provides the stimulus for bony change. It is tooth movement that produces compensatory changes in the alveolar bone.

Occlusal Forces

Once the tooth has erupted, it is repositioned through normal physiologic actions. Mastication is the most prominent force that alters tooth position, but other continuous forces (such as tongue pressure or thumb sucking) can move the tooth (or teeth) in the direction of the force over time. Occlusal force, the load applied between the upper and lower teeth, produces mechanical stimulation from the bone, through the PDL, to the alveolar bone. This mechanical stimulation induces collagen synthesis in the PDL and bone remodeling within the surrounding bone. In normal occlusion, there is a balance between normal bone resorption and deposition. Repositioning of teeth due to normal physiologic loading is gradual. When occlusal force is significantly increased for a period, such as in *bruxism* (clenching or grinding teeth), there is a potential for increased alveolar bone deposition and hypertrophy of the bone. Conversely, when the forces of occlusion are removed, one of two things may happen, depending on the method through which forces are removed. If the tooth is removed from the socket, due to extraction, trauma, or disease, then the tensile forces transmitted from the PDL to the bone are removed and normal bone resorption will be greater than the stimulus for bone formation. The net result is bone resorption. However, if the tooth is still within the socket and there is no occlusal force acting on the tooth, e.g. the opposing tooth is absent, then the tooth will erupt beyond the occlusal plane.

Orthodontic Tooth Movement

The intrinsic mechanism through which the teeth move in space is utilized in orthodontics to speed tooth movement and modify the oral cavity morphology. Effective orthodontic treatment strives to use the greatest amount of force with minimal deleterious effects. If tooth movement is not well designed, however, the translational or parallel movement of the tooth may not be uniform, and the coronal portion of the tooth may move one direction while the apical portion, i.e. the root, may move in the opposite direction, producing a tipping of the tooth (Fig. 11.9). Although orthodontic-induced tooth movement is based on the normal physiologic movement of teeth, the processes differ slightly because the forces used exceed those

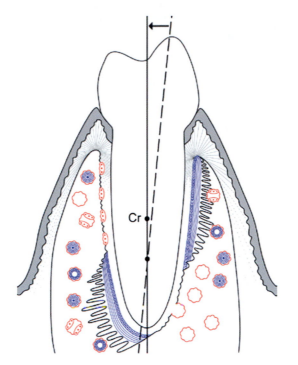

FIGURE 11.9 If tooth movement is not well designed, translational movement of the tooth may not be uniform, and the coronal portion of the tooth will move one direction while the apical portion, i.e. the root, may move in the opposite direction, producing a tipping of the tooth. This schematic drawing of a tooth tipping to the left shows the coordinated bone modeling and remodeling events associated with movement of the tooth relative to apical bone. Note that there is an elevated remodeling rate in bone supporting the moving tooth. The center of resistance (Cr) is occlusal to the center of rotation (intersection of the solid and dashed lines). *Reproduced with permission from Roberts, W. Eugene, et al. Seminars in Orthodontics. 12. 4. 2006.*

achieved during normal mastication. Unlike physiologic tooth movement, orthodontic tooth movement is usually a coupled bone resorption and formation process. When moving a tooth in a parallel fashion, the PDL is subjected to tension force on one side and compression force on the other side. Bone formation occurs adjacent to the PDL on the tension side and bone resorption on the compression side. However, this balanced bone formation-resorption couple only exists under light orthodontic force. When applying heavy force to the tooth, more bone resorption than formation occurs, which can cause resorption of the tooth root apex, leading to tooth root shortening.

SUTURE EXPANSION

The size of the maxilla and position of the mandible can be altered using orthodontic techniques that involve suture expansion and mandibular advancement. Sutures serve as growth sites that respond to brain expansion, leading to cranial bone growth and an enlarged cranial vault. The same mechanism can also be used clinically to stimulate bone growth in the nasomaxillary complex for increasing bone volume to correct midface deficiencies. The nasomaxillary complex can be expanded in transverse or anteroposterior directions. In the transverse direction, the treatment strategy of suture expansion is commonly referred as *palatal expansion*, although all the intermaxillary sutures and paired bones are expanded bilaterally. Palatal expansion can be used to widen the maxillary dental arch to alleviate dental crowding or treat patients with posterior cross bite. In the anteroposterior direction, suture expansion is often referred as *maxillary protraction*, which pulls the nasomaxillary complex forward to treat patients with midface deficiency usually accompanied with anterior crossbite (the lower incisors overlay the upper incisors).

Although many appliances can be used for suture expansion, the treatment principles and biological responses are similar. When a suture is expanded by mechanical force, tissue breakdown, acute inflammation, and cell death occur within a few hours. Within 3–4 days, preexisting and newly recruited osteoblasts deposit new lamellar bone along the suture margins. The collagen fibers and cells align transversely across the suture, depending on the amount of tension. In the subsequent 1–2 weeks, there is continued bone formation at the suture margins. Once the expansion forces cease, remodeling of the newly formed sutural bone continues until normal morphology has been established.

For palatal expansion, the most common approach is to attach an expander to the posterior teeth (Fig. 11.10). The jackscrew can be activated daily (rapid) or weekly (slow) to expand the midpalatal suture. Rapid palatal expansion accumulates up to approximately 90 N of force to widen the narrow maxilla over 1–3 weeks. This can sometimes cause discomfort or pain in patients. Slow palatal expansion only generates about 9 N of force but takes 8–13 weeks.

Effective suture expansion necessities achieving maximal bone formation with the minimal suture gap after expansion. Although studies indicate that either tensile or compressive cyclic loadings can stimulate suture growth with increased bone formation, most clinicians currently use continuous forces for suture expansion. For example, although a jackscrew palatal expander is usually turned or activated once per day, expansion force still continuously accumulates in the skeleton to open the midpalatal suture. On the other hand, wearing a face mask for more than 12 h/day to protract the maxilla is considered an intermittent force because it allows a few hours break per day.

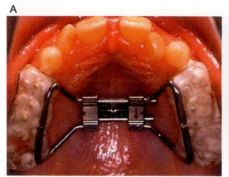
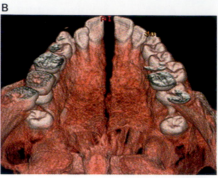

FIGURE 11.10 **Palatal expansion technique.** A jackscrew-type expander (A) is commonly utilized to expand the palate by adjusting the force over time. (B) computed tomography imaging can be used to track the expansion. *Reproduced with permission from Habersack, Karin et al., American Journal of Orthodontics and Dentofacial Orthopedics 131, 6 (2007): 776–781.*

Continuous force has been shown to be more effective for separating the suture and induces more bone formation, compared to intermittent forces. The amount of bone apposition during suture expansion is proportional to the amount of suture separation. When forces are intermittently applied, the periods of relief from the mechanical forces do not allow the collagen fibers to continue to appose the bone with reduced suture separation; this leads to less bone formation. A continuous force keeps fibers stretched and stimulates a greater response in bone formation than intermittent forces. This phenomenon apparently contradicts the concept in bone biology that sensitivities of bone cells become "saturated" with prolonged continuous (static) force application (see Chapter 9). The intermittent force described in this chapter is different to the *cyclic force* previously described. Intermittent force used by dentists is essentially a continuous force that allows long breaks (of several hours) between the loading regimens of continuous force. Cyclic force is usually loaded with much higher frequencies. However, it is important to keep in mind that the craniofacial skeleton forms through intramembranous ossification and in the regions of the sutures bone formation is caused by this process, and not by direct apposition of bone, as it is in the postcranial skeleton. These different bone-forming mechanisms can account for the apparent paradox.

Suture bone formation increases with higher levels of continuous expansion force. The rate of bone formation is highly correlated with suture separation but, because the skeletal resistance of the articulated bones limits the capacity for suture separation, the amount of both suture separation and bone formation reaches a plateau with even greater force. This occurs because DNA synthesis in fibroblasts and osteoblasts also plateau at higher forces, leading to less concomitant bone formation. With even higher force, fiber disruption and tissue damage occurs, along with evident cell death. Higher force also creates significantly wider suture gaps after expansion; these have a high relapse potential because stretched fibers tend to pull bones back to their original positions.

Treatment relapse often occurs after suture expansion. Without a proper retention protocol, treatment relapse can be as high as 45% of the amount of sutural separation. Higher expansion rates tend to produce higher relapse rates. At least 3–6 months of retention after rapid palatal expansion and 1–3 months after slow palatal expansion have been suggested to prevent potential relapse. Reducing relapse can also be achieved by accelerating the healing and bone formation processes within the gap. Clinical trials demonstrate that low-level lasers can facilitate suture opening of the midpalatal suture and speed bone regeneration toward the suture gap. Recombinant human BMP-2 (rhBMP-2) has also been used effectively in animal models to induce and accelerate bone formation while expanding sutures. Along with absorbable collagen sponge, rhBMP2 can be used as an alternative to bone grafts to augment bone in the floor of the maxillary sinus or repair large cranial defects. Although suture bone formation can be accelerated during suture expansion with very low concentrations of rhBMP-2, higher concentrations of rhBMP-2 can fuse the suture prematurely and prevent the suture from opening further.

Suture Compression

A suture can also respond to compression force to adapt to environmental changes. It is believed that compression force can induce bone resorption and reduce bone volume. In patients with a large dental overjet, clinicians sometimes instruct patients to wear headgear at least 12 h/day to compress the maxilla, with the assumption that the compression will displace the maxilla backward and bone resorption will occur adjacent to the sutures. However, most clinical studies agree that headgear can only inhibit maxillary growth and retract the upper dentition with intermittent heavy compression force. While some animal studies show bone displacement and histological bone resorption, others demonstrate that when a suture is compressed the adjacent bones become thickened rather than resorbed. Therefore, whether compression force

Phases of Treatment   **Factors introduced in each phase**

FIGURE 11.11 Diagram showing the three phases of dental implant treatment and the factors introduced in each phase that can affect bone-implant integration.

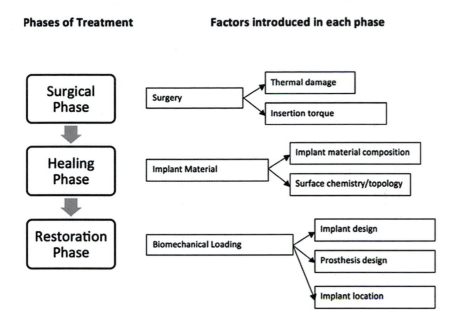

Mandibular Advancement

To treat patients with a large overjet, another treatment option is to position the mandible forward to stimulate condyle growth and increase mandibular length. In contrast to epiphyseal growth plates and synchondroses as growth centers, the condylar cartilage is considered as a specialized osteogenic growth site similar to a suture. When subjected to tension force, condyle cartilage allows an intramembranous bone formation response to mechanical stimulation. Using functional appliances to advance the mandible can accelerate condyle and mandibular growth to reduce overjet in the mixed dentition stage. While evidence suggests that increasing mandibular length is temporary and that the mature mandible size remains similar with or without mandibular advancement, reducing dental overjet still benefits patients because it improves their appearance and self-esteem. Similar to sutures, the condylar cartilage can also respond to compression force. However, reduction in cartilage thickness is transient, and using an appliance, such as a chin cup, to inhibit mandibular growth is not clinically effective.

IMPLANTS

Another important application of craniofacial bone biology is implant treatment. Since the late twentieth century, the development and application of dental implants have changed many treatment processes in clinical dentistry. From single tooth replacement to the restoration of a completely edentulous (absence of teeth) mouth, the use of dental implants has significantly improved potential treatment outcomes of restorative procedures. The success of the dental implant relies on an intimate integration between alveolar bone and the dental implant to provide a functional support to the occlusal loads placed on the implants. The integration between bone and implant is often considered in three phases: (1) surgical; (2) healing; and (3) restoration. In each phase, unique and important factors are introduced to the alveolar bone that can lead to the success or failure of integration between bone and implants (Fig. 11.11).

Surgical Phase

Implant placement starts with the surgical phase. In this phase, surgical wounds are created in both soft and hard tissue. Incisions are made in the oral mucosa using surgical blades and an osteotomy is created in the alveolar bone using a rotating drill under irrigation acting as a coolant. The osteotomy is first prepared using a small drill and is subsequently enlarged by sequentially larger drills until it reaches the size required for the implant. The implants are then inserted either by torquing or by tapping, depending on the implant design. The major impacts of this phase on alveolar bone are caused by (1) the thermal injury imposed during drilling and (2) the stresses applied to the bone-implant interface at insertion.

reduces bone volume remains controversial. The most significant treatment effects on overjet reduction are caused by dental movement rather than suture compression and bone resorption.

Thermal Damage

Heat associated with the drilling process can create a zone of devitalized bone about 1 mm or more around the implant. The devitalized bone will be resorbed and replaced by vital bone over time. However, excessive heat from drilling can significantly delay, or in extreme cases completely inhibit, the regenerative capacity of bone. The amount of heat generated during the osteotomy procedure is determined by the drilling speed,

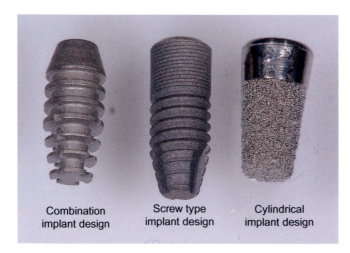

FIGURE 11.12 **Three basic types of implant body design.** Cylinder (Endopore by Sybron Dental Specialties; right), screw-type (OsseoSpeed by Astra Tech; middle), and combination (Bicon by Bicon Dental Implant; left).

sharpness of the drill, rate of irrigation, temperature of the irrigant, the size difference between subsequent drill bits, the size of osteotomy, and the bone density of the osteotomy site. A local temperature increase to just 40°C will start to cause bone cell death. At 50°C, bone practically loses all of its ability to regenerate, possibly due to cell death, decreased blood flow, and resorption of fat cells. The temperature during drilling can rise from the 37°C baseline to >41°C and will require 34–58 seconds for the temperature to return to baseline. Evidence has shown that overheating is one of the main reasons for early implant failure. To minimize thermonecrosis associated with the drilling procedure, it has been suggested that a minimum irrigation rate of 50 mL/min, a sharp drill, minimal size difference between subsequent drill bits, and a fast drilling speed should be used.

Stress on Insertion

Aside from excessive heat during osteotomy preparation, excessive stress placed on the bone at the time of insertion can cause cell death, bone necrosis, and early implant failure. The manner in which stress is applied to the alveolar bone during insertion is closely related to the design of the implant. Currently, there are three major types of implant designs: cylinder, screw, and combination design (Fig. 11.12).

The cylindrical design takes on the shape of a cylinder, with the surface of the implant usually coated with a rough material or sintered metal beads (Fig. 11.13).

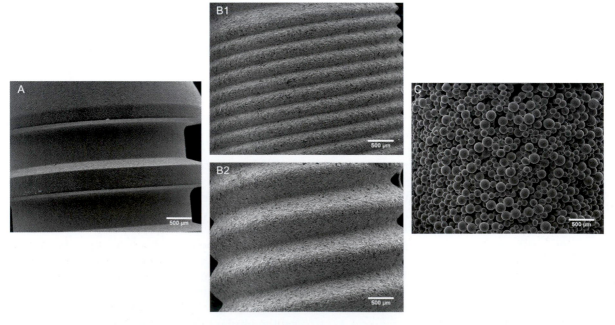

FIGURE 11.13 **Scanning electron microscope image showing the surface topology design of the three implants in Fig. 11.12 at 50×magnification.** (A) Bicon implant shows shelf-like macroretentive design. (B) OsseoSpeed implant shows screw-type macroretentive design with two screw-pitch levels: a microthread area toward the top of the implant (B1) and a macrothread area toward the bottom of the implant (B2). (C) Bicon implant shows macro-porous retentive design created by sintered beads.

These implants are inserted by pushing or gentle tapping into an osteotomy site. Screw-type implants show screw threads as the macroretentive elements. After the standard osteotomy preparation, tapping drills are sometimes used to cut grooves in the osteotomy site to receive the threads in the screw-type implant. Screw-type implants are inserted through a torqueing motion, but bone death can occur if excessive torque is used. Combination design implants have "shelves" along the long axis of the implant. Combination design implants are placed by tapping or pushing, as are cylindrical design implants. All three implant types place compressive stress on the bone-implant interface. Upon insertion, cylindrical implants in general apply a more even stress distribution along the implant surface, while screw-type implants induce stress concentration in the alveolar bone next to the screw threads.

Healing Phase

In this phase, the devitalized bones that result from drilling in the surgical phase are resorbed and replaced. This process continues for about 3–6 weeks after implant surgery, which represents the period when the bone-implant interface is the weakest. The resorbed bone is replaced by unorganized, less mineralized woven bone with eventual remodeling into lamellar bone. The mineral density around the implant will eventually reach about 70% of that of an intact bone by 4 months postimplant placement. During that time, a series of events occurs rapidly at the implant surface. Proteins, sugars, lipids, and mineral ions from serum absorb onto the implant surface. Proteins such as albumin, fibronectin, and immunoglobulin G (IgG) are found on the surface almost immediately. Blood clots forming around the implant bring fibrin, fibronectin, and hyaluronic acid, along with many platelet-derived growth factors, such as TGF-β and platelet-derived growth factor (PDGF). Through chemotaxis, mesenchymal cells migrate to the implant surface, interact with the absorbed species, and start producing extracellular matrix. This is followed by cell differentiation, matrix vesicle production, matrix vesicle maturation, and, finally, calcification on the implant surface. The two bone formation processes, one on the osteotomy surface and one on the implant surface, together establish the new bone-implant interface. The timing of these events and the strength of the bonding at the bone-implant interface are intimately related to the chemical composition and surface characteristics of the implant.

Influence of Implant Composition and Surface Characteristics on Healing

Current dental implants are predominantly made from either commercially pure (CP) titanium or Ti-6Al-4V titanium alloy. CP titanium is titanium with a purity level (98.9–99.6%), which contains trace amounts of other elements such as carbon, hydrogen, iron nitrogen, and oxygen. Ti-6Al-4V alloy is a titanium alloy containing 6 weight percent (wt%) of aluminum and 4 wt% of vanadium. These metals are used because of their biocompatibility, corrosion resistance, and favorable mechanical properties. Direct contact between bone and an implant made of these materials results in optimal integration.

For CP titanium, an oxide layer forms spontaneously on the surface when the metal is exposed to air; this produces a tightly adhered barrier to prevent further corrosion and oxidation. This tightly adhered oxide layer provides the required corrosion-resistance property for titanium and facilitates interactions between the implant and the biological system. In commercially produced CP titanium implants, a surface oxide layer of 3–6 nm naturally forms on the surface. Ti-6Al-4V alloy also naturally forms an oxide film on the surface when exposed to ambient atmosphere.

Oxide layer formation does not prevent the metal ions from leaching into the tissue adjacent to the implants, and higher concentrations of titanium ion are found in peri-implant tissue, lung, and lymph nodes after implant placement. In rare cases, the dissolved titanium ion can trigger allergic reactions in soft and/or hard tissues. Aluminum and vanadium ions from the Ti-6Al-4V can also have negative effects on bone cells.

Surface roughness and chemistry have a strong impact on the cellular and molecular events on the implant surface during the implant healing phase. They affect cell behavior on multiple fronts, including cell attachment, cell adhesion, cell proliferation, cell differentiation, matrix production, calcification, and local factor production. Current treatments used for modifying implants surfaces can be classified into subtractive and additive methods.

Subtractive surface treatments create surface roughness by removing surface materials through blasting, etching, grinding, or polishing. Blasting refers to the use of high-hardness particles to bombard the implant surface, thus creating pitted surface features. Commonly used particles include aluminum oxide and titanium oxide. Additive surface treatment creates surface roughness by adding materials to the surface through anodization, the sol-gel process, or thermal spraying. Anodization is an electrochemical process in which controlled oxidation occurs on the target surface to create thick, rough oxide layers. Combinations of treatment methods are often used in commercial dental implants (Fig. 11.14).

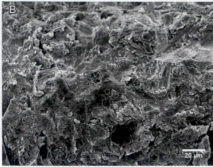

FIGURE 11.14 Scanning electron microscope image showing the surface texture from the three implants in Fig. 11.12. At 1000× magnification. (A) The surface of a Bicon implant is created by a proprietary grit-blasting and acid-etching process. (B) The OsseoSpeed surface is created by blasting the titanium surface with titanium dioxide particles, followed by a hydrofluoric acid etching procedure. (C) The Endopore surface is created by sintering Ti-6Al-4V particles in the size range of 44−15 μm to the solid titanium substrates.

Molecular and Cellular Events in Implant Healing

The surface layer is modified further when placed in a biological environment. Several physicochemical events can take place to modify the surface layer, including hydration of the oxide layer; dissolution and growth of the oxide layer; formation of electric charge and an oxide-fluid surface; interaction with inorganic ions; and interaction with biological molecules.

When the implant is exposed to an aqueous environment, the oxide layer becomes fully hydrated to form one of two kinds of hydroxide groups: a negatively charged acidic group ($[M-O]^-$) or a positively charged basic group ($[M-OH_2]^+$). At or close to physiologic pH, this hydroxylated surface carries a slight negative charge. The weakly charged surface is beneficial for osseointegration, since such a surface is less likely to induce an excessive material-protein interaction, which can potentially denature the protein.

The thickness of the oxide layer can increase in the biological environment found in certain implant locations. There is no significant change in the oxide layer on implants retrieved from the cortical bone. However, oxide layer thickness on samples retrieved from the bone marrow increase by threefold to fourfold. Several potential mechanisms have been proposed to explain this, including increased oxygen diffusion, formation of reactive oxygen metabolites, surface reaction with enzymatically produced peroxides, and assisted transport of titanium ions.

The oxide layer can absorb cations such as calcium (Ca^{2+}) and anions such as phosphate (PO_4^{3-}). When both Ca^{2+} and PO_4^{3-} are present in the oxide film, a thin layer of calcium phosphate precipitates in the form of amorphous or nanocrystalline apatite, or brushite (dicalcium phosphate dihydrate). This thin interfacial layer of calcium phosphate between the titanium oxide and bone cell is believed to serve as an important template on which new bone formation occurs.

Many studies have investigated the adsorption of single proteins (e.g. albumin, fibrinogen, and fibronectin) on titanium oxide films. When multiple proteins are present, the adsorbed protein can be displaced by another protein, a process called the *Vroman effect*. Initially adsorbed albumin can be displaced by IgG, which is in turn displaced by fibrinogen and then fibronectin. Fibronectin coating is known to enhance cell adhesion on biomaterial surfaces.

Cell attachment and adhesion to the implant surface is an important first step toward implant osseointegration. Many studies have demonstrated the effects of surface roughness on the attachment and adhesion of cells on titanium surfaces. Early studies showed that osteoblasts attach most tightly to surfaces with a roughness value (R_a) of about 1 μm. Surface roughness that stimulates differentiation usually inhibits their proliferation. Test implants with rough surfaces (profile height of 39.80 μm) created by blasting with alumina followed with plasma spray coating were found to stimulate higher levels of alkaline phosphatase activity, osteocalcin, prostaglandin E2 (PGE2), TGF-β1, and type I collagen production when compared to control implants without any treatment (profile height of 19.9 μm). Proliferation of the cells on the test implants, however, was significantly lower than in the control surface group. The same trend occurs on rough surfaces created by acid etching and sputter-deposited surfaces, where higher expression of osteocalcin and type I collagen are seen compared to on the smoother surfaces in the control group.

Tissue Level Events during Healing

Toward the end of the implant healing phase, a new bone-implant interface with sufficient bone-implant contact should form to provide a mechanical stability capable of bearing occlusal load. Many in vivo studies have evaluated the effects of implant

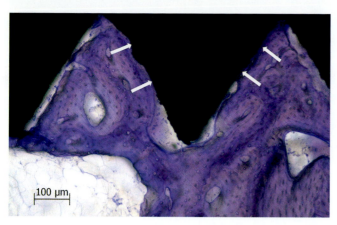

FIGURE 11.15 **Histology showing the bone-implant interface.** Following a period of healing, regions surrounding the implant can be processed for histology. The amount of contact between the implant and bone is measured by tracing the implant surface, and dividing it into sections that are or are not in contact with the bone. Section stained with toluidine blue; image taken at $10\times$ magnification. White arrows indicate areas of bone-implant contact.

surface treatments on the formation of the bone-implant interface. Two most commonly measured parameters are bone-implant contact (BIC) and removal torque (RTQ). In BIC measurements, a histology section showing a cross section of implant embedded in bone is evaluated microscopically (Fig. 11.15). BIC is calculated by measuring the length showing direct bone-implant contact along the implant threads and dividing it by the total length of the threads. In an RTQ measurement, a section of bone containing the implant is used. A torque gauge is first connected to the implant. The bone containing the implant is then subject to torsion in a direction opposite to the insertion direction (e.g. counterclockwise vs. clockwise) until the implant is loosened from the bone. RTQ is defined as the highest amount of torque resistance recorded during removal.

The effects of implant blasting on the quality of the formed bone-implant interface in the implant healing phase have been a main focus in dental implant research for many years. Blasted implants typically show higher RTQ and sometimes higher BIC compared to smooth-surface implants. The effects of acid etching on the in vivo performance of titanium implants are similar to those of blasting treatment. Many commercial implant manufacturers actually use combinations of blasting and etching treatments for their implants. Even rinsing and storage conditions can play significant roles in the in vivo performance of dental implants. Implants that are rinsed under nitrogen and then stored in saline show significantly higher BICs and RTQs when compared to implants with identical surface roughness that are rinsed without nitrogen protection and stored in air.

These results demonstrate the effects of surface treatments on the bone-implant interface properties (RTQ and BIC) in the implant healing phase. One can view this as a demonstration of the net positive effect of all cellular and molecular events that are influenced by surface roughness (with or without the associated chemical modification), surface cleaning, and calcium phosphate coating on the bone-implant interface.

Restoration Phase

In this phase, the prosthesis is attached to the implants that show successful osseointegration. The implant is subjected to occlusal loads through the force transmitted from the prosthesis during occlusion. The bone-implant interface formed in the implant healing phase is modified by this occlusal force. The mechanical loading transmitted to the bone-implant interface will induce further woven bone formation. This woven bone is called *reactive* woven bone and has similar structural and mechanical properties to the *repair* woven bone induced by surgical trauma. It is produced in the bone-implant interface to repair the microdamage in the alveolar bone created by the fatigue loading transmitted from the prosthesis to the implant-bone interface. Excessive force from occlusion can lead to bone resorption and thereby destabilize the bone-implant interface. A careful analysis of the stresses placed on the implant is therefore of paramount importance. Specifically, critical factors in determining the force transmitted to the bone-implant interface include (1) prosthesis design, (2) implant position, and (3) implant design. As the first two factors involve the theory of occlusion, which is beyond the scope of this chapter, we will only focus on implant design here.

Implant Design

The threads in screw-type dental implants are designed to maximize initial bone-implant contact in the implant healing phase and facilitate stress distribution in the restoration phase. Basic elements in a screw thread design include thread shape, thread pitch, and thread depth. After implant insertion, bone eventually grows to conform to the space between the threads. From this perspective, the thread design essentially determines the three-dimensional macroscopic contour of the bone-implant interface. The thread design therefore also determines how the axial occlusal load is modified and transmitted to the bone. There are basically four types of thread shapes: (1) V-shape, (2) buttress thread, (3) reverse buttress thread, and (4) square thread. Square design in general is better in modifying the axial occlusal load into a more favorable compressive load on alveolar bone. On the other hand, the axial occlusal load is modified into more shear load in the V-shape thread. This is important, since bones are stronger under compression and weaker in shear. Thread pitch is the distance between the two parallel threads on a screw. A smaller pitch translates to a higher number of threads and a larger surface area per unit distance. Higher implant surface area typically leads to higher bone-implant contact in the implant healing phase and better load transfer in the restoration phase. This is similar for pitch depth; the higher the pitch depth, the larger the surface of the implant becomes. Under loading, trabecular bone in the peri-implant region reorganizes to withstand the occlusal load. There is evidence to show that trabecular bone is aligned perpendicular to the threads after 18 months under occlusal loading.

Remodeling Around Implants

As mentioned earlier in the chapter, natural tooth connects to alveolar bone through the PDL. This thin layer of connective tissue functions as a "shock-absorber" and reduces the occlusal load transmitted to alveolar bone. Dental implants, on the other hand, connect to alveolar bone through a rigid bone-implant interface. Higher stresses transmitted to the alveolar bone can create microdamage in the peri-implant region. Long-term data indicate that bone-implant contact in the loaded implants remains relatively constant, in the range of 60% in humans. However, the bone turnover rate approaches 400% in the region adjacent to the implant, i.e. about 10 times higher than the 40% bone turnover rate in normal mandibular alveolar bone. It has been suggested that the high bone remodeling activity helps to prevent microdamage from accumulating in this high stress region.

In general, the survival rates for many current dental implant systems are high. Many studies show implant survival rates of 90% and above at follow-up periods of 5 years and longer. The general confidence toward the prognosis for dental implants is reflected in the suggestion that the success criteria for implant treatment should be set at an implant survival rate of 90% at 10 years. This is a testimony to the success in the development of dental implant systems.

STUDY QUESTIONS

1. Describe the biological response of alveolar bone during orthodontic tooth movement.
2. Describe biological response of peri-implant tissue during the dental implantation process.
3. Describe the importance of sutures in jaw development, and how can jaw growth be promoted by suture expansion?
4. Which clinical problems induce alveolar bone loss?
5. What are the main clinical factors that determine dental implant success?

Suggested Readings

Alan, R., 1985. Progress in Clinical and Biological Research, Vol. 187. Liss, Inc, New York, 259–267.

Björk, A., Skieller, V., 1972. Facial development and tooth eruption. An implant study at the age of puberty. Am. J. Orthod. 62, 339–383.

Boyan, B.D., Dean, D.D., Lohmann, C.H., Cochran, D.L., Sylvia, V.L., Schwartz, Z., 2001. The titanium-bone cell interface in vitro: the role of the surface in promoting osteointegration. In: Brunette, D. M., et al., (Eds.), Titanium in Medicine. Springer, New York, pp. 561–586.

Buschang, P.H., Martins, J., 1998. Childhood and adolescent changes of skeletal relationships. Angle Orthod. 68, 199–208.

Carlson, D.S., Buschang, P.H., 2012. Craniofacial growth and development: evidence-based perspectives. In: Garber, L.W. (Ed.), Orthodontics: Current Principles and Techniques, fifth ed. Elsevier Mosby, Philadelphia.

Davies, J.E., 2003. Understanding peri-implant endosseous healing. J. Dent. Educ. 67, 932–949.

Enlow, D.H., 1990. Handbook of Facial Growth. third ed. WB Saunders, Philadelphia.

Gasser, R.F., 1976. Early formation of the basicranium in man. In: Bosma, J.F. (Ed.), Symposium on Development of the Basicranium. DHEW Publication No. (NIH) 76-989, Washington D.C., pp. 29–43.

Liu, S.S., Kyung, H.M., Buschang, P.H., 2010. Continuous forces are more effective than intermittent forces in expanding sutures. Eur. J. Orthod. 32, 371–380.

Liu, S.S., Opperman, L.A., Kyung, H.M., Buschang, P.H., 2011. Is there an optimal force level for sutural expansion? Am. J. Orthod. Dentofacial. Orthop. 139, 446–455.

Misch, C.E., 2008. Contemporary Implant Dentistry. third ed. St. Louis, Mosby Elsevier.

Nanci, A., 2008. Ten Cate's Oral Histology: Development, Structure and Function. seventh ed. Mosby.

Proffit, W., 2013. In: Proffit, W.R., Fields, H.W., Sarver, D.M. (Eds.), "Chapter 2. Concepts of Growth and Development" Contemporary Orthodontics, fifth ed. Mosby Elsevier.

Solow, B., 1980. The dentoalveolar compensatory mechanism: background and clinical implications. Br. J. Orthod. 7, 145–161.

Van der Linden, F.P.G.M., 1985. Bone morphology and growth potential: a perspective of postnatal normal bone growth. In: Dixon, A.D., Sarnat, B.G. (Eds.), Normal and Abnormal Bone Growth: Basic and Clinical Research. Progress in Clinical and Biological Research, Vol 187. Alan R. Liss, Inc, New York.

Vignery, A., Baron, R., 1980. Dynamic histomorphometry of alveolar bone remodeling in the adult rat. Anat. Rec. 196, 191–200.

Whittaker, D.K., 1985. The effect of continuing eruption on the bone growth at the alveolar margin. In: Dixon, A.D., Sarnat, B.G. (Eds.), Normal and Abnormal Bone Growth: Basic and Clinical Research. Progress in Clinical and Biological Research, Vol. 187. Alan R. Liss, Inc, New York, pp. 259–267.

HORMONAL AND METABOLIC EFFECTS ON BONE

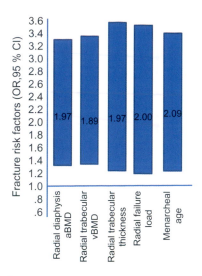

FIGURE 12.6 This Figure depicts the risk of fracture for a 1 SD decrease in radial bone mineral density (BMD), structural characteristics and strength variables of the distal radius in girls measured from 8 to 20 years of age. At 20 years of age, women with lower trabecular volumetric BMD (vBMD) and thickness of the distal radius were associated with reduced strength and high fracture risk. Columns represent odds ratio (OR) ± 95% confidence interval (CI), calculated by logistic regression. aBMD; areal bone mineral density. *Adapted from Chevalley et al., 2012. J Clin Endocrinol Metab, 97 (11):4174–4181.*

such as cortical thickness and cortical area in childhood are risk factors for both the risk of fracture and osteoporosis.

This developmental period between toddler and prepubertal years is one of the least studied periods. The rate of growth is not as rapid as that which occurs earlier or later during adolescence. Changes in bone size, geometry, and mineral accrual are directionally similar to the changes that occur during puberty, but they occur at a slower rate. Through prepuberty, bone growth and mineral accrual are faster in the legs and arms than in the spine. Gains in volumetric BMD are limited. Observational studies have shown positive associations between milk consumption and linear growth, and therefore height. Mean calcium retention is 161 mg/day at age 1–4 years on calcium intakes of 551 mg/day compared to calcium retention rates of 140–160 mg/day at age 7–8 years. Toddlers respond to gross motor exercise with increased tibial cortical cross-sectional area, but both calcium supplements and exercise result in increased BMD and cortical thickness, as well as bone size.

DEVELOPMENT OF PEAK BONE MASS

Peak bone mass assessed by BMD is an important determinant of adult bone health and is associated with an individual's fracture risk. Together with bone material properties and geometry, it determines bone strength. Peak bone mass occurs a few years after the closure of the epiphyseal growth plates and varies by sex and skeletal site (Fig. 12.1). Women typically attain peak bone mass at an earlier age than men; however, boys typically accrue a greater amount of mineral than girls, resulting in women having an overall lower ultimate peak bone mass. There is a tremendous amount of heterogeneity through the skeleton. When evaluating skeletal maturity it is crucial to examine both a child's age and stage of sexual maturity (see Box) as assessed by skeletal age and Tanner staging, respectively. The value for peak bone mass at about 30 years of age is used to derive T-scores for BMD, obtained by DXA for diagnosing osteoporosis. T-scores describe the difference in units of standard deviation within which an individual's BMD lies with respect to an average BMD in premenopausal women.

During the 2 years surrounding peak bone mineral accretion, approximately 26% of adult BMC is laid down. The higher rate of bone accrual in adolescent boys compared to girls is accomplished by a greater efficiency in calcium retention for any given calcium intake (Fig. 12.7). For both boys and girls, calcium retention increases with increasing calcium intake until a plateau is achieved. Additional calcium intake is excreted, so there is no further increase in calcium retention. Calcium retention achieves much higher rates than in prepuberty. If calcium intake increases from a typical intake of 800 mg/day to 1300 mg/day in an adolescent girl, then bone mass could increase by as much as 4% over the course of a year. In boys, serum insulin-like growth factor I (IGF-I) is as important as dietary calcium in regulating calcium retention. Serum IGF-I and male sex hormones are associated with periosteal bone expansion, resulting in greater bone size. In girls, estrogen suppresses periosteal bone expansion but also prevents remodeling from removing bone from the endocortical surface and widening the marrow cavity. Females have greater total trabecular bone area, which may facilitate mobilization of calcium and phosphate during pregnancy and lactation without compromising bone strength. Peak bone accrual rates occur at approximately 12.5 years in girls and approximately 14.1 years in boys (Fig. 12.8). Over the whole adolescent period, nearly half of peak bone mass is acquired.

Race is also an important determinant of peak bone mass. Asians have lower peak bone mass and blacks have higher peak bone mass on average, compared to whites. When BMC is adjusted for bone area or other measures of bone size, many racial differences disappear, showing the association of BMC with bone size. However, racial differences in cortical

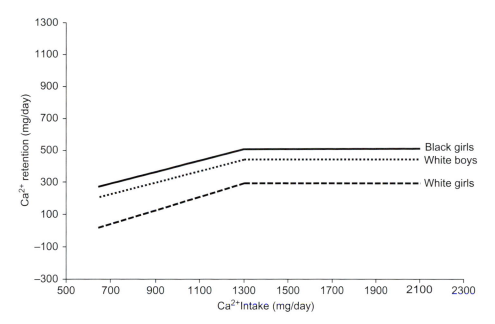

FIGURE 12.7 **Calcium is a threshold nutrient.** Calcium (Ca^{2+}) retention increases with intake up to a point of maximal retention, after which excess calcium is excreted. Higher rates of skeletal calcium accretion occurs in white boys (short dashed line) compared to white girls (long dashed line), and in black girls (solid line) compared to white girls during puberty.

and trabecular bone measures, including bone area and section modulus and sometimes BMC and BMD, persist when adjusted for size. Greater cross-sectional cortical bone dimensions confer greater bone strength, which helps explain sex and racial differences in vulnerability to fracture. Of course, lifestyle choices that differ by race and ethnicity can also influence peak bone mass. For example, Asians often have much lower calcium intakes than whites, but can accrete as much calcium as whites when calcium intakes are increased.

The attainment of peak height velocity precedes peak mineral accumulation by about 7 months (Fig. 12.8). At the age of peak height velocity, a girl has achieved approximately 90% of her final adult height, but only approximately 60% of her final total body BMC. The lag period due to an imbalance between bone size and mineral accumulation makes the skeleton more susceptible to fracture (Fig. 12.9). However, from a fracture perspective, impairments in the final size of a bone at the completion of growth are more indicative of future fracture risk than low bone mass. The normal heterogeneity in growth rates and timing of puberty present a challenge for clinical studies of bone outcomes because growth-related changes can overwhelm changes due to an intervention, along with difficulties in interpreting DXA scans in children. This can be minimized if subjects are matched for stage of sexual maturation, which can be achieved by categorizing using Tanner stages, or more accurately by determining peak height velocity (see box below). Although skeletal growth sites and timing are largely preprogrammed, lifestyle choices or interruptions of bone growth such as immobilization, loss of menstrual periods, or general illness can

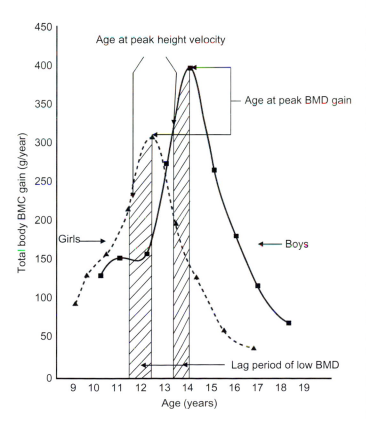

FIGURE 12.8 **Total body BMC gain in boys and girls from longitudinal DXA data.** Peak bone mineral content (BMC) gain in boys lags behind that in girls (14.0 vs. 12.5 years) and peak height velocity precedes peak BMC gain (11.8 years in girls vs. 13.4 years in boys). The lag period between peak height and BMC velocities of approximately 7 months represents a period of low bone mass for age. *Adapted from Bailey et al., 1999. J Bone Miner Res 14:1672–1679.*

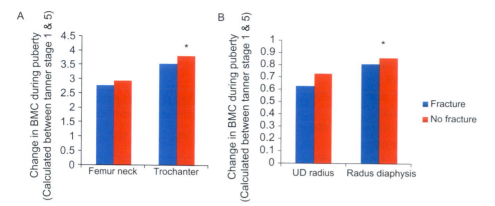

FIGURE 12.15 Fractures in childhood are linked to low childhood bone mineral content (BMC). Girls with a history of fracture gained less bone mass in the (A) trochanter and (B) ultradistal (UD) radius compared to girls with no fracture history. These data show changes in bone mass from prepuberty (Tanner stage 1) to pubertal maturity (Tanner stage 5). *, $P < 0.05$ versus fracture. *Adapted from Ferrari et al., 2006. Journal of Bone and Mineral Research 21(4): 501–507.*

At the same time there are no specific quantitative structural parameters [trabecular thickness (Tb.Th), spacing (Tb.Sp), and number (Tb.N)] assessed by pQCT that give a clear picture of a child's fracture risk. These standards may be developed at some point; the use of pQCT to evaluate bone structure has only recently been included as a method to evaluate bone tissue in both children and adults.

Nutritional Considerations

Adequate nutrition is a key building block for optimal bone health across all ages. Children who avoid dairy products and have suboptimal vitamin D levels have lower bone mass, smaller stature, and a higher incidence of fracture. Insufficient intake of calcium has been found to correlate with vBMD of the distal radius, a common site for childhood fractures. This is particularly important during the period of rapid growth surrounding puberty when there is a mineralization lag. Researchers report children who sustain multiple fractures have approximately 66% less calcium intake than their peers with no history of fracture.

Carbonated beverages and sweetened drinks have been associated with an increased fracture risk. One reason for this relationship is the displacement of milk and other dairy foods, the primary source of calcium and other nutrients needed for bone. The high acidic value and large amount of phosphoric acid (found in caffeinated beverages) and citric acid (found in noncaffeinated beverages) contained in carbonated beverages are linked to lower bone mass and a higher risk of fracture in children. There is conflicting data in the literature on the impact of caffeinated versus noncaffeinated soft drinks on bone modeling and remodeling. The consumption of adequate calcium has been shown to be protective against fractures despite an intake of carbonated beverages. However, children who avoid dairy products typically fail to choose alternative sources of calcium in sufficient amounts to meet their requirements.

The management of pediatric fractures should include dietary modifications to optimize calcium and vitamin D intake, along with engaging in moderately intensive physical activity. The literature is conflicting on whether adequate calcium intake combined with vitamin D or regular participation in weight-bearing physical activity is more important for optimizing bone health as a means to prevent fracture.

Race and Ethnic Differences

Children of African American descent are typically found to have higher bone mass and reduced fracture incidence than both white and Asian children. The basis for some of this is that black girls use calcium more efficiently than white girls given similar calcium intakes (Fig. 12.7). At similar calcium intakes, black girls excrete about half the amount of calcium in the urine as do white girls. They also have much higher rates of bone formation, but only slightly higher rates of bone resorption, than white girls. Interestingly, Asian adolescent girls have a greater efficiency of calcium utilization on low calcium intakes relative to white and black adolescent girls. The usually very low calcium intakes of Asian adolescents are probably responsible for their lower BMC. Investigators report that children of white descent have the highest risk for fracture when compared to blacks, Hispanics, and Asians. Fig. 12.16 depicts an example of fracture risk in children of European descent (whites) compared to nonwhites (including blacks, Hispanics, and Asians). The greatest number of fractures is typically found to occur during the pubertal years, a period of time when mineralization lags behind changes in bone size. Despite racial and ethnic differences, blacks share similar traits associated with increased fractures risk including increased adiposity and lower vitamin D intake.

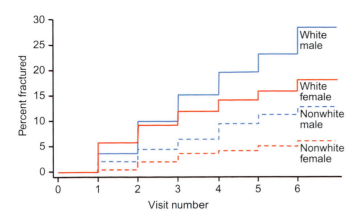

FIGURE 12.16 **Prevalence of fracture over 6 years in boys and girls from different racial backgrounds.** The Figure is plotted as percentage of fractures across 6 years (baseline mean age approximately 10 years) between white and nonwhite (Hispanic, black, and Asian) boys and girls. *Adapted from Wren et al. The Journal of Pediatrics, 161 (6): 1035–1040.*

The reasons for racial differences found in both children and adults are not clearly understood at this time. Differences in vitamin D metabolism, heritable differences in bone size, and daily activity levels may account for differences in fracture risk. It is important to note that although children of African American descent typically have greater bone mass than white children, African Americans who are vitamin D deficient actually have a higher incidence of fractures than those with higher vitamin D status. Thus, vitamin D deficiency increases an individual's risk for fracture despite ethnic origin. Ethnic differences in bone strength and structure may also play an important role in fracture risk. African Americans and Hispanics are found to have larger bones and greater bone strength at the distal radius compared to Caucasian children.

Role of Adiposity in Fracture Risk

Obese children have greater fracture risk compared to normal and overweight children. Repeated forearm fractures have been found in children with a combination of low bone mass and high body weight, along with a greater percentage of fat to lean mass. There are some reports of a protective effect of higher body mass on fracture risk. Children who have higher body weight with a greater percentage of lean mass sustain fewer fractures than do children with more adipose tissue. This points to the importance of lean mass as an important factor in fracture risk. Overweight and obese children typically have higher absolute bone mass (BMC) than their normal weight peers, but when corrected for bone size (aBMD) these children may in fact have lower relative bone mass

than normal weight peers. Children who are overweight may have a greater risk for fracture due to poor balance and coordination and reduced muscle strength when compared to normal weight children. Thus, fall prevention strategies are important for children who are overweight. In addition, ensuring adequate calcium and vitamin D intake combined with weight-bearing exercise, and reducing body weight are all important to maximize bone mass and structure to protect the skeleton from excessive loads during a fall.

The relationship between obesity and increased rate of forearm fractures during childhood is of concern, particularity given the increased prevalence of obesity worldwide. Approximately 16–19% of children between the ages of 2 and 19 years are classified as obese and current trends suggest this number will continue to increase. Childhood obesity is associated with an increased number of both upper and lower extremity injuries, and a greater number of injuries is associated with a higher number of fractures in this population. Increased body weight and BMI have been linked to a higher number of lower extremity injuries, including fractures of the foot, ankle, leg, and knee.

Children who are overweight or obese have an increased risk not only of fracture but also of a multitude of comorbidities such as diabetes, cardiovascular disease, and of reduced physical function in adulthood. These comorbidities also are linked to lower bone mass, with a higher chance of developing osteoporosis in adulthood.

STUDY QUESTIONS

1. Explain briefly how maternal diet may impact skeletal health during fetal growth, infancy, and childhood.
2. Describe how and why Ca^{2+} regulation and Ca^{2+} needs change during pregnancy and lactation.
3. Discuss how IGF-I and GH deficiency can affect skeletal health during growth.
4. Fracture risk depends on several factors. Describe how the incidence of fracture may vary across the following parameters:
   a. Age
   b. Sex
   c. Ethnicity
   d. Skeletal site
   e. Body weight
5. Explain the consequences of an inadequate Ca^{2+} diet during adolescence. How will this impact skeletal health later in life? If a rapid intervention occurs (i.e. Ca^{2+} supplementation or switching to a higher Ca^{2+} diet) explain the possible pathways in which catch-up growth can occur.

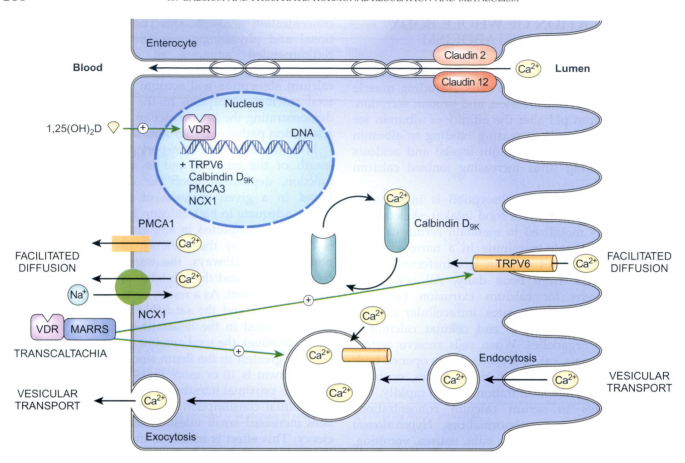

FIGURE 13.3 **Intestinal calcium absorption.** Passive paracellular transport following the concentration gradient involves claudin-2 and claudin-12. Active transport may occur through multiple mechanisms: facilitated diffusion, vesicular transport, and transcaltachia. Facilitated diffusion (middle right, middle left) uses the transient receptor potential cation channel subfamily V member 6 (TRPV6), calbindin-D_{9K}, basolateral transport via plasma membrane Ca^{2+} ATPase 1 (PMCA1), and the sodium-calcium exchanger 1 (NCX1). Gene transcription for expression of these proteins is upregulated by 1,25(OH)$_2$D. Transcaltachia (lower left) refers to the action of 1,25(OH)$_2$D on either a membrane form of the vitamin D_3 receptor (VDR) or membrane-associated, rapid-response steroid-binding protein (MARRS), triggering a rapid increase in transport through TRPV6 or through lysosomal cycling, independent of gene transcription. Vesicular transport (lower right, lower left) occurs through endocytosis or entry of cytoplasmic calcium into vesicles for transport and basolateral exocytosis.

their acceptance. The *vesicular transport model* predicts that cycling of calcium-containing lysosomes in the intestine is necessary for calcium absorption (as in Fig. 13.3). 1,25(OH)$_2$D increases lysosomal calcium in enterocytes, as well as increasing the number of lysosomes. Models of uptake into lysosomes after transport into the cell by TRPV6, and also of endocytosis, have been proposed. Transient receptor potential cation channel subfamily V member 5 (TRPV5) and TRPV6 may be present in some vesicular structures, which may facilitate transport, and the presence of calbindin-D_{28K} in chick enterocyte vesicles is reported. However, it is not yet clear whether this accumulation of calcium in vesicles is specific to regulation of transcellular calcium transport in mammals.

An alternative model is known as *transcaltachia* (Fig. 13.3). This is a model for a rapid, 1,25(OH)$_2$D-stimulated increase in calcium transport that does not require gene transcription. During transcaltachia, the process of transportation across the cell may still involve vesicular transport, while the effects of 1,25 (OH)$_2$D on molecules involved in the *facilitated diffusion model* require transcription. In ex vivo perfused chick intestine, exposure to 1,25(OH)$_2$D for 14 min dramatically increases calcium transport across enterocytes, detectable as increased calcium in the perfusate. Transcaltachia appears to be mediated by a basolateral membrane receptor: either by a unique role for the vitamin D_3 receptor (VDR) at the basolateral surface, by a novel membrane vitamin D receptor called the membrane-associated, rapid response steroid-binding protein (MARRS), or by PTH-PTHrP. However, MARRS knock-out mice do not have disrupted transcellular calcium absorption or whole body calcium metabolism.

A third alternative model involves regulated paracellular movement of calcium through tight junctions.

Production of the tight junction proteins claudin-2 and claudin-12 is increased in response to 1,25(OH)$_2$D and decreased in *Vdr*-null mice; thus, this may provide vitamin D-dependent regulation of paracellular calcium transport. However, vitamin D regulates active calcium absorption in the proximal small intestine, as opposed to the ileum, where claudin-2 and claudin-12 expression is highest. Finally, the voltage-dependent L-type calcium channel subunit alpha-1D (also known as voltage-gated calcium channel subunit alpha Cav1.3) may also contribute to intestinal calcium absorption. However, the gene for this protein is not regulated by vitamin D and neither calcium nor bone metabolism is strongly disrupted in Cav1.3 knockout mice.

Currently, no single model fully explains observations of intestinal calcium transport, and the timing of responses to 1,25(OH)$_2$D. Data supports both transcriptional and more rapid responses of calcium transport to 1,25(OH)$_2$D; these more rapid increases in calcium transport suggest that mechanisms such as vesicular transport may be important. Redundancy in this system may enable greater control and efficiency of calcium absorption, given that relative dietary deficiency is common. Further studies are necessary to delineate the relative contributions of these various mechanisms.

Renal Calcium Reabsorption

The basic functional unit structure of the kidney is the nephron. Blood is filtered at the first segment of the nephron, the glomerulus, and the filtered fluid and contents proceed along the course of the nephron: through the proximal convoluted tubule, the loop of Henle, and the distal convoluted tubule and connecting tubule. At multiple steps along the nephron, calcium reabsorption may occur (Fig. 13.4A). Ionized calcium (about 45% of the total plasma calcium) enters the glomerular filtrate. Most calcium (about 65%) is reabsorbed, passively and paracellularly, in the proximal convoluted tubule (Fig. 13.4B). In the proximal tubule, calcium reabsorption occurs due to *solvent drag*, where transport happens primarily due to calcium following the reabsorption of water rather than to the action of ion channels.

In the thick ascending limb of the loop of Henle, about 20−25% of the filtered calcium becomes reabsorbed. The thick ascending limb uses both passive and active transport as it contains claudin-16 (or paracellin-1), a protein expressed in tight junctions that is critical for calcium and magnesium transport, as well as ion channels (Fig. 13.4C). While a PTH-responsive sodium-calcium exchanger (NCX1) and a PMCA are present in the thick ascending limb (discussed in the distal convoluted tubule below), most of the calcium transport in this region is paracellular. Calcium reabsorption in this region is driven by an electrochemical gradient as part of nonselective cation reabsorption. On the apical (luminal) surface, both the ATP-sensitive inward rectifier potassium channel/renal outer medullary potassium channel (ROMK) and the solute carrier family 12 member 1, also known as the sodium-(potassium)-chloride cotransporter 2 (NKCC2), together with the basolateral sodium/potassium-transporting ATPase (NaK-ATPase), maintain a positive electrochemical potential in the lumen, which facilitates paracellular reabsorption of the cations calcium, magnesium and sodium.

On the basolateral (interstitial) side of the renal tubular epithelial cells in the thick ascending limb, the G protein-coupled extracellular calcium-sensing receptor (CaSR, encoded by the *CASR* gene), detects high extracellular calcium, increasing phospholipase C (PLC), and inhibiting the renal outer medullary potassium channels, which inhibits potassium transfer into the nephron lumen, and indirectly limits NKCC2 activity, which ultimately impairs generation of the lumen positive voltage potential. Thus activation of the CaSR limits paracellular calcium (and magnesium) transportation in this region of the nephron.

Active transcellular transport of calcium against an electrochemical gradient occurs in the late distal convoluted tubule and the connecting tubule of the nephron. Apically, calcium enters the cell through the TRPV5 channel, binds to calbindin-D$_{28K}$, and is transported on this protein to the basolateral membrane (Fig. 13.4D). Here, calcium is extruded from the cell via transporter proteins, including the sodium-calcium exchangers NCX1 and PMCA.

TRPV5 is sometimes termed the gatekeeper for distal tubular calcium reabsorption. It is a six-transmembrane domain protein that assembles into homotetramers to provide the channel for calcium reabsorption. TRPV6 in the intestine is highly homologous to TRPV5, consistent with similar biological functions in these two tissues. TRPV5 is regulated in a variety of ways, including glycosylation and protein kinase C (PKC)-mediated phosphorylation. TRPV5 phosphorylation is induced by PTH and inhibits internalization of TRPV5 from the membrane. This promotes calcium reabsorption. Free intracellular calcium also can inhibit TRPV5 activity. The receptor Klotho is expressed in the distal convoluted tubule and connecting tubule, as well as being secreted into the urine and into the circulation. Klotho cleaves oligosaccharide chains from TRPV5, increases membrane TRPV5 retention, and increases TRPV5 activity.

Once calcium enters renal tubular cells, its intracellular transport and basolateral export involve similar mechanisms to those necessary for intestinal calcium

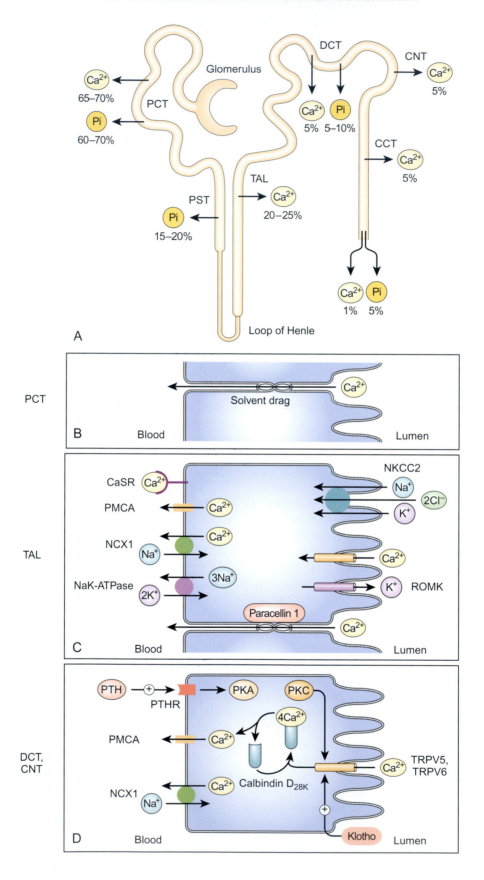

absorption. In the distal convoluted tubule and connecting tubule, calbindin-D_{28K} interacts with TRPV5 to increase its activity under low calcium conditions. Calbindin-D_{28K} also binds to calcium and shuttles the cation to the basolateral transport systems. Additional models of distal renal calcium transport indicate some lysosomes also facilitate calcium transport to the basolateral surface.

The final step in calcium reabsorption in the distal convoluted tubule and connecting tubule involves ATP-dependent processes that transport calcium against the electrochemical gradient. PMCA1b hydrolyzes ATP and transports calcium out of the cell with high affinity, with its highest activity in the distal convoluted tubule. The NaK-ATPase is also expressed at the basolateral membrane, where it functions to transport sodium extracellularly, a process that is also stimulated by Klotho. This facilitates the activity of the NCX1 protein, which transports three sodium ions intracellularly while extruding one calcium molecule. The bulk of the calcium transport in the connecting tubule is mediated by NCX1.

HORMONES CONTROLLING CALCIUM METABOLISM

Vitamin D and its Metabolites

Vitamin D plays a critical role in calcium metabolism. Unlike true vitamins that must be obtained in small quantities from dietary intake but cannot be synthesized by the body, vitamin D can be synthesized from cholesterol, and is therefore actually a steroid hormone. Vitamin D is metabolized into its most bioactive hormone form, $1,25(OH)_2D$, through a series of hydroxylation steps (Fig. 13.5).

Vitamin D is known as the "sunshine vitamin," since it can be produced from 7-dehydrocholesterol in skin exposed to ultraviolet B (UVB) light in the range of 295−300 nm. In the presence of adequate sunlight exposure, there is no dietary requirement for vitamin D. However, there are many conditions under which dermal vitamin D production is limited. UVB exposure decreases with distance in latitude from the equator. This effect is exaggerated during winter, when the angle of the sun changes and the atmosphere filters more UVB rays. Skin production of vitamin D is also reduced by increases in skin melanin; with aging, as the quality of skin changes and people are more likely to be homebound; and whenever the skin is covered by sunscreen, clothing, or shelter. Since these conditions are common, for many people vitamin D intake as a dietary nutrient is necessary to maintain adequate serum concentrations.

While all endogenously produced vitamin D is vitamin D_3, intestinally absorbed vitamin D can either consist of vitamin D_2 (ergocalciferol, from plant sources) or vitamin D_3 (cholecalciferol, from animal sources). In the gut lumen, dietary vitamin D is incorporated into fat-containing particles (micelles). Diseases that decrease fat absorption are, therefore, associated with poor vitamin D uptake from the gut. Vitamin D then passively diffuses across the tall, columnar enterocytes of the small intestine. Once absorbed, about 40% of circulating vitamin D is packaged into a different type of fatty particles (called *chylomicrons*). The remaining 60% circulates bound to vitamin D-binding protein (DBP).

Whether from endogenous production or exogenous intake, vitamin D is then transported through the circulation to the liver, where it is hydroxylated on its side-chain carbon-25 to form 25-hydroxyvitamin D [25(OH)D]. This hydroxylation is mediated by a cytochrome P450 enzyme family member commonly known as vitamin D 25-hydroxylase (cytochrome P450 2R1). Multiple cytochrome P450 enzymes can catalyze this reaction but the two most common are cytochrome P450 2R1 and mitochondrial sterol 26-hydroxylase/cytochrome P450 27A1. This step has some feedback

FIGURE 13.4 **Calcium and phosphate transportation in the nephron.** (A) Calcium (Ca^{2+}) and inorganic phosphate (Pi) are filtered from the blood at the glomerulus into the nephron luminal fluid. The filtered fluid proceeds along the nephron through the proximal convoluted tubule (PCT), the loop of Henle, up the thick ascending limb of the loop of Henle (TAL), through the distal convoluted tubule (DCT), the connecting tubule (CNT), and the cortical collecting duct (CCD) to join with the contents of other nephrons in the renal collecting system. Most of the filtered calcium and phosphate is reabsorbed at multiple nephron segments indicated by the arrows. Most of this reabsorption occurs in the proximal tubule. Phosphate transport is further described in Fig. 13.10. (B) After filtration at the glomerulus, 65−70% of calcium is reabsorbed paracellularly in the PCT following water reabsorption. (C) In the TAL, about 20−25% of the filtered calcium is reabsorbed. Paracellular transport occurs across tight junctions using paracellin 1. Active transcellular calcium transport involves multiple proteins. A positive electrochemical potential is maintained in the lumen via apical renal outer medullary potassium channels sodium-potassium-chloride cotransporters (NKCC2), and the basolateral sodium/potassium-transporting ATPase (NaK-ATPase), which facilitates paracellular reabsorption of calcium, magnesium, and sodium. In addition, the extracellular calcium-sensing receptor (CaSR) and parathyroid hormone (PTH) receptors in the TAL regulate the sodium-calcium exchanger (NCX1) and a plasma membrane calcium-transporting ATPase (PMCA) for transcellular transport of calcium. (D) In the DCT and CNT, transient receptor potential cation channel subfamily V member 5 (TRPV5) and TRPV6 channels transport calcium into the cell. TRPV5 is activated by Klotho. Calbindin-D_{28K} transports calcium to the basolateral membrane, to be transferred out of the cell via NCX1 and PMCA. PTH stimulates protein kinase A (PKA) and protein kinase C (PKC) to phosphorylate TRPV5 to increase uptake of calcium. PTH upregulates calbindin-D_{28K}.

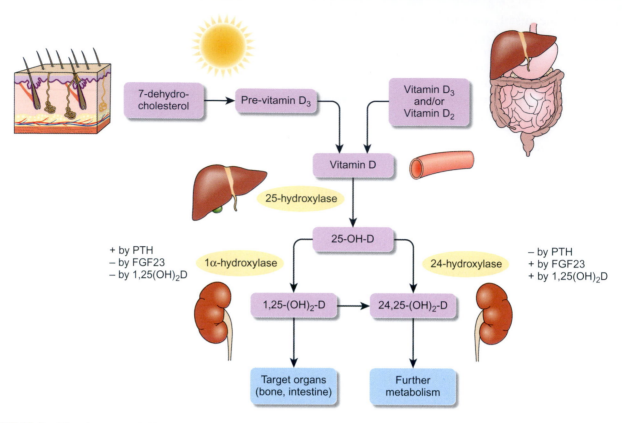

FIGURE 13.5 **Vitamin D metabolism.** Vitamin D can either be absorbed from dietary intake (D$_2$ or D$_3$) or produced endogenously (D$_3$). In the skin, 7-dehydrocholesterol is converted to Vitamin D$_3$ in steps catalyzed in part by ultraviolet light (UVB) and heat. Vitamin D is transported by vitamin D-binding protein. In the liver, 25-hydroxylase modifies it to the storage form 25-hydroxyvitamin D [25(OH)D]. In the kidney, 1α-hydroxylase activates 25(OH)D to the most potent form 1,25(OH)$_2$D, which then circulates to have actions in various tissues. Renal 1α-hydroxylase is upregulated by parathyroid hormone (PTH), and downregulated by both fibroblast growth factor 23 (FGF-23) and 1,25 (OH)$_2$D. 24-hydroxylase begins the process of degradation of 25(OH)D or 1,25(OH)$_2$D, producing 24,25(OH)$_2$D or 1,24,25(OH)$_3$D. 24-hydroxylase is downregulated by PTH and upregulated by FGF-23 and by 1,25(OH)$_2$D.

regulation, but in the setting of excessive vitamin D ingestion, sufficient conversion still occurs to cause vitamin D toxicity with high levels of 25(OH)D.

25-Hydroxyvitamin D Metabolism

25(OH)D is a very stable metabolite of vitamin D. Its serum concentration is used clinically to assess vitamin D status. 25(OH)D$_2$ (derived from vitamin D$_2$) has a lower affinity than 25(OH)D$_3$ for DBP, and therefore has a shorter half-life in circulation. Most assays measure 25(OH)D$_2$ and 25(OH)D$_3$ [hereafter referred to simply as 25(OH)D]. 25(OH)D has a biological half-life of 2–3 weeks. A recent report from the Institute of Medicine identified a serum 25(OH)D value of >50 nM (20 ng/mL) as adequate to support optimal bone and mineral metabolism. Many experts argue that higher concentrations (>80 mM or 32 ng/mL) may have beneficial effects (particularly extraskeletal effects), and the optimum level for health remains controversial. When 25(OH)D is measured in human populations, it is very common for a majority to be assessed as "deficient" if a cut-off of 32 ng/mL is used.

Like vitamin D, 25(OH)D is primarily transported through the serum by DBPs. After renal filtration, 25 (OH)D-binding protein complexes are actively reabsorbed by cell membrane receptor complexes consisting of cubilin and megalin in the proximal tubule. Once DBP binds to the receptor complex, receptor-mediated endocytosis occurs. Since this process is very efficient, little 25(OH)D is excreted in the urine. Several studies have shown that polymorphisms in the gene encoding DBP influence the serum concentration of 25(OH)D, probably by influencing 25(OH)D-binding affinity and renal vitamin losses.

1,25-Dihydroxyvitamin D Regulation and Metabolism

At normal physiologic concentrations, 25(OH)D has limited biological activity and must be hydroxylated at carbon-1 to form the potent hormone 1,25(OH)$_2$D.

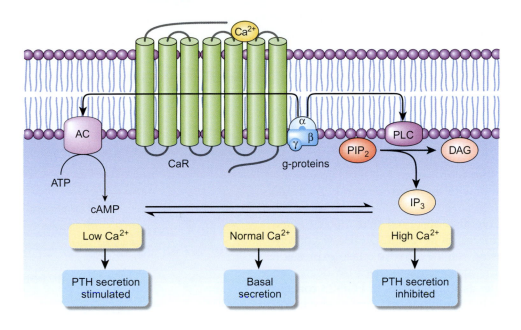

FIGURE 13.6 **Calcium-sensing receptor in the parathyroid chief cell.** The extracellular calcium-sensing receptor (CaSR) is a seven-transmembrane glycosylated protein that interacts with G proteins (α, β, and γ). Calcium interaction with this receptor activates phospholipase C (PLC), stimulating release of diacylglycerol (DAG) from phosphatidylinositol-4,5-bisphosphate (PIP2), thus producing inositol-1,4,5-trisphosphate (IP3). IP3 leads to release of calcium from the endoplasmic reticulum to the cytoplasm, and subsequent suppression of stored parathyroid hormone (PTH) secretion from secretory granules and of *PTH* gene expression. The CaSR also activates inhibitory G-protein signaling, and inhibits adenylate cyclase, reducing cAMP and suppressing PTH production. In contrast, decreases in extracellular calcium decrease signals through the CaSR, leading to increased cAMP and increased production and secretion of PTH.

This conversion is mediated by mitochondrial 25-hydroxyvitamin D-1 alpha hydroxylase/cytochrome p450 27B1 (commonly known as 1α-hydroxylase; encoded by *CYP27B1*). In contrast to the production of 25 (OH)D, the renal synthesis of 1,25(OH)$_2$D is highly regulated (largely by PTH and FGF-23) in order to control calcium and phosphate metabolism.

Decreases in serum calcium are sensed at the parathyroid gland through CaSR, leading to the increased production and release of PTH into the circulation (Fig. 13.6). PTH is a strong stimulator of *CYP27B1* expression. A feedback loop limits overproduction of 1,25(OH)$_2$D by 1,25(OH)$_2$D-mediated suppression of the *CYP27B1* and *PTH* genes.

Circulating 1,25(OH)$_2$D acts as an endocrine hormone that stimulates bone resorption, renal calcium reabsorption in the distal convoluted tubule, and active calcium and phosphorus absorption in the proximal small intestine. Concentrations of 1,25(OH)$_2$D can be measured clinically. However, since they primarily reflect acute changes in PTH signaling in response to circulating calcium concentrations, rather than overall vitamin D stores, 1,25(OH)$_2$D measurements are not useful for assessments of vitamin D status; instead, 25

(OH)D concentrations are used. Signaling by 1,25 (OH)$_2$D is attenuated within each target tissue by mitochondrial 1,25-dihydroxyvitamin D 24-hydroxylase (commonly known as 24-hydroxylase; encoded by *CYP24A1*), which hydroxylates 1,25(OH)$_2$D to the short-lived metabolite 1,24,25(OH)$_3$D (Fig. 13.5). Furthermore, 24-hydroxylase expression is induced by 1,25(OH)$_2$D. The 24-hydroxylase enzyme can also act upon 25(OH)D to form the metabolite 24,25(OH)$_2$D. Both 1,24,25(OH)$_3$D and 24,25(OH)$_2$D are substrates for other enzymes that lead to degradation and inactivation of 1,25(OH)$_2$D.

Although the classic model of vitamin D action is an endocrine system that relies solely upon the renal production of 1,25(OH)$_2$D, several studies indicate that tissues other than the kidney have the capacity to produce 1,25(OH)$_2$D. A variety of cells, including inflammatory cells, parathyroid cells, and osteoblasts express 1α-hydroxylase and this expression confers local ability to produce 1,25(OH)$_2$D.

Transcription of *CYP27B1* is suppressed by 1,25 (OH)$_2$D both at the kidney and at extrarenal sites. However, *CYP27B1* gene expression is strongly induced by PTH in the kidney, but not at nonrenal sites. This suggests that calcium homeostasis could be controlled by 1,25(OH)$_2$D at two levels: (1) as an endocrine system sensitive to PTH, which is critical for maintaining serum calcium concentrations at times of inadequate dietary calcium intake; and (2) as an autocrine or paracrine system sensitive to 25(OH)D concentrations [i.e. production of 1,25(OH)$_2$D is driven by the availability of 25(OH)D]. In this working model, the autocrine/paracrine function reduces the need for the endocrine system, a scenario that is supported by rat data showing that high serum 25(OH)D concentrations are associated with reduced serum 1,25(OH)$_2$D concentrations.

Investigators have also observed that dietary phosphate restriction increases serum 1,25(OH)$_2$D concentrations by increasing renal 1α-hydroxylase expression and activity. This is at least partly because decreased phosphate intake leads to decreases in FGF-23 concentrations, thus relaxing inhibition of renal 1α-hydroxylase by FGF-23.

Vitamin D Actions

Within the cells of vitamin D target tissues is a 1,25(OH)$_2$D receptor, VDR. VDR is a member of the steroid hormone receptor superfamily of ligand-activated transcription factors. Two lines of evidence demonstrate that VDR is crucial for calcium economy. Humans with recessively inherited inactivating mutations of the *VDR* gene have a phenotype that includes severe rickets and hypocalcemia. Deletions of the *Vdr* gene in mice lead to an identical phenotype. Interestingly, the phenotype of rickets can be reversed in mice by normalizing serum calcium through the administration of very high-calcium diets or by intestine-specific transgenic expression of VDR. Very high calcium intake is also necessary to treat hypocalcemia in patients with recessive *VDR* mutations. This demonstrates that one of the most important roles of 1,25(OH)$_2$D signaling through the VDR is the control of intestinal calcium absorption.

VDR is found in both the cytoplasm and nucleus of vitamin D target cells. Binding of 1,25(OH)$_2$D to VDR promotes an association between VDR and the retinoic acid receptor/retinoid X receptor (RXR). This heterodimerization is required for migration of the RXR-VDR-ligand complex from the cytoplasm to the nucleus. Once in the nucleus, the 1,25(OH)$_2$D-VDR-RXR complex regulates gene transcription by interacting with specific vitamin D response elements (VDREs) in vitamin D-responsive genes. These VDREs can be found within the promoter region, within introns, and far distal from the transcription start site of a gene. The steps leading to vitamin D-mediated gene transcription are summarized in Fig. 13.7. Multiple proteins involved in renal calcium transport are regulated by 1,25(OH)$_2$D,

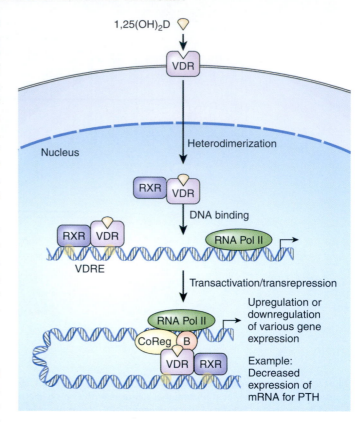

FIGURE 13.7 **The vitamin D receptor.** The vitamin D$_3$ receptor (VDR) is present in the cytoplasm and nucleus of vitamin D target cells. 1,25(OH)$_2$D binds to the VDR and facilitates the interaction of VDR with the retinoic acid receptor/retinoid X receptor (RXR). This heterodimer of VDR:RXR with bound vitamin D translocates to the nucleus. The 1,25(OH)$_2$D-VDR-RXR complex interacts with specific vitamin D response elements (VDREs) in vitamin D-responsive genes. Various coregulating proteins may also be recruited to this site. VDREs may be within the promoter region, within introns, or far distal from the transcription start site. VDR activation upregulates some genes and downregulates others. For example, *PTH* mRNA is downregulated by activation of VDR.

including calbindin-D$_{28K}$, TRPV5, and NCX1. Increased production of these proteins induced by the action of 1,25(OH)$_2$D results in increased renal calcium reabsorption.

1,25(OH)$_2$D increases gut calcium absorption by increasing the maximal capacity of saturable calcium transport, suggesting that it increases transporter production. Intestinal calcium absorption efficiency is reduced by >75% in vitamin D-deficient animals and in dialysis patients with compromised renal function and low circulating 1,25(OH)$_2$D concentrations. In patients with normal hepatic and renal function, the deficit in calcium absorption caused by vitamin D deficiency can be restored by giving either vitamin D (enterally) or 1,25(OH)$_2$D (enterally or parenterally). In patients with compromised 25-hydroxylase or 1α-hydroxylase activity, 1,25(OH)$_2$D must be used.

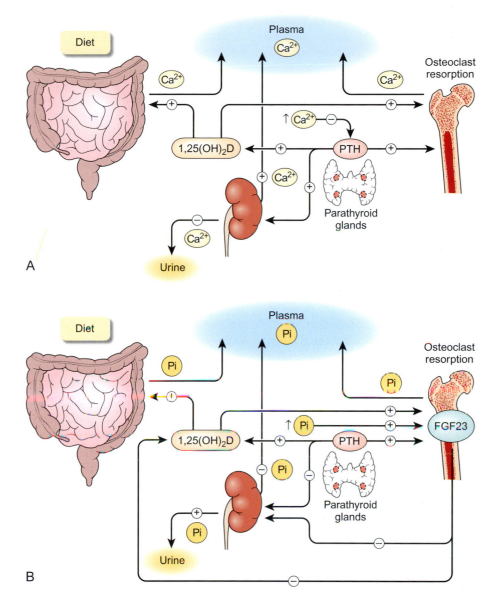

FIGURE 13.8 **Hormonal regulation of calcium and phosphate.** (A) The parathyroid gland senses extracellular calcium level and secretes parathyroid hormone (PTH) in response to low plasma calcium concentrations. High calcium suppresses PTH secretion. PTH stimulates resorption of calcium from the bone, as well as reabsorption of calcium in the kidney. PTH also stimulates production of 1,25(OH)$_2$D, which increases osteoclastic resorption of calcium from bone, as well as intestinal absorption of calcium. 1,25(OH)$_2$D also inhibits PTH production. (B) Fibroblast growth factor 23 (FGF-23) is produced in osteocytes. PTH and FGF-23 both inhibit renal phosphate reabsorption. FGF-23 also suppresses both renal 1,25(OH)$_2$D production and PTH. The net effect of FGF-23 is to decrease plasma phosphate concentration. 1,25(OH)$_2$D stimulates intestinal phosphate absorption. Both phosphate and 1,25(OH)$_2$D stimulate FGF-23 production.

Parathyroid Hormone and Parathyroid Hormone-Related Peptide

Active PTH is a short-lived 84-amino-acid polypeptide hormone secreted by the parathyroid glands. Its biological activity is conferred by the first 34 amino acids, which are highly conserved across species. PTH is derived from 115-amino-acid preproPTH made in the ribosomes. PreproPTH is converted to proPTH hormone in the endoplasmic reticulum after cleavage of a 25-amino-acid signal sequence. In the Golgi apparatus, an additional six amino acids are cleaved off, leaving the mature 1-84 polypeptide, which is packaged into secretory granules.

PTH plays a critical role in the regulation of blood calcium concentrations (Fig. 13.8A). PTH and ionized Ca^{2+} concentrations are inversely related to each other along a steep sigmoidal curve. PTH synthesis and secretion are stimulated when serum calcium concentrations fall. Conversely, increases in ionized Ca^{2+} concentrations signal at the CaSR and rapidly inhibit PTH secretion. PTH works to increase bone calcium release both by mobilizing calcium adsorbed to the bone surface and by stimulating osteoclastic resorption of bone mineral and matrix. In the kidney, PTH increases renal calcium reabsorption (to decrease renal losses) and stimulates $1,25(OH)_2D$ via renal CYP27B1 expression.

A second calcium-regulating hormone, PTHrP, is produced as a paracrine/autocrine factor by a wide variety of fetal and adult tissues. It has close homology with PTH at its N-terminus. PTHrP has multiple functions, the most important of which is regulation of endochondral bone formation and mineralization. Animals that lack PTHrP have major developmental abnormalities, including significant prenatal hypocalcemia and lethal skeletal dysplasia characterized by premature endochondral bone mineralization. PTHrP is also produced by the lactating breast, has a critical role in breast development, and is abundant in breast milk. During lactation, PTHrP promotes calcium mobilization from bone and regulates transport of calcium into breast milk.

Control of PTH Production and Release

Serum calcium, phosphorus, and $1,25(OH)_2D$ all influence parathyroid gland production and release of PTH. Changes in serum calcium concentrations are sensed by the CaSR (Fig. 13.6), a dimeric glycoprotein that belongs to the superfamily of G protein-coupled receptors. CaSR resides on the cell surface of many cell types including the chief cells of the parathyroid gland and the cells along the kidney tubules that are involved in mineral ion homeostasis. Upon binding of calcium to CaSR, several intracellular signaling pathways are activated (see Chapters 3 and 15). CaSR interacts with G proteins (α, β, and γ). Calcium interaction with this receptor activates PLC, stimulating release of diacylglycerol from phosphatidylinositol-4,5-bisphosphate (PIP2) and producing inositol-1,4,5-trisphosphate (IP3). IP3 triggers intracellular release of calcium from the endoplasmic reticulum, and the rise in intracellular calcium suppresses secretion of stored PTH in secretory granules. PTH gene expression is also suppressed. CaSR also activates inhibitory G proteins and inhibits adenylate cyclase, thus reducing cAMP and suppressing PTH production. In contrast, decreases in extracellular calcium decrease CaSR signaling, leading to increased cAMP and increased production and secretion of PTH.

Changes in serum calcium can alter three aspects of PTH biology in the parathyroid gland: secretion (with responses within seconds), intracellular degradation (with responses within 30 min), and gene expression (with responses within hours). In response to hypocalcemia, PTH secretion increases rapidly. Next, suppression of intracellular PTH degradation into C-terminal fragments leads to an increase in the amount of intact PTH(1−84) available for secretion. Finally, both transcriptional regulation of the PTH gene and stabilization of PTH mRNA lead to increased PTH synthesis. This is mediated in part by a decrease in protein binding to the 3′ untranslated region of PTH mRNA. When serum calcium is high, the CaSR signals and the parathyroid glands work in the opposite manner to decrease PTH production and secretion. During frank hypercalcemia, most secreted immunoreactive PTH is present as biologically inactive C-terminal fragments.

Serum phosphorus levels affect binding of cytosolic proteins to the PTH mRNA transcript. In conditions of hypophosphatemia, there is less binding and PTH mRNA is more rapidly degraded; in hyperphosphatemia, there is an increase in binding of proteins to the 3′ untranslated region of the PTH mRNA and increased transcript stability. However, the effect of a high-phosphorus diet to increase PTH may also be partly mediated by slight decreases in serum calcium that occur following dietary phosphorus intake. The ability of high serum phosphate concentrations to increase PTH concentrations suggests that high dietary phosphate may promote bone loss by increasing PTH-mediated bone resorption. However, short-term consumption of a daily diet containing 2000 mg of elemental phosphate does not increase urinary markers of bone resorption in young men.

As described in the section on vitamin D, PTH secretion leads to activation of the renal 1α-hydroxylase and increases the production of $1,25(OH)_2D$. In a feedback mechanism, $1,25(OH)_2D$ reduces PTH concentrations in two main ways: first, the influence of $1,25(OH)_2D$ on intestinal calcium absorption leads to increases in serum calcium, which is sensed by the CaSR; and second, $1,25(OH)_2D$ directly influences the parathyroid gland. Thus, $1,25(OH)_2D$ reduces PTH secretion, reduces transcription of the PTH and CASR genes, and regulates the rate of parathyroid cell proliferation. The effects of $1,25(OH)_2D$ on PTH requires binding of the $1,25(OH)_2D$-VDR-RXR complex to two VDREs in the promoter of the PTH gene. It is likely that these interactions recruit corepressors with histone deacetylase activity that maintain the PTH gene in a transcriptionally repressed state.

PTH activates both cellular and paracellular pathways of calcium reabsorption in the thick ascending limb of the loop of Henle. In addition, PTH also stimulates mechanisms controlling calcium reabsorption in the distal convoluted tubule. PTH stimulates expression of calcium transporters, calbindin-D_{28K}, NCX1, PMCA1b, and TRPV5. PTH-activated calcium uptake in the distal tubule depends upon the activation of both PKC and protein kinase A (PKA) and requires voltage-gated calcium channel family members. PTH stimulates PKC phosphorylation of TRPV5, resulting in increased cellular uptake of calcium. However, pharmacologic inhibition of calcium channel transport (through TRPV5) decreases the expression of calbindin-D_{28K} and NCX1, and blocks PTH-induced stimulation of the encoding genes, suggesting that the effect of PTH on these genes is at least partly mediated by cellular calcium uptake from the renal tubule lumen.

Estrogen

The primary estrogen in humans is estradiol. Estradiol and other estrogens bind to estrogen receptors (primarily ERα) in bone and the kidney. These receptors are ligand-activated transcription factors that work through classical *genomic* pathways to regulate gene transcription, as well as *nongenomic* mechanisms that produce cellular responses. Subsequently, they affect bone by mediating changes in expression of multiple regulatory factors. For example, estrogen increases osteoblast production and the release of osteoprotegerin, suppresses the membrane form of macrophage colony-stimulating factor 1 (M-CSF), and suppresses cytokine production by osteoblasts, monocytes, and T cells that normally promote osteoclastogenesis [e.g. interleukin-1 (IL-1) and tumor necrosis factor (TNF-α)]. Estrogen also mediates the impact of mechanical strain on bone, with low estrogen concentrations resulting in loss of mechanotransduction and accelerated bone loss.

Estrogen is thought to affect calcium homeostasis through the intestine and kidney in several ways. First, when estrogen levels drop during the early phase of menopause, bone remodeling increases, leading to increased calcium release from bone. This leads to PTH suppression, and a subsequent reduction in renal production of 1,25(OH)$_2$D. By affecting PTH and 1,25(OH)$_2$D, estrogen deficiency indirectly decreases the efficiency of intestinal calcium absorption and increases renal calcium excretion. This results in net urinary calcium loss at the expense of bone. Second, once the system has later adapted to the loss of estrogen and bone loss has slowed, the previous decrease in calcium intestinal absorption efficiency coupled with enhanced renal calcium loss results in subsequent increases in PTH.

Estrogen deficiency may also disrupt intestinal vitamin D signaling. For example, oophorectomy reduces both basal and 1,25(OH)$_2$D-induced intestinal calcium absorption in young women. This can be reversed by estrogen repletion. Impaired response to vitamin D during estrogen deficiency is related to reduced expression of VDR, which is regulated in part by estrogen.

Estrogen may also directly regulate other genes encoding proteins involved in calcium absorption and excretion. Intestinal *TRPV6* mRNA levels are lower in *Esr1* (encoding ERα)-null mice, and pharmacologic treatment with estradiol increases duodenal *TRPV6* mRNA in both normal and *Vdr*-null mice, indicating that vitamin D-independent mechanisms exist. Estrogen administration increases expression of calbindin-D_{28K}, NCX1, PMCA1b, and TRPV5, promoting calcium reabsorption in the kidney; deficiency of estrogen thus contributes to increased urinary calcium losses.

Calcitonin

Calcitonin is primarily made by the parafollicular cells (also called C cells) of the thyroid gland, although some is made in mammary and placental tissues. Calcitonin acts to decrease calcium concentrations and may play a role in regulating serum magnesium. Serum calcitonin concentrations are higher in fetuses than in adults. However, calcitonin does not appear to play a large role in calcium homeostasis in adult humans, as neither excess nor deficiency causes calcium disorders. For example, calcium concentrations do not change appreciably after thyroidectomies unless the parathyroid glands are also damaged or removed. Calcitonin secretion is regulated by binding of serum calcium to CaSRs on C cells. High serum calcium concentrations lead to CaSR-mediated cellular depolarization, activation of voltage-dependent calcium channels, and calcitonin secretion. Calcitonin may play a role in limiting maternal skeleton resorption in pregnancy. Mechanistically, calcitonin inhibits osteoclast activity by binding to a cell surface receptor on osteoclasts. When bone turnover is very high, calcitonin-mediated suppression of osteoclast activity can lead to hypocalcemia, and this phenomenon is utilized in the treatment of hypercalcemia with calcitonin. After calcitonin-induced hypocalcemia, compensatory activation of the PTH-vitamin D axis occurs to normalize serum calcium.

Somatotropin/Growth Hormone

GH and IGF-1 are important during growth; they are critical for both attainment of adult bone size and accrual of peak bone mass. IGF-1 is a peptide hormone, whose production is regulated in part by GH and which serves as a physiologic paracrine mediator

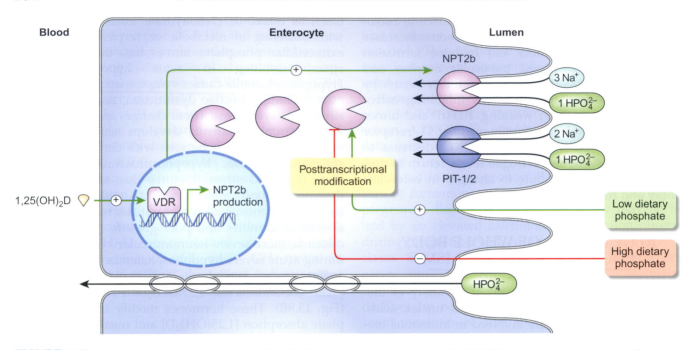

FIGURE 13.9 **Phosphate absorption in the intestine.** Sodium-phosphate cotransporter 2b (NPT2b) is the major transporter in the intestine. 1,25(OH)$_2$D binding to the vitamin D receptor (VDR) leads to increased expression of NPT2b, increasing absorption. Dietary phosphate content results in posttranscriptional modification altering NPT2b protein expression at the brush border, independent of transcription and 1,25 (OH)$_2$D. PIT-1 and PIT-2 are ubiquitously expressed phosphate transporters but these are not regulated. Passive paracellular absorption also occurs.

phosphate most efficiently in the ileum (distal small intestine), while rats and humans absorb phosphate more efficiently in the proximal segments of the small intestine (duodenum and jejunum). In the intestinal epithelium, multiple phosphate transporters are expressed. However, NPT2b accounts for 90% of active phosphate transport, and about half of total phosphate transport (Fig. 13.9). NPT2b in the intestine, and the related transporters NPT2a and NPT2c in the kidney, are membrane glycoproteins with eight membrane-spanning domains and long intracellular N-terminal and C-terminal domains. NPT2b also has roles in other tissues, which are beyond the scope of this chapter, including liver, lung, mammary gland, and testis. NPT2b transports sodium and phosphate in a 3:1 molar ratio (3Na$^+$:1HPO$_4^{2-}$), consequently generating an intestinal lumen negative voltage potential. At high dietary levels of phosphate, NPT2b internalization occurs to limit phosphate absorption. NPT2b apical surface expression can be downregulated by endocytosis, and this process is regulated by PKC-mediated phosphorylation. Mechanisms for transport of phosphate across the intestinal epithelial cell and for extrusion on the basolateral side are not currently understood.

Periods of low- or high-phosphate dietary intake alter expression of both 1,25(OH)$_2$D and NPT2b to upregulate or downregulate phosphate absorption, respectively. As with calcium, restricting dietary phosphate stimulates processes to improve absorption efficiency. Feeding pigs a 0.25% phosphate diet increases phosphate absorption by 90% compared to a 0.4% phosphate diet.

Adaptation to the phosphate content of diet is mediated by changes in *SLC34A2*/NPT2b mRNA and protein levels. However, the mechanisms for this type of NPT2b regulation are unclear. Phosphate restriction decreases FGF-23 (see below) and increases renal 1α-hydroxylase activity, leading to higher circulating 1,25(OH)$_2$D concentrations. Both vitamin D deficiency and hypoparathyroidism-associated deficits of 1,25 (OH)$_2$D impair phosphate absorption. Treatment with 1,25(OH)$_2$D doubles the maximal capacity of the active (saturable) phosphate transport but does not change the passive nonsaturable component. This effect is demonstrated by treating vitamin D-deficient mice with 1,25(OH)$_2$D, which increases surface expression of NPT2b protein in the intestinal brush border. This suggests that 1,25(OH)$_2$D directly regulates the *SLC34A2* gene. However, *SLC34A2* does not contain a classical VDRE. Additionally, as mice age, they lose the ability to respond to 1,25(OH)$_2$D by increasing *SLC34A2* mRNA expression. However, adult mice do increase intestinal phosphate transport in response to 1,25(OH)$_2$D, suggesting that this effect is not mediated by new gene transcription, but may be due to enhanced translation or redistribution of existing transporter proteins to the apical membrane.

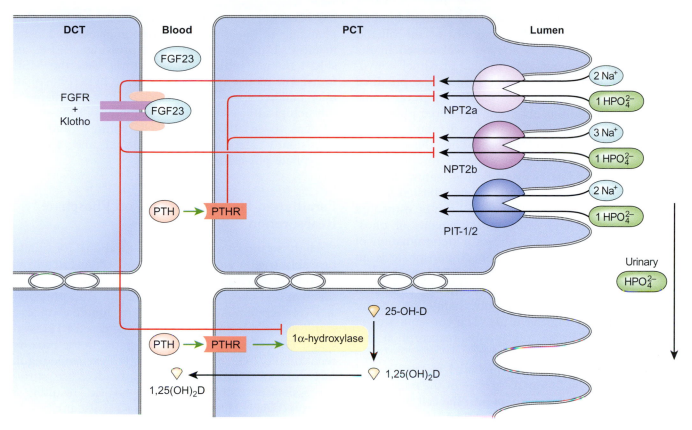

FIGURE 13.10 **Phosphate reabsorption in the proximal convoluted tubule.** Phosphate is reabsorbed primarily in the proximal convoluted tubule (PCT) by the action of sodium-dependent phosphate transport protein 2A (NPT2a) and NPT2c. Parathyroid hormone (PTH) action at the PTH receptor (PTHR) inhibits apical expression of NPT2a and NPT2c, and stimulates expression of 1α-hydroxylase to increase 1,25(OH)$_2$D. Fibroblast growth factor 23 (FGF-23) also inhibits the apical expression of NPT2a and NPT2c, but inhibits expression of 1α-hydroxylase to decrease 1,25(OH)$_2$D. FGF-23 signals through FGF receptors (FGFR) type 1, 3, and 4, but requires Klotho as a cofactor. There is controversy regarding the location of action for FGF-23 receptors. Klotho is primarily expressed in the distal convoluted tubule (DCT), and some studies indicate that FGF-23 signaling begins in the distal tubule. Other investigators report direct signaling in the PCT. Basolateral phosphate transport is not well understood.

Furthermore, independently of 1,25(OH)$_2$D, a low-phosphate diet can upregulate brush-border NPT2b and phosphate transport; this was confirmed in the *Vdr*-null mouse. In these mice, intestinal phosphate absorption is reduced by 30−70% compared to wild-type mice. *Vdr*-null mice have a reduction in NPT2b protein but not in mRNA, which may indicate that 1,25(OH)$_2$D requires the classical VDR for regulation of phosphate transport, but that this effect is not primarily mediated by *SLC34A2* gene transcription. Interestingly, even in the absence of 1α-hydroxylase, or with vitamin D deficiency, there is still an increase in intestinal phosphate absorption efficiency after dietary phosphate restriction. Collectively, this demonstrates that while vitamin D signaling through the VDR is important for normal intestinal phosphate absorption, phosphate restriction stimulates both vitamin D-dependent and vitamin D-independent mechanisms to increase phosphate absorption.

Multiple additional factors may also interfere with phosphate absorption, most importantly the ingestion of divalent cations such as calcium, magnesium, and aluminum which are used as antacids or specifically as phosphate binders during treatment of hyperphosphatemia. Other phosphate-binding agents include sevelamer, used especially to limit phosphate absorption from the diet in the setting of impaired excretion due to chronic kidney disease.

Dietary phosphate content does alter FGF-23 expression. While some studies have indicated that injections of FGF-23 may inhibit intestinal phosphate absorption in mice, this effect may be partially confounded by changes in 1,25(OH)$_2$D. Notably, FGF-23 does not suppress jejunal phosphate absorption in *Vdr*-null mice. Thus, an FGF-23 contribution to intestinal phosphate absorption is probably mediated through the effects of FGF-23 on renal 1α-hydroxylase activity.

Renal Phosphate Reabsorption

Regulation of renal phosphate handling is the greatest contributor to overall phosphate metabolism. About 90% of plasma phosphate is freely filtered at the renal glomerulus, limited in part by protein binding and calcium complexes. The majority of the filtered phosphate (about 85%) is actively reabsorbed in the proximal tubule, about 70% in the proximal convoluted tubule section, and about 15% in the proximal straight tubule section prior to the loop of Henle (Fig. 13.4A). In addition about 5% of filtered phosphate is reabsorbed at the distal convoluted tubule. Due to the greater percentage of reabsorption, the proximal tubule is the most important segment for phosphate regulation. Phosphate reabsorption is saturable, and the transport maximum for phosphate adjusted for the glomerular filtration rate (TmP/GFR) can be estimated by fasting serum and urine measurements of phosphate and creatinine. This allows a determination of whether the percent reabsorption of phosphate is an appropriate response to the serum phosphate level. Dietary changes and some diseases alter the TmP/GFR to appropriately or inappropriately change the amount of phosphate entering the urine. The consequences of abnormalities in renal phosphate reabsorption are more fully discussed in Chapter 16.

Phosphate reabsorption is mediated by NPT2a and NPT2c at the apical (luminal) surface of the renal tubular cell. Transport of phosphate into the cell by NPT2a is electroneutral ($2Na^+:1PO_4^{2-}$), while NPT2c transport is electrogenic ($3Na^+:1PO_4^{2-}$). This is a saturable transporter system, which is regulated according to states of dietary phosphate intake. The ubiquitous sodium-dependent phosphate transporters (PiT-1 and PiT-2) are also present, although these are not regulated to control total body phosphate balance. Once in the renal tubular cell, phosphate must be transported across the basolateral side, although this process remains poorly understood. Evidence suggests that basolateral transport involves a phosphate-specific transporter that is independent of sodium transport but is potentiated when there is a high intracellular phosphate concentration, and is facilitated by an intra- to extracellular electrical potential.

Both PTH and FGF-23 regulate proximal renal phosphate reabsorption, while they have reciprocal effects on 1,25(OH)$_2$D metabolism. However, PTH has a primary role in regulating serum calcium, while FGF-23 is a more important regulator of serum phosphate. Other so-called *phosphatonins* have been proposed, including FGF-7, matrix extracellular phosphoglycoprotein (MEPE), and secreted frizzled-related protein 4 (sFRP-4), many of which induce renal phosphaturia when injected. However, of these, PTH, FGF-23, and FGF-7 are the only ones clearly associated with clinical disorders of hypophosphatemia. Both MEPE and sFRP4 are expressed along with FGF-23 in tumors, causing hypophosphatemic osteomalacia, and FGF-7

has been implicated as the mechanism for hypophosphatemia in one case report of tumor-induced osteomalacia. Data on mice overexpressing MEPE and sFRP4 do not indicate phosphaturia, and the MEPE transgenic mouse is actually prone to hyperphosphatemia, making these two molecules less promising as physiologically relevant phosphatonins. Of these, FGF-23 has been the most well studied and most consistently associated with normal and abnormal phosphate physiology in humans.

Parathyroid Hormone Effects on Phosphate Reabsorption

High serum phosphate concentrations (or intake) probably trigger small decreases in serum calcium via binding of calcium and phosphate. Furthermore, high doses of oral phosphate impair calcium absorption, and as PTH secretion is very sensitive to minute changes in extracellular calcium, this triggers increased PTH secretion. PTH acts at PTH/PTHrP receptors in the proximal renal tubule, interacting with Na^+/H^+ exchange regulatory cofactor NHERF1 and PLC to decrease surface expression of both NPT2a and NPT2c, resulting in decreased phosphate reabsorption. PTH also inhibits distal tubular phosphate reabsorption, although this represents a much smaller proportion of phosphate transport. PTH stimulates expression of renal 1α-hydroxylase resulting in increased 1,25(OH)$_2$D, while inhibiting expression of 24-hydroxylase, which is the enzyme responsible for converting vitamin D to its less active forms. Thus, PTH effects on the kidney indirectly stimulate intestinal calcium and phosphate absorption. However, the net effect of PTH on phosphate is to lower serum phosphate.

Fibroblast Growth Factor 23 is an Important Hormone in Phosphate Reabsorption and Vitamin D Regulation

FGF-23 is a glycosylated peptide hormone produced primarily in osteocytes. Under normal physiologic conditions, some of the FGF-23 produced is cleaved into inactive fragments before secretion into the circulation. FGF-23 glycosylation by polypeptide N-acetylgalactosaminyltransferase 3 (GalNac-T3) decreases its susceptibility to cleavage and is necessary for adequate secretion of intact FGF-23. Mutations in *FGF23* that impair this cleavage cause autosomal dominant hypophosphatemic rickets due to FGF-23 excess. Deficiency of either phosphate-regulating gene with homologies to endopeptidases on the X chromosome (PHEX) or dentin matrix acidic phosphoprotein

1 (DMP-1) also results in increased FGF-23 expression and consequent syndromes of excess FGF-23 biological activity (e.g. X-linked hypophosphatemia and autosomal recessive hypophosphatemia). The mechanisms by which these two proteins affect FGF-23 expression are not completely understood. Emerging evidence indicates that subtilisin-like protein convertase 2 (PC2) enzyme activity is decreased in the presence of PHEX deficiency due to decreased amounts of a chaperone protein necessary for PC2 activity. In vitro studies show that lack of PC2 activity increases FGF-23 expression and protein levels.

Intact FGF-23 interacts with FGFR-1, FGFR-3, and FGFR-4, requiring klotho as a critical coreceptor. There is controversy regarding the location of action for FGF-23 receptors. Klotho is primarily expressed in the distal convoluted tubule, and some studies indicate that FGF-23 signaling begins in the distal tubule. In support of this, distal tubule Klotho knockout mice have hyper-phosphatemia. However, the canonical effects of FGF-23 occur in the proximal renal tubule (Fig. 13.10). The mechanism explaining this discrepancy is uncertain, but probably occurs through an unknown signal transferred from the distal renal tubule to the proximal tubule cells. However, there is conflicting information about whether or not Klotho is active in the proximal tubule and some investigators have reported Klotho expression and direct FGF-23 signaling in the proximal convoluted tubule.

Using multiple methodologies, FGF-23 has been demonstrated to trigger intracellular signaling via MAPK and extracellular signal-regulated kinase (ERK) phosphorylation. FGF-23 inhibits surface expression of NPT2a and NPT2c in the proximal renal tubule, similarly to PTH, leading to decreased phosphate reabsorption. However, contrary to the effect of PTH on $1,25(OH)_2D$ metabolism, FGF-23 inhibits proximal tubule 1α-hydroxylase expression and stimulates 24-hydroxylase expression. Thus, FGF-23 decreases both serum phosphate and $1,25(OH)_2D$ concentrations.

Diseases resulting in increased FGF-23 production result in hypophosphatemia due to excessive renal phosphate losses and cause osteomalacia and rickets. Conditions of decreased FGF-23 function, either due to increased cleavage of FGF-23 into inactive fragments [due to mutations in *GALNT3* (encoding polypeptide N-acetylgalactosaminyltransferase 3 (GalNac-T3) or *FGF23*], or insufficient FGF-23 activity (due to mutations in the coreceptor *Klotho* or to kidney disease) cause hyperphosphatemia and result in ectopic soft tissue and vascular calcifications.

FGF-23 and PTH also interact. The parathyroid glands express klotho, and cell culture data indicate that FGF-23 treatment suppresses PTH production. This process is blocked by a MAPK inhibitor. In the setting of chronic kidney disease, parathyroid gland klotho expression decreases, and the response of the uremic parathyroid gland to FGF-23 declines. Conversely PTH receptor activation stimulates osteocyte production of FGF-23.

Limited data suggest that other hormones may also interact with FGF-23. Studies in ovariectomized mice with chronic kidney disease indicate that restoring estrogen levels increases FGF-23 expression. This was confirmed in an osteoblast cell line. Additionally, human studies of X-linked hypophosphatemia indicate that single dose treatment with calcitonin increases $1,25(OH)_2D$ concentrations and improves serum phosphate. In one study, this correlated with a decrease in serum FGF-23 concentrations.

Changes in Dietary Phosphate Intake Alter Fibroblast Growth Factor 23 Concentration

Since the Western diet is very high in phosphate, low phosphate intake in persons consuming standard diets is rare. Rather, low phosphate intake is most often seen in the contexts of phosphate-binding agent ingestion or general states of malnutrition. In these settings both $1,25(OH)_2D$ and NPT2b expression increases, leading to a greater percentage of ingested phosphate being absorbed. During low dietary phosphate intake, FGF-23 protein production decreases in both mouse models and humans. This allows renal phosphate reclamation via NPT2a and NPT2c, and is permissive for increased $1,25(OH)_2D$ production, which increases NPT2b-mediated intestinal absorption. Conversely, in response to chronic high phosphate intake, FGF-23 concentrations increase in both humans and mice, resulting in decreased 1α-hydroxylase and lower $1,25(OH)_2D$ concentrations. FGF-23 action at the kidney leads to decreased phosphate reabsorption that clears excess phosphate and reestablishes phosphate homeostasis.

STUDY QUESTIONS

1. Describe the role of FGF-23 in phosphate and vitamin D metabolism.
2. What is the primary role of PTH? How does it function to control both calcium and phosphate metabolism?
3. Calcium levels are very tightly regulated. Describe two mechanisms utilized to increase serum calcium levels in situations of hypocalcemia.
4. What roles do vitamin D and its metabolities play in both phosphate and calcium metabolism?
5. Describe the three mechanisms of active calcium absorption in the intestine.

TABLE 14.1 Essential and Nonessential Bone-Related Nutrients

Nutrient	Function in Bone Health	Dietary Sources
ESSENTIAL MACRONUTRIENTS		
Protein	10—15% of bone weight Dietary protein increases serum IGF-I levels	Animal foods, nuts, and seeds
Fat	5—10% of bone weight	Fats and oils
ESSENTIAL MICRONUTRIENTS		
Calcium	36% of bone ash Reservoir for maintaining serum calcium Suppresses parathyroid hormone release and consequent bone resorption Secondary messenger in cell signaling Primary messenger in PTH release	Dairy, dark green leafy vegetables
Phosphorus	17% of bone ash Acid-base buffer Energy currency of cells	Dairy, meats, processed foods, colas
Magnesium	0.8% of bone ash Improves quality of bone by controlling size of hydroxyapatite crystals	Dark green leafy vegetables, nuts, whole grains, and dairy
Potassium	Produces an alkaline ash to help maintain normal pH	Fruits and vegetables
Zinc	Stimulates osteoblast activity and collagen synthesis	Animal foods, pulses, nuts, and seeds
Iron	Cofactor for enzymes in collagen synthesis and 25-hydroxycholecalcierol hydroxylase	Pulses and animal foods
Copper	Cofactor in lysyl oxidase, essential for cross-linking of collagen	Nuts and legumes
Manganese	Cofactor for enzymes	Nuts, seeds, whole grains, fortified foods
Vitamin D	Vitamin D-dependent calcium-binding proteins facilitate transcellular calcium absorption	Fortified milk, fatty fish/seafood
Vitamin K	Cofactor of vitamin K-dependent gamma-carboxylase, functional in carboxylation of glutamic acids residues of osteocalcin	Vegetables, liver, milk, eggs
Vitamin C	Cofactor in hydroxylation of lysine and proline for cross-linking collagen fibers May affect bone resorption	Fruits, vegetables
Vitamin A	Essential for bone remodeling	Fortified milk, fruits, vegetables, and liver
NONESSENTIAL NUTRIENTS		
Strontium	May play a role in increasing bone formation and reducing bone resorption	
Boron	May play a role in bone remodeling through interactions with vitamin D and mineral metabolism	Fruits and vegetables

PTH, parathyroid hormone; IGF-1, insulin-like growth factor 1.

The influence of phosphorus intake on bone and cardiovascular health is an emergent topic in the bone and mineral field.

Magnesium is the third most abundant mineral in bone. Magnesium has a smaller ionic radius than calcium, so it seeds crystal formation as it introduces a different shape and acts as an interfering agent in the same way that corn syrup seeds smaller crystal formation in the making of crystalline candies. This prevents hydroxyapatite crystals from becoming too large and brittle. Magnesium also influences mineral metabolism through its indispensable role in the metabolism of ATP and as a cofactor for over 300 enzymes.

Magnesium intakes are on average less than current recommendations by about one-third in women and one-quarter in men. One concern is that the magnesium content of plant foods has decreased over time due to the practice of adding lime to acidic soils. The calcium from lime competes with magnesium for uptake by the roots of plants, and therefore the plants we eat are often deficient in magnesium. Magnesium deficiency has been associated with osteopenia and bone fragility, but most of the evidence for a role of magnesium in bone is from animal studies. Magnesium intakes around the current requirements have been associated with greater bone mineral

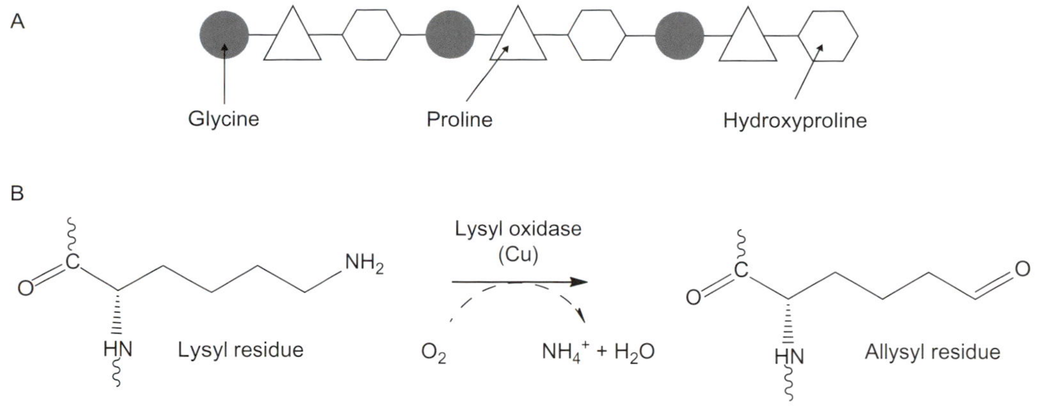

FIGURE 14.1   Role of amino acids and copper in collagen synthesis and fibril cross-linking.

density (BMD), and a few magnesium supplementation studies have found a beneficial effect of supplementation in postmenopausal women.

The role of other micronutrients and amino acids is mainly involved with connective tissue synthesis and maturation, as depicted in Fig. 14.1. Zinc stimulates osteoblastic bone formation, collagen synthesis, and alkaline phosphatase activity, and inhibits osteoclastic bone resorption. However, excessive zinc consumption limits the size of hydroxyapatite crystals. Overt zinc deficiency results in stunting and underdevelopment of sexual organs in children, and low zinc levels have been associated with osteoporosis in adults. Zinc deficiency is especially common in countries with low intakes of animal foods and unleavened bread. These diets are both low in zinc content and the zinc in grain is largely unavailable due to complexing by phytic acid. In cultures that consume leavened bread, the phytases in yeast release the mineral cations during fermentation. High calcium intakes can also impair zinc absorption. Supplementation can correct the defects associated with zinc deficiency in children and zinc supplementation in postmenopausal women has been shown to be beneficial in women with low dietary zinc intakes but not in women with high dietary zinc intakes. Similar to phosphorous, the take home message is that adequate but not excessive intake is optimal for bone health.

Other essential minerals have also been shown to play a role in bone health. Iron deficiency has a detrimental effect on bone morphology. Iron is a cofactor in enzymes responsible for bone collagen synthesis and cross-linking. Iron deficiency also leads to increased porosity in bone. Iron deficiency was very common in prehistoric populations, and is associated with characteristic pitting of the roof of the orbit (cribra orbitalia). Copper is needed in smaller amounts to assist in normal cell function, and is also a cofactor in enzymes responsible for bone collagen synthesis and cross-linking. Copper deficiency reduces insulin-like growth factor I (IGF-I) and decreases bone strength in rats; in humans, there is evidence of a benefit of copper supplementation (along with other trace minerals, i.e. minerals required in trace amounts) in lessening postmenopausal bone loss.

Fluoride is helpful in strengthening teeth enamel to reduce risk of cavities. At one time, sodium fluoride was considered as a treatment for osteoporosis. It can be anabolic and stimulates osteoblastic bone formation. Its replacement of hydroxyl groups in the hydroxyapatite crystal also increases BMD. Animal studies showed that these combined effects increase bone strength. However, incorporating fluoride into hydroxyapatite also makes the crystal—and bone—more brittle, causing it to break more easily, even though it is denser and stronger (recall in Chapter 6 that strength is defined by maximum load but is not necessarily related to fracture resistance). Because of this, it was discarded as a potential treatment for osteoporosis. Nevertheless, fluoridation of drinking water has been one of the most significant public health measures by enhancing mineralization of tooth enamel and greatly reducing the dental caries, and appropriately fluoridated drinking water is also associated with higher BMD and lower risk of fractures. However, there is current controversy over water fluoridation based largely on unfounded claims that fluoridation is associated with low intelligence quotient and health problems like Alzheimer disease, cancer, and even acquired immunodeficiency syndrome (or AIDS). While these claims are unsubstantiated, a true concern is the risk of fluorosis, which results in discoloration and pitting of tooth enamel and occurs with overfluoridation of drinking water or overingestion of fluoride products like toothpaste. In addition, overfluoridated drinking water has been associated with bone have high rates of bone loss, but cross-cultural comparisons have many confounders such as physical activity and genetics. Within cultures, fruit and vegetable intakes are weakly correlated with BMD. Controlled feeding studies have shown mixed effects of dietary protein, fruit and vegetables intake, and acid or base on calcium retention, partly related to dose and type of intervention. Nevertheless, for many health advantages it is prudent to eat a diet rich in fruits, vegetables, and dairy products and not to overindulge in meat.

DIETARY BIOACTIVES AND THEIR MECHANISMS

Osteoporosis is one of many chronic diseases now thought to have an inflammatory component to its pathology. Thus, a number of foods and isolated constituents with postulated anti-inflammatory properties are being evaluated for their effect on bone. Among the most promising dietary sources for bone are those with anti-inflammatory properties such as blueberries and plums, and hesperidin in orange. Many of these have polyphenolic structures called flavonoids that are typically abundant in plant foods. Most of these foods and bioactive constituents have been tested in vitro and in animal models. Soy isoflavones, a flavonoid subgroup, have been studied extensively in humans. The major isoflavones in soy, daidzein and genistein, have chemical structures similar to estrogen and bind weakly to estrogen receptors. Clinical studies show less reduction in postmenopausal bone loss than do in vitro and animal studies, which may be related to the dose and their short duration relative to changes in bone. In contrast to many randomized controlled trials (RCTs), epidemiological studies show that in countries with high soy intakes, soy consumption is associated with a reduced incidence of hip fracture. There are too many differences in design between the positive epidemiological studies and the negative RCTs of soy isoflavones to interpret their conflicting results. The epidemiological studies took place in Asian populations who consume whole soy foods, typically throughout life. The RCTs were of white women who do not habitually consume soy and they used isolated soy isoflavones for the relatively short duration of less than 3 years. Therefore, the conflicting findings could be due to the food versus supplement, the dose or duration of feeding, or genetic differences.

A number of potential mechanisms have been proposed to account for the bone benefits of flavonoids. The alkaline salts in plant-rich diets do not explain why specific flavonoid-rich sources are most effective. In fact, it is some of the phenolic acids such as those in blueberries that confer the most benefit. Bone benefits also cannot be adequately explained by bone-protective nutrients such as vitamin K or boron, which are high in plums and, therefore, have been speculated to partly explain the bone anabolic capacity of plums. Current evidence suggests that the mechanism of action of polyphenolic compounds is through modulating cell signaling pathways that influence bone turnover (Fig. 14.8). Bioactive metabolites cause antioxidant responsive elements in cells to be activated, which stimulates gene expression of many cytoprotective enzymes that reduce oxidative stress and inflammation. Some dietary bioactives suppress bone resorption by inhibiting osteoclast differentiation, which protects against ovariectomy-induced bone loss similar to the way estrogen acts. Remarkably, bioactives from some sources such as plum and berries also increase bone formation through the Wnt-β-catenin and bone morphogenic protein (BMP) signaling pathways regulated by RUNX2. These bioactives can also influence redox statues and oxidative stress, which stimulate osteoclast differentiation through RANK ligand [tumor necrosis factor ligand superfamily member 11/receptor activator of the NF-κB ligand (RANKL)] expression and osteoblast apoptosis and senescence. It is not always the bioactive compounds that exist naturally in the food that are the bioactive compounds; rather, it is often the metabolites of the flavonoids that influence inflammatory pathways that are produced by metabolism in the intestine and liver or by the gut microbiome of the host. As this has become understood, chemical measures of antioxidant capacity are being abandoned in favor of measuring the biological activity of the antioxidant that is of interest to health.

One metabolite that has received attention recently is equol. Equol is a metabolite produced by certain gut microbiota from the substrate daidzein, a soy isoflavone. Fewer than half of individuals have gut microbial communities that produce equol. Equol is a more potent phenolic than any of the natural soy isoflavones. While equol has some benefits in reducing bone loss, it also has weak uterotropic activity. Much work remains to be done to identify the bioactive constituents and their effective doses, as well as to completely understand their mechanisms of action. Few clinical trials have been performed to validate the benefits to humans.

EFFECTS OF OBESITY AND WEIGHT LOSS ON BONE

Obesity is a disease that most often results from poor nutrition, namely excess energy intake, and inadequate physical activity. Overweight and obesity rates are increasing in the United States and around the

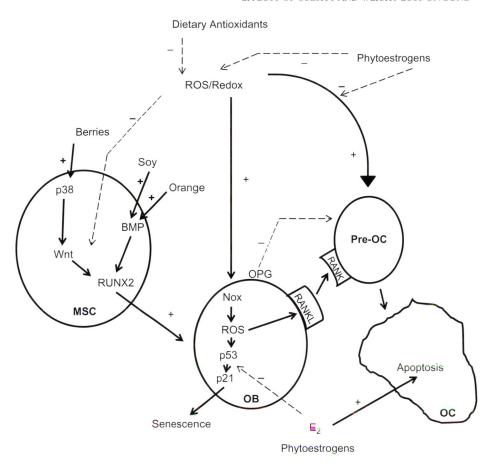

FIGURE 14.8 Bioactive-rich foods such as dietary antioxidants and phytoestrogens that impact cell signaling pathways to control bone turnover through precursor mesenchymal stem cells, osteoblasts for bone formation, osteoclast activation to stimulate bone resorption or removal of activated osteoclasts. BMP, bone morphogenic protein; E_2, 17β-estradiol; MSC, mesenchymal stem cell; NEFA, nonesterified free fatty acids; Nox, NADPH oxidase; OB, osteoblast; OC, osteoclast; OJ, orange juice; OPG, osteoprotegerin; ROS, reactive oxygen species.

world. Based on a BMI (>30), approximately one-third of Americans are overweight and another third are obese. Obesity is associated with many comorbidities such as cardiovascular disease, diabetes mellitus, dyslipidemia, hypertension, metabolic syndrome, pulmonary diseases, rheumatologic diseases, and stroke.

Obesity, Osteoporosis, and Fracture Risk

In adults, obesity is generally considered to be protective to bone due to increased bone mass. Overweight and obese adults have higher BMD and a lower incidence of osteoporosis compared with healthy weight adults. This is probably a result of greater mechanical stimulation to the skeleton (see Chapter 9), but may also be due to effects of adipokines such as leptin, which promotes bone formation (see Chapter 15), or to the storage and aromatization of estrogen in fat.

Adults with a healthy BMI or who are slightly overweight seem to have some protection from fracture compared to those who are underweight. However, fracture risk for adults with a BMI in the obesity range is less clear. This may be attributed to the presence of potential harmful factors (including decreased

stability, greater risk of falls, greater force during falls, comorbid degenerative joint diseases, diabetes, and inflammation) counteracting the positive effects of fat mass and increased mechanical stimulation on bone mass. There is evidence that the location of excess fat may affect fracture risk, where excess subcutaneous fat is protective and excess visceral fat is harmful. Moreover, fracture risk in obesity may vary by site, where obese individuals may be protected from hip and vertebral fractures but have increased risk of wrist and ankle fractures.

Other comorbid conditions in obesity such as type 2 diabetes mellitus (T2DM) can adversely affect bone health and fracture risk. Obese adults are at high risk for developing T2DM. Although patients with T2DM generally have normal or high BMD (usually associated with higher body weight), their fracture risk is approximately twice that of nondiabetics, indicating that bone quality is impaired in these individuals. A current hypothesis is that bone quality is impaired in diabetes due to increased nonenzymatic cross-linking of type I collagen by advanced glycation end products (AGEs) in the presence of high blood glucose, thus decreasing the integrity of the collagen matrix and making the tissue more brittle. Therefore, although osteoporosis is less common in obesity, bone health

Hormonal Effects on Bone Cells

Teresita Bellido[1] and Kathleen M. Hill Gallant[2]

[1]Roudebush Veterans Administration Medical Center, Indianapolis, Indiana, USA [2]Department of Anatomy and Cell Biology, Indiana University School of Medicine, Indianapolis, Indiana, USA

INTRODUCTION: DIRECT VERSUS INDIRECT EFFECTS OF HORMONES ON BONE CELLS

Systemic hormones can affect bone either directly or indirectly. Direct action occurs through receptors expressed in bone cells. Indirect action occurs when a hormone modulates mineral homeostasis through regulation of calcium and phosphate absorption by the intestine and excretion or reabsorption by the kidney. The goal of this chapter is to discuss the current knowledge about the direct effects of hormones on the skeleton.

PARATHYROID HORMONE

Parathyroid hormone (PTH) is a peptide hormone that controls the minute-to-minute level of ionized calcium in the circulation and extracellular fluids. PTH is secreted by the chief cells of the parathyroid gland in response to low levels of calcium in the blood. The two main target tissues of PTH are bone and kidney. By binding to receptors in cells of these tissues, PTH induces responses leading to an increase in blood calcium concentrations. This increase in circulating calcium, in turn, feeds back on the parathyroid gland to reduce PTH secretion.

Actions of Parathyroid Hormone on Bone

The primary effect of PTH on the skeleton is to induce bone resorption with the goal of liberating calcium from the mineralized matrix and increasing its concentration in the blood and extracellular fluids.

PTH has profound effects on the skeleton at the tissue level. Elevated circulating levels of the hormone can generate both catabolic and anabolic effects on bone, depending on the temporal profile of its increase. Continuous (or chronic) elevations in PTH, as in primary or secondary hyperparathyroidism, increase the rate of bone remodeling, and can result in loss of bone. In contrast, intermittent increases of PTH in the blood, as achieved by daily injections of the pharmaceutical agent teriparatide [recombinant human PTH; rhPTH(1−34)], results in bone gain.

The high bone remodeling rates and bone loss resulting from chronic PTH elevation are associated with excessive production and activity of both osteoclasts and osteoblasts. The enhancement of osteoclast activity outpaces that of osteoblasts and thus results in a negative basic multicellular unit (BMU) balance (see Chapter 4, Fig. 4.11). Conversely, the primary effect of intermittent PTH elevation is a rapid increase in the number and activity of osteoblasts and in bone formation, leading to net bone gain. The mechanism of this anabolic effect is attributed to the ability of PTH to promote proliferation of osteoblast precursors, inhibit osteoblast apoptosis, reactivate lining cells to become matrix synthesizing osteoblasts, or a combination of these effects (see below). In humans, intermittent PTH administration stimulates bone formation by increasing the bone remodeling rate and the amount of bone formed by each BMU in a process named *remodeling-based formation*. PTH also stimulates bone formation not coupled to prior resorption, referred to as *modeling-based formation*. The latter mechanism appears to be more evident in rodents.

Parathyroid Hormone Receptors and Downstream Signaling

PTH binds with high affinity to the parathyroid hormone/parathyroid hormone-related peptide and that deletion of RANKL from osteocytes leads to osteopetrosis. Moreover, RANKL expression, osteoclast number, and bone resorption are elevated in transgenic mice with constitutive activation of the PTH1-R in osteocytes. These findings raise the possibility that at least part of the effects of PTH on osteoclast differentiation and resorption are due to osteocytic RANKL regulation.

SEX STEROIDS

In the 1940s, Fuller Albright made the association between women's loss of estrogen at menopause and bone loss. For decades, this association was believed to be indirect, until the discovery in the late 1980s that estrogens bind directly to bone cells, indicating a direct effect of estrogen on the skeleton. In men, the gradual reduction in androgen secretion with aging is associated with bone loss. Some of the effects of androgens are due to their conversion to estrogen. However, bone cells express receptors that specifically bind androgens and mediate their biological effects independently of estrogens. This section addresses the general and sex-specific effects of the main sex steroid hormones affecting skeletal tissue: androgens and estrogens.

Sex Steroid Production

Sex steroid hormone synthesis begins by hydrolysis of cholesterol esters and uptake of cholesterol by the mitochondria of target tissue cells. Cholesterol is metabolized to pregnenolone, which is further metabolized to produce all sex steroid hormones. Estrogens are sex steroids secreted by the ovaries in women and to a small extent by the testes in men. Over 80% of estrogen in men is produced through peripheral conversion of androgens to estrogens by cytochrome P450 aromatase. Adipose tissue is the main tissue for estrogen production in men and for extraovarian estrogen production in women.

Androgens are sex steroids secreted by the testes in men, the ovaries in women, and the adrenal glands in both men and women. Testosterone, the main androgen in men, is secreted primarily by the testes (approximately 95% of total testosterone). In women, only about 25% of testosterone comes from the ovaries; another 25% comes from the adrenal glands, but half of total testosterone in women comes from conversion of other sex steroids, such as dehydroepiandrosterone (DHEA) and androstenedione, by peripheral tissues such as adipose tissue.

Most testosterone is bound to proteins in the circulation. Approximately half is bound with high affinity to steroid hormone-binding globulin, with the other half bound with low affinity to albumin. Only 1−2% of testosterone is free (unbound) in the circulation. Bioavailable testosterone refers to both free testosterone and albumin-bound testosterone. Free testosterone diffuses passively through cell membranes and binds to the androgen receptor. Testosterone can be metabolized in peripheral tissues to the potent androgen, dihydrotestosterone by 5-alpha-reductase, or to 17β-estradiol by cytochrome P450 aromatase.

Sex Steroid Receptor Signaling

Sex steroid signaling occurs through genotropic and nongenotropic signaling pathways (Fig. 15.3). Genotropic signaling occurs when the sex steroid ligands bind to the sex steroid receptors, which then dimerize and translocate to the nucleus to initiate gene transcription. Dimerized sex steroid receptors can bind directly to

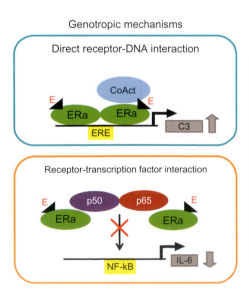

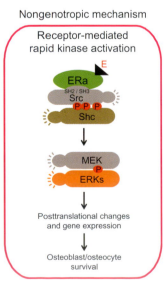

FIGURE 15.3　**Signaling pathways activated by sex steroids.** Estrogens (ERs; depicted in the figure) and androgens (not shown) activate genotropic and nongenotropic pathways. CoAct, coactivator; E, estradiol. See text for details.

response elements in promoters of the target genes [e.g. estrogen response elements (EREs), for estrogens]. Alternatively, receptor monomers can directly interact with transcription factors, and these complexes subsequently bind to promoters in the target genes through response elements for the particular transcription factors. Major transcription factors that participate in sex steroid signaling include nuclear factor-kappa-B (NF-κB) and activator protein 1 (AP-1).

Sex steroids also activate rapid, kinase-mediated signaling by binding to membrane-bound receptors. Rapid signaling is initiated by binding of the ligand to the receptor at the cell membrane. Signaling is amplified through the interaction of receptors with scaffolding proteins and culminates with activation of kinases, including Akt, proto-oncogene tyrosine-protein kinase Src, extracellular signal-regulated kinases (ERKs), PI3-K, PKA, and protein kinase C (PKC). This mechanism has been termed nongenotropic because it does not involve direct binding of the receptor to DNA. However, it is important to note that kinase signaling leads not only to posttranslational changes in proteins (such as phosphorylation) but also to transcriptional changes that involve alterations of gene expression mediated by kinase-activated transcription factors.

Sex Steroids during Growth

At puberty, boys and girls experience a period of rapid height gain followed by a period of rapid bone mineral accrual. Girls experience these growth spurts on average a year and a half before boys, but boys achieve higher peak height velocity and peak bone mineral velocity. Subsequently, boys are taller and have greater bone mass than girls by the end of puberty, and ultimately have higher peak bone mass in adulthood.

The sexual dimorphism of the skeleton during growth is attributed to a general stimulatory effect of androgens and inhibitory effect of estrogens on bone growth. Androgens appear to be stimulatory for periosteal bone expansion, which is greater in boys than girls during puberty and throughout the years of peak bone mass acquisition. Androgen deficiency in males results in a reduction in periosteal bone expansion. Conversely, estrogens are inhibitory of periosteal expansion, as estrogen-deficient females have a drastic increase in periosteal bone expansion. On the endocortical surface, estrogens promote and androgens suppress bone formation during growth. As a result, girls at puberty have cortical thickening from endosteal contraction with little periosteal expansion, whereas boys have cortical thickening mostly from periosteal expansion being greater than endosteal expansion (Fig. 15.4). Estrogen signaling through ERβ appears to be responsible for the effects of estrogen on the periosteal and endosteal surfaces. Thus, female ERβ knockout mice have bones that resemble wild-type males, with greater periosteal and endosteal circumferences and greater cross-sectional diameter.

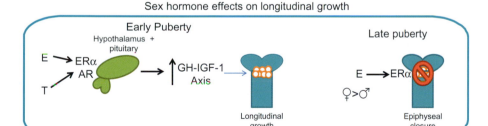

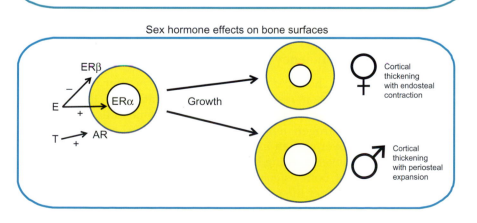

FIGURE 15.4 Concept model of the effects of sex hormones during growth. Early in puberty, low levels of estrogen and testosterone stimulate longitudinal bone growth. In both sexes late in puberty, estrogen stimulates epiphyseal closure. Estrogen is stimulatory to bone formation at the endosteal surface and inhibitory at the periosteal surface of bone, whereas testosterone is stimulatory at the periosteal surface. AR, androgen receptor; GH, somatotropin/growth hormone; E, estrogen; ERα/β, estrogen receptor α/β; IGF-I, insulin-like growth factor I; T, testosterone. See text for details.

TABLE 15.1   Bone Loss in Men and Women with Aging and Sex-Steroid Loss

Life Stage	Compartment	Rate of Loss	Amount of Loss
5–10 years postmenopause ♀	Cancellous;	4–6%/years;	Cancellous > cortical
	Cortical	1–2%/years	
Older age ♀	Cancellous;	1–2%/years;	Cortical > cancellous
	Cortical	1–2%/years	
Older age ♂	Cancellous;	1–2%/years;	Cortical > cancellous
	Cortical	1–2%/years	

At the beginning of puberty, both estrogen and testosterone activate the somatotropin/growth hormone (GH)-IGF-I axis to stimulate longitudinal bone growth. The effects of estrogen during growth are dependent on the stage of development. Early in puberty, estrogen (at relatively low levels in girls compared with later puberty) signaling through ERα in the hypothalamus and pituitary is necessary for GH secretion, which acts directly and indirectly through IGF-I to increase longitudinal bone growth by stimulating proliferation of growth plate cartilage. At the end of puberty, estrogen levels are high and act directly through ERα signaling in growth plate chondrocytes to slow and then cease longitudinal bone growth. Estrogen signaling through ERα is responsible for epiphyseal closure in both sexes; however, higher estrogen in girls explains the shorter period of longitudinal bone growth and ultimate bone length in girls compared with boys (Fig. 15.4). The necessity of estrogen signaling for epiphyseal fusion has been demonstrated by the lack of epiphyseal fusion in men with aromatase deficiency and ERα loss-of-function mutations.

Loss of Sex Steroids in the Adult Skeleton

At menopause, the ovaries cease to produce estrogens, thus making peripheral production of estrogen, mainly through conversion of adrenal androgens in adipose tissue, the primary source of estrogen in postmenopausal women. In men, total testosterone gradually declines by approximately 1% per year starting by the third decade of life. In addition, the levels of sex hormone-binding proteins are markedly increased with age in men, thus reducing the amount of bioavailable testosterone.

The difference in bone loss between women and men during aging is mainly attributable to the rapid bone loss in women in the years immediately following menopause. In the first 5–10 years following menopause, women lose cancellous bone at a rate of approximately 4–6% per year and cortical bone at a rate of approximately 1–2% per year. After this

TABLE 15.2   Effects of Sex Steroid Deficiency on Bone Cells

Cell Type	Number	Birth	Death
Osteoclasts	Increased	Increased	Decreased
Osteoblasts	Increased	Increased	Increased
Osteocytes	Unknown	Unknown	Increased

Supply of osteoclasts exceeds demand.
High rate of bone remodeling.

period, women lose bone at a slower rate of 1–2% per year in both compartments, which is similar to the rate of bone loss in men (Table 15.1). The rapid loss in the years immediately after menopause leaves women with lower bone mass, but also with lower trabecular connectivity and number, which makes their bone more susceptible to fracture. Men also have approximately three times greater periosteal expansion than females during aging, which produces stronger bone geometry. Decreased estrogen signaling also impairs bone's response to mechanical loading, which contributes to bone loss.

The mechanism for the slow rate of bone loss in aging men is similar to that of the slow bone loss phase in women. Testosterone deficiency has some estrogen-independent effects on calcium absorption and bone cell functions; however, much of the effect of testosterone deficiency on bone loss in men is related to the resulting estrogen deficiency and its consequences.

Changes in Bone Cells Induced by Estrogen Deficiency

The rapid rate of bone loss early after menopause in women results from a combination of the effects of loss of estrogens on different bone cells (Table 15.2). Estrogen deficiency leads to increased rate of bone turnover and an imbalance in focal remodeling at the BMU level favoring bone resorption. There is overproduction of both osteoclasts and osteoblasts, accompanied by a longer life span of osteoclasts and a shorter life span of osteoblasts. The longitudinal extent of the BMU (which

TABLE 15.3 Effects of Estrogens on Bone Cells

Cell Type	Effect of Estrogen	Mechanism
Osteoclasts	Induction of apoptosis (Fas ligand and ERK/JNK activation)	Genotropic and nongenotropic
Stromal/osteoblastic cells and T lymphocytes	Inhibition of pro-osteoclastogenic cytokine production (IL-1, IL-6, and TNFα)	Genotropic, mediated by receptor-transcription factor interaction
Osteoblasts and osteocytes	Inhibition of apoptosis (ERKs and PI3-K)	Nongenotropic

ERK, extracellular signal-regulated kinase; IL-1/6, interleukin-1/6; JNK, c-Jun N-terminal kinase; PI3-K, phosphoinositide 3-kinase.

is related to the lifetime of the BMU) is determined by the supply of osteoclast and osteoblast precursors, whereas the depth of the BMU's erosion lacunae depends on the timing of apoptosis in mature osteoclasts. In estrogen deficiency, the supply of osteoclast precursors is enhanced, resulting in the origination of more BMUs per unit bone area (i.e. a higher activation frequency), and there are more osteoclasts and osteoblasts contributing to extend the progression of each BMU. Moreover, osteoclasts live longer, resulting in deeper resorption pits and delayed BMU reversal to the formation phase. Furthermore, osteoblast apoptosis is increased and thus bone formation is disproportionately lower compared to resorption, contributing to a negative balance within each remodeling cycle and leading to bone loss. The prevalence of osteocyte apoptosis is also increased, adding to the bone fragility that characterizes conditions of loss of sex steroids.

Effects of Estrogens and Androgens on Osteoclasts

Consistent with the increase in osteoclasts and bone resorption induced by sex steroid deficiency, estrogens and androgens decrease the number of osteoclasts in vivo and in vitro (Table 15.3). The cellular mechanism of reduction of osteoclasts involves inhibition of osteoclast generation combined with induction of osteoclast apoptosis. Estrogens decrease the production of interleukin-1 (IL-1,) IL-6, and tumor necrosis factor (TNF-α) in cells that support osteoclast formation, resulting in inhibition of proliferation and the differentiation of osteoclast precursors toward mature osteoclasts. The inhibitory effect of estrogens on cytokine production is mediated by an interaction between the estrogen receptor and NF-κB and regulation of gene expression mediated by this transcription factor (Table 15.3). Androgens exert similar effects as estrogens on the production of pro-osteoclastogenic cytokines. In addition, estrogens induce apoptosis in mature osteoclasts by acting directly on these cells. Current evidence indicates that estrogens induce osteoclast apoptosis by activating proapoptotic pathways, including the mitogen-activated protein kinase (MAPK)-c-Jun N-terminal kinase (JNK) and TNF ligand superfamily member 6 (Fas ligand) pathways.

Effects of Estrogens and Androgens on Osteoblasts and Osteocytes

In contrast to their proapoptotic effect on osteoclasts, estrogens and androgens inhibit apoptosis in osteoblasts and osteocytes (Fig. 15.3; Table 15.3). The mechanism of this survival effect involves rapid activation of survival kinases ERKs and PI3-K. This is followed by phosphorylation of the proapoptotic protein Bad, which leads to inactivation of the apoptotic properties of the protein, and phosphorylation and activation of the transcription factors ETS domain-containing protein (Elk) and CCAAT/enhancer-binding protein beta (C/EBP β), with subsequent changes in gene expression. These kinase-mediated posttranslational and transcriptional effects are required for estrogen-induced survival of osteoblasts and osteocytes.

GLUCOCORTICOIDS

Glucocorticoids are produced and released by the adrenal glands in response to stress. They regulate numerous physiologic processes in a wide range of tissues. Among several effects, these hormones exert profound immunosuppressive and anti-inflammatory actions and induce apoptosis in many cell types, including T lymphocytes and monocytes. Because of these properties, exogenous glucocorticoids are extensively used for the treatment of immune and inflammatory conditions, the management of organ transplantation, and as components of chemotherapy regimens for hematological cancers. However, long-term use of glucocorticoids is associated with severe adverse side effects in several organ systems. In particular, prolonged use of exogenous glucocorticoids leads to a dramatic loss of bone mineral and strength, similar to endogenous elevation of glucocorticoids in Cushing disease.

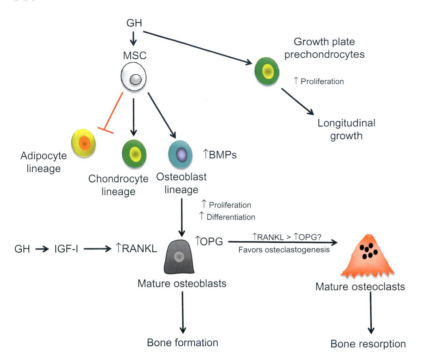

FIGURE 15.6 **Effects of growth hormone on bone cells.** Growth hormone (GH) directs mesenchymal stem cells toward chondrocytic and osteoblastic lineages and away from the adipocyte lineage. GH increases osteoprotegerin (OPG) production, but through IGF-I also increases production of the RANK ligand [tumor necrosis factor ligand superfamily member 11/receptor activator of the NF-κB ligand (RANKL)], generally favoring osteoclastogenesis. GH can also directly affect longitudinal growth by stimulating growth plate prechondrocyte proliferation. BMP, bone morphogenetic protein; IGF-I, insulin like growth factor-I; MSC, mesenchymal stem cell; RANKL, tumor necrosis factor ligand superfamily member 11/receptor activator of the NF-κB ligand.

may influence bone metabolism indirectly through its actions on PTH, $1,25(OH)_2D_3$, and phosphate handling. GH helps to maintain PTH secretion and circadian rhythm and increases the production of $1,25(OH)_2D_3$ by increasing 1α-hydroxylase and inhibiting 24-hydroxylase. GH also increases phosphate retention by increasing the renal maximal reabsorption threshold for phosphate. Together, these actions of GH favor bone formation.

INSULIN

Insulin and IGFs are highly homologous, as are their receptors and their functions. The effects of IGFs on bone are discussed in Chapter 3. In this chapter, the more direct effects of insulin on bone cells will be discussed.

Insulin is a peptide hormone secreted by pancreatic beta cells in response to increased concentrations of glucose in blood. Insulin increases glucose uptake into target tissues and inhibits the release of stored energy. Insulin (and IGFs) signals through the insulin receptor (IR), a cell surface tyrosine kinase receptor present in two isoforms: α and β. IRs exist as either homodimers of the same IR isoform, or as heterodimers of IRα and IRβ or an IR with insulin-like growth factor-1 receptor (IGF-IR). Signaling transduction occurs by conformational changes upon ligand binding, which result in autophosphorylation, followed by increased kinase activity of the receptor, and phosphorylation of a number of substrate proteins that serve as effector molecules.

Establishing the importance of insulin for bone independent of IGFs is difficult due to their overlapping functions. However, type 1 diabetes mellitus (T1DM) patients, who are insulin deficient due to loss of pancreatic beta cell mass and function, have lower bone mass and are at increased risk for early onset osteoporosis and increased fracture risk. In addition, animal models of T1DM show that bone formation is reduced, providing evidence for a relationship between insulin and bone, although these animals also have low circulating IGF-I.

IRs have been identified in osteoblasts, and treating osteoblasts with insulin increases collagen synthesis and ALP activity. Global IR knockout mice are not viable past the early postnatal period, but studies of cell-specific IR and IGF-IR deletion in osteoblasts have been informative about the individual roles of insulin signaling versus IGF-I signaling. These studies show that diminished insulin signaling in osteoblasts results in reduced cancellous bone volume, with no defects in mineralization but reduced osteoblast number. On the other hand, diminished IGF-I signaling in osteoblasts results in reduced cancellous bone volume and under-mineralized bone, but with a normal number of osteoblasts. Cultured IR-deficient osteoblasts exhibit impaired proliferation and differentiation, whereas wild-type osteoblasts treated with insulin have increased proliferation and differentiation (Fig. 15.7A).

More recently, insulin signaling in osteoblasts has been implicated in controlling whole body glucose

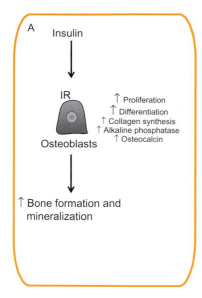

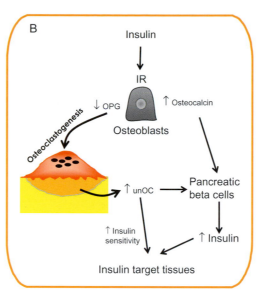

FIGURE 15.7 **Effects of insulin on osteoblasts and its proposed role in glucose metabolism.** Insulin increases proliferation and differentiation of osteoblasts (A) and increases collagen synthesis, bone formation, and mineralization. Insulin may also act through an osteocalcin-mediated mechanism to regulate whole body glucose homeostasis (B). IR, insulin receptor; OPG, osteoprotegerin; unOC, undercarboxylated osteocalcin.

metabolism through an osteocalcin-dependent mechanism. Insulin signaling in osteoblasts increases the production of osteocalcin, which in turn acts on the pancreas to increase insulin production. Additionally, insulin signaling in osteoblasts decreases OPG and thus increases osteoclastic bone resorption. During bone resorption, undercarboxylated osteocalcin, which is considered the active hormonal form of osteocalcin regarding glucose metabolism, is liberated from the bone matrix. This provocative animal experimentation demonstrates a novel metabolic function of bone. However, the relative importance of insulin signaling in bone to overall glucose metabolism and the validity of the hypothesis in humans remains to be determined (Fig. 15.7B).

1,25-DIHYDROXYVITAMIN D$_3$

1,25-Dihydroxyvitamin D$_3$ [1,25(OH)$_2$D$_3$ or cholecalciferol] is a steroid hormone derived from vitamin D in the diet or from subcutaneous synthesis. Vitamin D undergoes hydroxylation in the liver to produce 25 (OH)D$_3$, the serum indicator of vitamin D status, and a second hydroxylation in the kidney to produce 1,25 (OH)$_2$D$_3$, the hormonally active vitamin D metabolite. 1,25(OH)$_2$D$_3$ signals by binding to the vitamin D receptor (VDR), which is a member of the superfamily of nuclear receptors. VDR knockout mice develop hypocalcemia, secondary hyperparathyroidism, and rickets, indicating a role for 1,25(OH)$_2$D$_3$ in bone mineralization. However, a diet high in calcium and phosphate rescues the abnormal mineral biochemistries and bone phenotype in the VDR knockout mouse, indicating that the main effects of 1,25(OH)$_2$D$_3$ on bone are to provide sufficient calcium and phosphate for normal mineralization, particularly by mediating intestinal calcium and phosphate absorption. The role of 1,25 (OH)$_2$D$_3$ on mineral homeostasis is discussed in Chapter 13. Here, the direct effects of 1,25(OH)$_2$D on bone cells are discussed.

VDR is present in cells of the osteoblastic lineage, including osteoblast progenitor cells, osteoblast precursors, and mature osteoblasts. 1,25(OH)$_2$D$_3$ signaling in osteoblastic cells increases production of RANKL and decreases the production of OPG, thus increasing RANKL-RANK-mediated osteoclastogenesis. This action of 1,25(OH)$_2$D$_3$ is consistent with the actions of PTH and 1,25(OH)$_2$D$_3$ to increase serum calcium by liberating calcium from bone mineral.

1,25(OH)$_2$D$_3$ signaling can also directly affect bone formation. 1,25(OH)$_2$D$_3$ increases production of RUNX2, an essential transcription factor for osteoblast differentiation. Transgenic mice that overexpress VDR in osteoblastic cells have increased bone formation. Though the main role of 1,25(OH)$_2$D$_3$ in promoting bone mineralization is through increasing intestinal calcium and phosphate absorption (as evidenced by the high-calcium/phosphate rescue diet in the VDR knockout mice), 1,25(OH)$_2$D$_3$ has also been shown to have direct effects on osteoblasts to increase production of osteocalcin and osteopontin, proteins involved in bone mineralization (Fig. 15.8). Conversely, studies have shown that high dose 1,25(OH)$_2$D$_3$ actually inhibits osteoblastic bone mineralization. Therefore, the direct effects of 1,25(OH)$_2$D$_3$ on bone are diverse, and can affect both bone resorption and formation processes. Its beneficial effects occur within a defined window, and either high or low levels can be detrimental to bone.

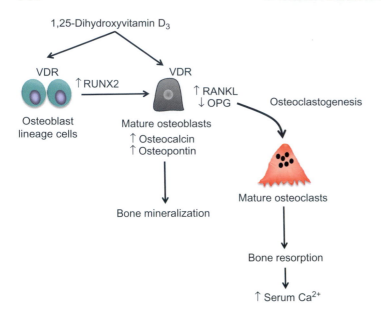

FIGURE 15.8 **Effects of 1,25-dihydroxyvitamin D₃ on osteo-blast lineage cells and osteoclasts.** In addition to its effects on intestinal calcium and phosphorus absorption (not shown), 1,25-dihydroxyvitamin D₃ [1,25(OH)₂D₃] increases bone mineralization by driving differentiation of osteoblast lineage cells toward mature osteoblasts and by increasing osteoblast production of osteocalcin and osteopontin. Conversely, 1,25(OH)₂D₃ increases bone resorption by increasing the RANK ligand/tumor necrosis factor ligand superfamily member 11 (RANKL):osteoprotegerin (OPG) ratio, thus promoting osteoclastogenesis. Ca^{2+}, calcium; VDR, vitamin D receptor.

LEPTIN

Leptin is a peptide hormone produced and secreted mainly by adipocytes. Leptin plays a role in energy homeostasis, appetite, neuroendocrine function, immune function, reproduction capacity, and bone metabolism. Leptin exerts its effects by binding to leptin receptors, which are members of the class I cytokine receptor superfamily. Leptin receptors are expressed throughout the central nervous system and in peripheral tissues. Leptin signaling in the hypothalamus is important for energy homeostasis. Congenital leptin deficiency results in obesity in both animal models and humans. Leptin deficient (*ob/ob*) mice exhibit an obese phenotype that is rescued upon administration of leptin. Paradoxically, leptin excess is observed in obese individuals, apparently due to hypothalamic leptin resistance. Because the main source of leptin is adipose tissue, circulating leptin is highly correlated with body fat mass, particularly subcutaneous adiposity.

The effects of leptin on bone are complex, emergent, and dependent on dual-effects of central and peripheral leptin signaling pathways (Fig. 15.9). A high bone mass phenotype has been characterized in leptin-deficient *ob/ob* mice, and intracerebroventricular infusion of leptin in both *ob/ob* and wild-type mice reduces bone mass, suggesting that leptin decreases bone mass through central mechanisms. However, the effect of leptin appears to vary by skeletal site. Thus, *ob/ob* mice have greater bone density and cancellous in the lumbar vertebrae, but they exhibit lower cortical bone density and volume in the femur. Due to the high contribution of cortical bone to total bone mass, *ob/ob* mice have reduced total body bone mass compared with wild-type mice. Therefore, central leptin signaling

appears to have dual effects on bone. Indeed, central leptin effects mediated by β2-andrenergic receptors decrease osteoblast activity and bone formation and increase remodeling of cancellous bone through increased RANKL, but central leptin effects mediated by β1-andrenergic receptors or the GH/IGF-I axis stimulate bone formation, particularly at cortical sites (Fig. 15.9).

In contrast to intracerebroventricular infusion, peripheral administration of leptin increases bone mass in *ob/ob* mice. BMSCs, osteoblasts, and osteoclasts express leptin receptors. Leptin increases the expression of osteogenic genes in BMSCs, leading to preferential differentiation into the osteoblast lineage over adipocytes. In addition, leptin signaling in osteoblasts increases the expression of OPG and decreases the expression of RANKL, leading to decreased osteoclastogenesis (Fig. 15.9).

Similar to adipocytes in peripheral body fat, adipocytes in the bone marrow also secrete leptin. Local effects of leptin produced by bone marrow adipocytes add another layer of complexity to the leptin-bone relationship. In contrast to the peripheral effects of leptin on bone cells discussed above, higher concentrations of leptin stimulate bone marrow stromal cell apoptosis and bone resorption, and decrease bone formation (Fig. 15.9). This suggests that a higher local concentration of leptin from increased marrow adiposity may contribute to bone loss, a concept that is consistent with the positive association between marrow adiposity and osteopenia.

Body weight is positively associated with bone mass. This is commonly attributed to influences of mechanical stimulation from increased load-bearing. Leptin signaling on bone is also a potential

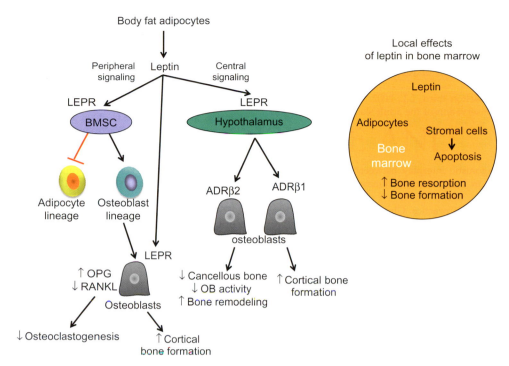

FIGURE 15.9 **Concept model of central and peripheral effects of leptin on bone and local effects of leptin in bone marrow.** Leptin secreted by body fat adipocytes increases bone formation through peripheral signaling and has dual-effects on bone through central signaling. Leptin produced locally by adipocytes in the bone marrow increases stromal cell apoptosis, increases bone resorption, and decreases bone formation. BMSC, bone marrow stromal cell; LEPR, leptin receptor; OB, osteoblast; OPG, osteoprotegerin; RANKL, tumor necrosis factor ligand superfamily member 11/receptor activator of the NF-κB ligand.

contributor to the higher bone mass observed with increased body weight, as body weight is associated with bone mass even at "non-weight-bearing" sites. However, the high correlation between circulating leptin and body fat in humans (accounting for more than 80% of the variation in body fat) makes distinguishing associations between leptin and bone from associations between body fat and bone challenging. Both positive and negative associations between circulating leptin and BMD have been reported, especially when adjusted for body composition. Leptin receptor polymorphisms have been associated with bone mass in humans, but these associations may be largely mediated through leptin effects on energy homeostasis and body size.

STUDY QUESTIONS

1. Describe how deficiencies of androgen and estrogen affect bone at the cellular and structural levels. Compare and contrast these effects on young and old bone.
2. Describe how PTH can cause anabolism and catabolism of bone.
3. Describe PTH signaling and its role in osteoblasts, osteoclasts, and osteocytes.

4. Describe why hypothyroidism and hyperthyroidism are both associated with fracture risk.
5. What are the roles of GH and IGF-I in the growing skeleton?
6. How does obesity affect skeletal structure and function?

Suggested Readings

PTH

Bellido, S., Divieti, in press. Effect of PTH on osteocytes. Bone.

Dempster, D.W., et al., 1993. Anabolic actions of parathyroid hormone on bone. Endocr. Rev. 14, 690–709.

Jilka, R.L., 2007. Molecular and cellular mechanisms of the anabolic effect of intermittent PTH. Bone. 40, 1434–1446.

Jilka, R.L., et al., 2008. Apoptosis of bone cells. In: Bilezikian, J.P., Raisz, L.G., Martin, T.J. (Eds.), Principles of Bone Biology, third ed. Academic Press.

Neer, R.M., et al., 2001. Effect of parathyroid hormone (1–34) on fractures and bone mineral density in postmenopausal women with osteoporosis. N. Engl. J. Med. 344, 1434–1441.

Obrien, C.A., 2010. Control of RANKL expression. Bone. 46, 911–919.

Sex Steroids

Kousteni, S., Bellido, T., et al., 2001. Nongenotropic, sex-nonspecific signaling through the estrogen or androgen receptors: dissociation from transcriptional activity. Cell. 104, 719–730.

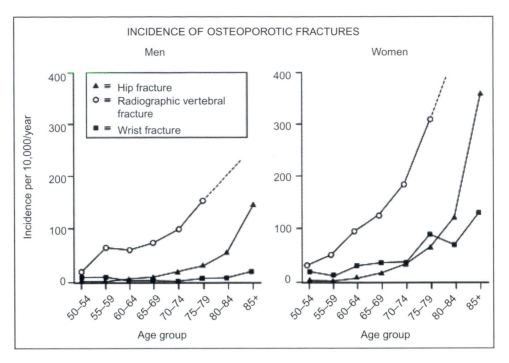

FIGURE 16.4   Age-related increase in fractures occurs in women and men. In women, the incidence of hip and spine fractures increases following menopause, but wrist fractures remain stable until about age 70. In men, the rise in hip and spine fractures begins about 10 years after that in women. In men there is not an age-related increase in wrist fractures. *Reproduced with permission from Lane NE and Sambrook PN. Osteoporosis and the Osteoporosis of Rheumatic Diseases. Elsevier; 2006. p. 2 [Chapter 1].*

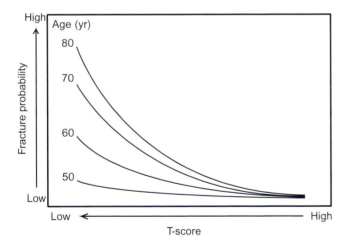

FIGURE 16.5   The 10-year probability of fracture is related to both bone mineral density (BMD T-score) and age. Older patients at any given BMD have higher fracture risk than younger patients. *Adapted from Kanis et al. 2001; 12: 989—995.*

thinning of trabeculae (see Chapter 1). It turns out that most of the bone loss in women occurs through loss of trabeculae, with very little attributed to thinning of trabeculae, whereas male bone loss is the opposite. This has the added effect in women of exacerbating the architectural deterioration of the cancellous structure, and leads to more rapid decline in strength and stiffness. Therefore, the skeletal differences between men and women involve bone size, as well as differences in the way that bone is lost and the resulting effects on cancellous architecture.

Diagnosis

DXA-based BMD measurement is the standard tool for diagnosing osteoporosis. As outlined in Chapter 5, DXA can provide BMD for the total body or for individual regions such as the vertebra or hip. A diagnosis of osteoporosis is made if any one of these sites has a sufficiently low BMD value. Although, on a population basis, BMD is lower in women who have osteoporotic fracture than in those who don't, there is a large overlap in BMD between those who fracture and those who don't (Fig. 16.1, inset). This makes it very difficult to predict which individuals are likely to fracture and which are not. Because of this clinical problem, a computer-based algorithm, the Fracture Risk

offset these behavioral risk factors). By the age of 65, although the risk for fracture in men is less than that in women, the rate of increase in fractures is about the same. However, women and men lose trabecular bone in different ways, and the way that women lose bone puts them at additional risk for fracture. We earlier learned that residual strength and stiffness were more compromised by loss of trabecular number than by

Assessment Tool (FRAX) was developed by the WHO to calculate the 10-year probability of any major osteoporotic fracture (hip, spine, humerus, and wrist), separately of hip fracture, from clinical risk factors. This tool, though still not a perfect predictor, can help physicians decide whether to treat or simply follow an individual patient. The utility of the FRAX tool is unclear. Several studies have examined how FRAX compares to BMD alone for predicting fractures. In general, FRAX does better than BMD alone but this is mostly due to the inclusion of age in the FRAX analysis; the other parameters that go into the algorithm appear to contribute less additional information.

Pathogenesis

Many osteoporoses develop as a consequence of both inherited traits (e.g. small body size) and contributions from nonheritable risk factors (e.g. inadequate calcium intake). Although not everyone becomes osteoporotic by the definitions used here, after reaching peak bone mass everyone loses bone due to age. As explained above, this rate of bone loss is influenced by genetic and lifestyle factors. In this section, we define acquired osteoporoses as those for which a specific known genetic mutation is not responsible. This includes postmenopausal, age-related, and glucocorticoid-induced osteoporosis. Other acquired forms of osteoporosis are listed in (Table 16.1); some of these are described elsewhere in this book.

Postmenopausal Osteoporosis

Estrogen-deficiency osteoporosis, usually occurring as a function of menopause, is the most common form of osteoporosis. Estrogen suppresses bone remodeling on trabecular and endocortical surfaces of bone, and maintains bone mass. Estrogen loss increases the initiation of new remodeling events—measured histomorphometrically by the activation frequency—and results in an imbalance between bone resorption and bone formation at sites of remodeling. The imbalance is not caused by a deficit in osteoblast number or activity at the level of the individual remodeling unit, but rather by an inability of osteoblasts to keep up with the acceleration in resorption. Although there is an imbalance within each bone metabolic unit (BMU), on a global scale bone formation is increased and there may be as much or more bone formation in the early postmenopausal skeleton as in the premenopausal one. The imbalance leads to a 1—2% or greater loss of bone each year. The increase in remodeling rate is responsible for the most rapid changes, although the imbalance between resorption and formation is responsible for the longer-term effects.

Estrogen normally suppresses the production of tumor necrosis factor ligand superfamily member 11/receptor activator of the NF-κB ligand (RANKL) by bone marrow stromal cells and by osteoblasts. At the same time, it increases osteoprotegerin/tumor necrosis factor receptor superfamily member 11B (OPG) production by cells in this lineage. Therefore, loss of estrogen alters the balance between RANKL and OPG in favor of RANKL, consequently increasing the differentiation of osteoclasts and stimulating their activation on bone surfaces, while removing estrogen's proapoptotic effects on osteoclasts. In addition, estrogen normally limits production of various cytokines, including interleukin 1 (IL-1), IL-6, tumor necrosis factor-α (TNF-α), and macrophage colony-stimulating factor 1 (M-CSF1); loss of estrogen stimulates these and promotes osteoclastogenesis. On the formation side, estrogen normally promotes osteoblastogenesis and limits apoptosis of osteoblasts, effects that are lost when estrogen levels decline. Estrogen loss may also impede the release of growth factors that would normally have a positive effect on bone formation. Finally, estrogen prevents osteocyte apoptosis, and the death of osteocytes at menopause may stimulate remodeling and reduce bone quality.

Postmenopausal bone loss leads to reduced cancellous bone volume and loss of trabecular connectivity. The architectural deterioration adds a significant component to the simple loss of bone mass, as thinner trabeculae without the support of trabecular cross-struts can more easily fail by buckling. The loss of cancellous bone is severe, partly because the greater surface area lends itself to more extensive remodeling, which increases the rate of loss. Consequently, sites rich in cancellous bone, such as the spine, femoral neck, and distal radius, are prime sites for increased fracture risk in postmenopausal osteoporosis (PMO). However, significant amounts of cortical bone are also lost. In absolute terms, it is possible that more cortical bone than cancellous bone is lost, although a greater percentage of the total cancellous bone mass is lost in PMO. Endocortical loss causes cortical thinning, which some now believe is equally important to the more noticeable loss of cancellous bone. This is partly compensated by periosteal apposition, which increases the overall diameter of the bone and can ameliorate the loss of strength from endocortical and cancellous bone loss (see Chapters 6 and 9), but the small amount of new bone periosteally does not completely offset the larger loss that occurs endocortically in women. Fig. 16.6 clearly demonstrates thinning of the cortical shell even in the lumbar vertebra, leaving the anterior part of the shell nearly as thin as a single trabecula. Cortical thinning and increased porosity of the superior femoral neck also occurs

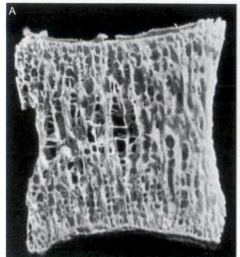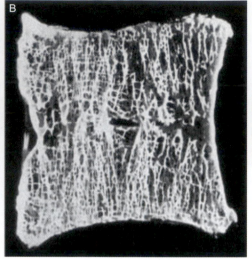

FIGURE 16.6   Although loss of cancellous bone is often the focus of bone loss in the vertebra, there is significant thinning of the cortical shell to the point that portions of the shell can be as thin as an individual trabecula. About 50% of the bone in a vertebra by volume is cortical bone, and thinning of the anterior (to the right in images above) cortex can lead to an increased risk for anterior wedge fractures, as well as compression fractures of the vertebra. *Reproduced with permission from Messent et al. 2007; Osteoarthritis Cartilage 15, 179–186.*

(Fig. 16.7), and is thought to contribute to the increased fragility at this site.

Other factors contribute to either the rate or the significance of bone loss during the postmenopausal period; bone loss is not completely the result of a decline in estrogen levels. As the rate of bone loss can be quite consistent among different individuals following the menopause, those with low bone mass at the start of menopause, whether from genetic causes, poor diet, physical inactivity, or other diseases, are likely to sooner reach a low level of bone mass at which their risk of fracture is greatly increased (this is sometimes called the *fracture threshold*). This is partly the reason that African American women are less susceptible to osteoporosis-related fractures; their bone mass tends to be higher at the start of menopause. Most women lose bone postmenopausally at similar rates, but there is a subpopulation of "rapid losers," and these women are also at greater risk for eventual fracture. Accelerated bone loss at the menopause can occur for genetic reasons, but certain diseases and/or drug treatments, poor nutrition, and inactivity can also accelerate loss.

Age-Related Osteoporosis

Everyone loses bone with age. Age-related loss of bone is sometimes called senile (not a very nice term) osteoporosis. Pathophysiologically, rather than being related to hormonal declines, age-related bone loss is more likely to be caused by calcium and vitamin D deficiency, as intestinal absorption of calcium declines with age and a reduced mesenchymal stem cell population decreases the number of cells available for bone

formation. Furthermore, with aging increased reactive oxygen species and decreased Wnt signaling contribute to decreased osteoblast activity. In age-related osteoporosis, the rate of remodeling may actually slow down, but an imbalance between resorption and formation still exists because of the failure of osteoblasts to completely refill erosion cavities.

Because the essential biological basis for bone loss in PMO and in age-related osteoporosis is different, the timing and site of skeletal loss is different in each pathology. Loss of vertebral bone mass occurs earlier in women than in men because this is the primary site of postmenopausal loss. Vertebral bone loss also occurs in men, but begins later—the seventh or eighth decade, rather than the fourth or fifth, as in women. The wrist fracture rate also increases at menopause in women, but does not increase significantly with age in men, indicating that loss at the radius is more closely related to hormonal deficits than to age effects. This may be one reason why loss of bone at the wrist is highly correlated to loss at the spine—both mostly relate to hormonal changes. Hip fractures, on the other hand, are probably more closely related to age and dietary considerations than to hormonal changes. They tend to occur at about the same age in both men and women, and later than the onset of vertebral fractures.

Glucocorticoid-Induced Osteoporosis

Glucocorticoid excess is most commonly caused by steroid treatment for various diseases, including inflammatory disorders. Less commonly it is the result of endogenous cortisol overproduction by the adrenal

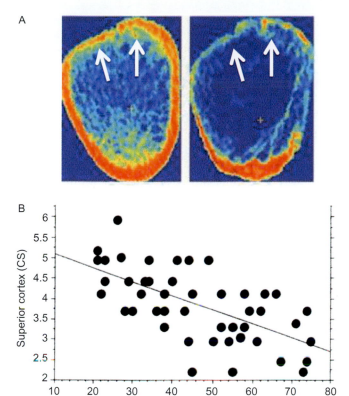

FIGURE 16.7 Cortical bone on the superior aspect of the femoral neck is normally thin, but becomes substantially thinner through endocortical bone loss with age. This is now considered to contribute to the risk of fracture from a sideways fall, which would generate substantial tensile stresses on this surface. (A) Cross-sectional images of the femoral neck from a 20-year-old (left) and 80-year-old (right). Arrows indicate the superior cortex. (B) Graph shows data indicating a consistent thinning of the superior cortex from the third decade to the eight decade of life. *Part (A) is reproduced with permission from Mayhew, P.M., et al Lancet 2005;366, 129–135. The Lancet, 2005;366 (9480):9–15.*

glands. Glucocorticoids affect mineral metabolism and all bone cell populations (Fig. 16.8), causing a rapid loss of bone mass and strength. On the one hand, steroid excess is associated with both a reduction in OPG and an increase in RANKL, which increases differentiation and activity of osteoclasts, while decreasing osteoclast apoptosis, thus causing more osteoclasts to be available for bone resorption. On the other hand, glucocorticoids decrease the differentiation of precursor cells to fully functional osteoblasts, and also increase the rate of apoptosis in fully differentiated osteoblasts. Osteoblast production of type 1 collagen is also impaired. Thus, glucocorticoids both increase resorption, especially early during glucocorticoid excess, and reduce coupled formation, especially with prolonged glucocorticoid excess. In addition, glucocorticoids also increase apoptosis of osteocytes, which can alter intercellular signaling and also appears to compromise bone

tissue quality, leading to osteonecrosis, in which some regions of bone are devoid of living osteocytes.

Glucocorticoids have effects on bone quality, independent of the loss of bone that also occurs in glucocorticoid-induced osteoporosis (GIOP; illustrated by the fracture rates in Fig. 16.9). For a given BMD T-score ≤ −2.0, the risk of hip fracture is about four times greater in those receiving glucocorticoids than in those who are not. Although this may in part be related to other aspects of the medical conditions that require the use of steroids, it nevertheless demonstrates that the risk of fracture in GIOP is not totally determined by low bone density and is probably partly due to high remodeling rates and to stress-risers in trabeculae caused by the incomplete filling of resorption cavities.

Glucocorticoid excess can also decrease intestinal calcium absorption while increasing urinary calcium excretion, leading to a systemic calcium deficiency that further stimulates resorption (via increases in PTH) and leads to bone loss. Additionally, glucocorticoids can reduce the concentrations of sex steroids (either estrogen or testosterone), which increases bone resorption and loss. Glucocorticoids also increase fracture rates through non-bone-related effects such as muscle weakness that increases the risk of falls.

Inherited and Other Conditions Associated with Osteoporosis

A variety of inherited conditions are associated with low bone density and fracture risk, through many different mechanisms. Some, such as cystic fibrosis, cause low bone density as a complication of systemic illness, inflammation, malnutrition, and the drugs such as glucocorticoids used to treat them. Others such as muscular dystrophy cause low bone density due in part to muscle weakness. Various lipid and glycogen storage diseases alter bone cell functions and may variously be associated with neuromuscular abnormalities, inflammatory cytokines, and a variety of effects on bone cells. Galactosemia contributes to low bone density due to ovarian failure and the necessary dietary restriction of dairy products (lactose) in the treatment of the disorder.

The osteoporosis pseudoglioma syndrome causes a primary bone metabolic defect due to mutations in the *LRP5* gene [encoding low-density lipoprotein receptor-related protein 5 (LRP5)] that affect Wnt signaling. Abnormalities in connective tissue proteins also cause osteoporotic phenotypes in osteogenesis imperfecta (OI; type 1 collagen), Ehlers-Danlos syndrome (various collagen genes), and Marfan syndrome (fibrillin-1). Inability to adequately mineralize bone matrix due to mutations in the *ALPL* gene [encoding tissue nonspecific alkaline phosphatase (TNSALP)] causes a range of skeletal disease ranging from a lethal inadequacy of bone formation, to rachitic features, to adult low bone density.

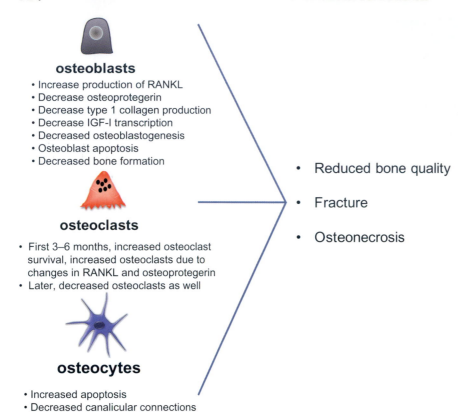

FIGURE 16.8 Glucocorticoids have multiple effects on osteoblasts, osteoclasts, and osteocytes, leading to osteoporotic fracture and also to osteonecrosis. IGF-I, insulin like growth factor-I; RANKL, tumor necrosis factor ligand superfamily member 11/receptor activator of the NF-κB ligand.

osteoblasts
- Increase production of RANKL
- Decrease osteoprotegerin
- Decrease type 1 collagen production
- Decrease IGF-I transcription
- Decreased osteoblastogenesis
- Osteoblast apoptosis
- Decreased bone formation

osteoclasts
- First 3–6 months, increased osteoclast survival, increased osteoclasts due to changes in RANKL and osteoprotegerin
- Later, decreased osteoclasts as well

osteocytes
- Increased apoptosis
- Decreased canalicular connections

- Reduced bone quality
- Fracture
- Osteonecrosis

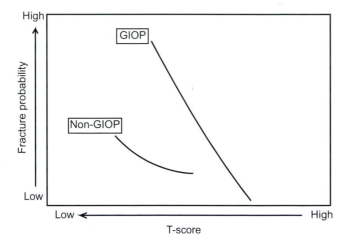

FIGURE 16.9 Among patients receiving glucocorticoids, there is a steeper slope between bone mineral density (BMD) and fracture rate than among those without glucocorticoid exposure, indicating that steroids increase fracture risk independently of BMD. GIOP, glucocorticoid-induced osteoporosis. *Adapted from Van Staa et al. Arthritis and Rheumatism, 2003: 48:3224.*

In addition, many other inherited and acquired diseases cause secondary abnormalities in mineral metabolism and/or function of bone cells, resulting in low bone density. Another condition involves transient, but idiopathic, juvenile osteoporosis with a propensity to fracture and low bone density that resolves after puberty.

As seen in Table 16.1, a large variety of other conditions and medications can contribute to secondary osteoporosis due to hormonal deficits or excesses, malabsorption, and inflammatory processes. While glucocorticoids are the major cause of drug-induced osteoporosis, other medications can also play a role. Aromatase inhibitors block the production and thus the effect of estrogen, resulting in pathology similar to PMO. In addition, adults or children treated with anticonvulsants develop osteoporosis and, in some cases, osteomalacia. The mechanisms whereby anticonvulsants lower BMD and increase fracture risk are not well understood, but have been postulated to involve effects on vitamin D metabolism, though the evidence for this is inconsistent; other effects such as inhibition of calcium absorption and direct osteoblastic inhibition are suggested by animal and in vitro studies.

Many inherited or acquired conditions present with fractures and low bone density in childhood. However, the adult bone density criteria for osteoporosis (a T-score of ≤ −2.5) are not appropriate for children. Children have not yet reached their peak bone density, and using T-scores (comparing to the mean BMD for young adults at peak bone density) will falsely label normal children as abnormal. Rather, children should be compared with age, race, and sex normative data, with adjustments for body size (height). Even then, densitometry is complicated by individual variation in maturation and growth rates across different ages. Appropriate comparisons generate a "Z-score" for age, and bone density more than 2 SD below the mean for age, race, and sex is defined as "low for chronological age." Additionally, children are not labeled osteoporotic based on bone density criteria alone. In children, a diagnosis of osteoporosis is made only after documentation of a combination of low bone density for age plus the occurrence of fragility fracture.

Osteoporosis Therapies

Available osteoporosis treatments address aspects of the bone remodeling cycle. Most of the currently available osteoporosis treatments inhibit osteoclastic bone resorption, as increased resorption is the most common mechanism for developing osteoporosis postmenopausally. Bisphosphonates are the most widely used antiresorptive agents. Other antiresorptive agents include calcitonin, estrogen, selective estrogen receptor modulators, and denosumab (antibodies to RANKL). Because bone remodeling is a coupled process, treatment with

antiresorptive agents also suppresses bone formation. Conversely, bone formation by osteoblasts is stimulated by teriparatide [recombinant human PTH(1–34)]. However, the cornerstone of osteoporosis treatment is encouraging weight-bearing physical activities and adequate (but not excessive) intake of calcium and vitamin D, so that adequate mineral substrate for bone mineralization is present. A detailed discussion of the mechanism of action and effectiveness of various osteoporosis therapies may be found in Chapter 17.

OSTEOGENESIS IMPERFECTA

Definition

OI is the name for a group of rare genetic diseases caused by disruptions in the amount or structure of type 1 collagen. The primary manifestations of OI are increased bone fragility with reduced BMD and ligamentous laxity. OI is highly variable clinically, and has been divided into *types* (currently numbered from I to

TABLE 16.2 Osteogenesis Imperfecta

Type	Inheritance	Phenotype	Gene Defect	Protein Affected
CLASSICAL TYPES				
I	Autosomal Dominant	Mild	Null *Col1A1* allele	Type 1 collagen
II	Autosomal Dominant	Lethal	*Col1A1* or *Col1A2*	Type 1 collagen
III	Autosomal Dominant	Progressively deforming	*Col1A1* or *Col1A2*	Type 1 collagen
IV	Autosomal Dominant	Intermediate	*Col1A1* or *Col1A2*	Type 1 collagen
OTHER ETIOLOGIES				
V	Autosomal Dominant	Hypertrophic callus formation, interosseous membrane calcification, distinctive histology	*IFITM5*	Interferon-induced transmembrane protein 5
MINERALIZATION DEFECT				
VI	Autosomal Recessive	Mineralization defect, distinctive histology	*SERPINF1*	Pigment epithelium-derived factor
3-HYDROXYLATION DEFECTS				
VII	Autosomal Recessive	Lethal (null), Severe (hypomorphic)	CRTAP	Cartilage-associated protein
VIII	Autosomal Recessive	Lethal to severe	*LEPRE1*	Prolyl-3-hydroxylase 1
IX	Autosomal Recessive	Lethal to moderate	*PPIB*	Peptidyl-prolyl isomerase B/cyclophilin B
CHAPERONE DEFECTS				
X	Autosomal Recessive	Lethal to severe	*SERPINH1*	Serpin H1/47 kDa heat shock protein
XI		Progressive deforming	*FKB10*	FK506-binding protein 10
XII	Autosomal Recessive	Intermediate	*SP7*	Specificity protein 7/osterix

Modified from Forlino et al. Endocrinol 2011;7:540–557.

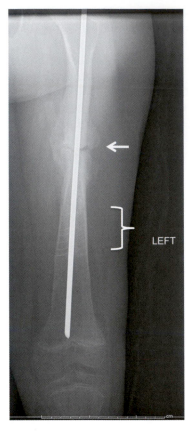

FIGURE 16.10 **Osteogenesis Imperfecta.** Left femur films from 8-year-old child with type V osteogenesis imperfecta. Bone is very hypomineralized. Note mid-transverse fracture and exuberant callus formation with heterotopic ossification (arrow). Fracture was treated with open reduction and intramedullary rod placement. Distal to fracture site, transverse dense lines are visible (to left of bracket), consistent with prior therapies with intravenous bisphosphonate.

XII) based upon the order they were discovered (Table 16.2). OI has an estimated prevalence of 1 in every 10,000 persons. This means that over 30,000 Americans have the disease. It is also important when OI is suspected in very young children to differentiate the disorder from child abuse (inflicted injuries) as well as from other conditions that can cause bone fragility or deformity, including some rachitic disorders.

Persons with mild (type I) OI may only have a few or even no fractures over their lifetimes, whereas other persons with severe forms of OI may have severe deforming skeletal disease with innumerable fractures over their lifetimes (Fig. 16.10). At the extreme end of severity is a lethal perinatal form. Growth impairment is also a common manifestation. Dentinogenesis imperfecta is present in about a third of patients, manifesting as tooth discoloration (opalescent yellow/gray and/or brown), a tendency for enamel cracking with attrition of the exposed dentin, and narrow pulp chambers. Other manifestations of OI can include basilar invagination of the skull, blue sclerae, and progressive hearing impairment. Persons

with OI can also have associated cardiovascular problems, including easy bruising, mitral and aortic heart valve disorders, and aortic root dilation. Some experience abnormal temperature regulation, including intraoperative hyperthermia. Abnormal collagen can also affect platelet-endothelial cell interactions, leading to defective platelet aggregation and a bleeding diathesis. Severe skeletal deformity can lead to disability, limit mobility, and in extreme cases cause respiratory failure due to abnormal thoracic shape.

Pathogenesis

Most cases are inherited in an autosomal dominant manner, although recessive forms have also been described. The majority of OI cases are caused by abnormalities in the amount or structure of type 1 collagen. Type 1 collagen is an important structural protein in bones, tendons and ligaments, skin, and the sclerae of the eyes. Type 1 collagen is a long and rope-like protein, composed of two α1 chains and one α2 chain twisted into a left-handed helix. *COL1A1* encodes the α1 chains and *COL1A2* encodes the α2 chain. Most OI patients have an autosomal dominant mutation in one of these two genes. Mutations that lead to structural collagen abnormalities generally lead to more severe disease than those that result in functional null alleles. This is because collagen molecules must interact to form the helical structure. A null allele results in less total collagen, but the collagen that is formed is normal in structure, usually resulting in a milder disease. An abnormally shaped collagen protein can work in a dominant-negative fashion to disrupt the helical shape and disorganize and weaken the extracellular matrix. In addition, mutations in *COL1A1* are often more severe than those in *COL1A2*, since the α1 chains make up two-thirds of the collagen fibril. However, despite the identification of more than 1500 dominant collagen mutations, phenotypes cannot be predicted from genotypes, and many patients cannot be satisfactorily classified by "type."

Several autosomal recessive mutations causing moderate to severe OI have been found in genes encoding proteins involved in the posttranslational modification or folding of type 1 collagen (Table 16.2). Genes involved directly in posttranslational collagen prolyl—3-hydroxylation are *CRTAP* (encoding cartilage-associated protein), *LEPRE1* (encoding prolyl 3-hydroxylase 1) and *PPIB* (encoding Peptidyl-prolyl cis-trans isomerase B). Peptidyl-prolyl cis-trans isomerase FKBP10/FK506-binding protein 10 (encoded by *FKBP10*) is a molecular chaperone found in the endoplasmic reticulum and involved in type 1 procollagen folding.

Other OI-causing mutations are found in genes encoding proteins involved in bone formation and turnover. Recently the mutation responsible for OI type V, characterized by bone fragility and hypertrophic callus formation has been identified in the *IFITM5* gene [encoding interferon-induced transmembrane protein 5 (IFITM5)]. IFITM5 is expressed only in skeletal tissue and is thought to be important in osteoblast maturation and bone formation. Other mutations that cause OI have been described in *SERPINF1* [which encodes pigment epithelium-derived factor (PEDF)] and *SERPINH1* (which encodes serpin H1/47 kDa heat shock protein). PEDF serves as an angiogenesis inhibitor. Persons with *SERPINF1* mutations have a characteristic mineralization defect on histologic examination and are classified as OI type VI. Serpin H1 is thought to function as a collagen chaperone-like protein, and persons with mutations in *SERPINH1* are now classified as having OI type X. Another recent discovery has been that OI type XII is caused by mutations in the *SP7* gene, which encodes the zinc finger transcription factor Sp7 (also known as osterix) that regulates bone cell differentiation. However, the causative mutation is unknown in 5–10% patients with clinical OI.

Therapies

Physical therapy and rehabilitation for fracture or deformity form the mainstay of OI therapy. After fracture, immobilization should be limited as much as is feasible to prevent further disuse osteoporosis. Calcium and vitamin D intake should also be optimized for age. In more severe cases, intramedullary rods may be placed in the long bones (femurs, tibiae, and humeri) to help stabilize them. Spinal rodding may be done for patients with severe scoliosis.

There are no currently US Food and Drug Administration (FDA)-approved therapies for OI. Bisphosphonates (primarily intravenous pamidronate, but also alendronate and zoledronic acid) have become a standard therapy for children with moderate to severe OI. Bisphosphonate treatment is generally reserved for those with multiple long bone and/or vertebral compression fractures. Therapy during childhood is usually continued until the end of linear growth, with treatments at an interval judged to optimize therapy during growth.

Mesenchymal stem cell transplant or gene therapy to correct the genetic defects are also potential therapies for OI. Gene therapy for humans with OI remains theoretical, but might involve antisense therapies designed to suppress expression of the mutant type 1 collagen allele while allowing expression of the other normal allele. This approach could convert a moderate or severe form of OI into a milder form of the disease by removing the dominant-negative effect of the abnormally shaped collagen.

Therapy for persons with OI should also include hearing assessments, starting in later childhood and adolescence. For persons with dentinogenesis imperfecta, dental care by a dentist with experience with this condition is also important.

HYPERPARATHYROIDISM

Definition

Normally PTH secretion maintains serum ionized calcium concentrations in a highly regulated manner. Hyperparathyroid disorders are characterized by high serum concentrations of PTH in the setting of high or low serum calcium, or PTH concentrations that are inappropriately normal in the setting of high serum calcium. Hyperparathyroidism can be characterized as primary, secondary, or tertiary.

Primary hyperparathyroid disorders are characterized by autonomous secretion of PTH by the parathyroid glands, leading to high or normal PTH concentrations accompanied by high serum calcium. Causes of primary hyperparathyroidism are outlined in Table 16.3.

Secondary hyperparathyroidism is found when PTH is elevated as an adaptive response to depressed serum calcium, decreased 1,25-dihydroxyvitamin D [1,25(OH)$_2$D] or 25-hydroxyvitamin D [25(OH)D] concentrations (in the context of vitamin D deficiency or renal failure), poor calcium intake, excessive renal calcium losses or hyperphosphatemia (especially in the context of renal failure). With prolonged secondary hyperparathyroidism, the parathyroid glands can respond by enlarging and starting to autonomously produce PTH, a condition known as tertiary hyperparathyroidism.

Demographics

Primary hyperparathyroidism is one of the most common endocrine disorders. The incidence of primary hyperparathyroidism in 1974 was estimated to be 16 in every 100,000. Once measurement of serum calcium concentrations on multielectrolyte biochemistry analyzers became routine, diagnosis of this condition improved and the incidence increased to 112 in every 100,000. During the 1990s, an estimated 22 in 100,000 cases per year occurred in Rochester, Minnesota. More current estimates, however, suggest a much lower incidence of 4 in 100,000, perhaps reflecting decreasing population exposures to ionizing radiation and/or improved vitamin D status. Most cases occur after age 45 and the incidence increases with age. The peak occurrence is in women in their 60s. Although male and female incidences are similar before age 45 years, after this age women are twice as likely to be affected as men.

TABLE 16.3 Parathyroid Diseases

Disorder	Cause	Description
Primary hyperparathyroidism	Adenoma (90% of cases)	95% of adenomas are single; 5% double
	Glandular hyperplasia (10% of cases)	All four glands enlarged
	Parathyroid carcinoma (approximately 1% of cases)	—
	Familial hyperparathyroidism	Found in Multiple Endocrine Neoplasia syndromes, although can be isolated; Also Familial Hypocalciuric Hypercalcemia due to inactivating mutations in the *CASR* (autosomal dominant, but homozygous or compound heterozygous states cause severe neonatal hyperparathyroidism)
	Lithium Therapy	—
Secondary hyperparathyroidism	Vitamin D deficiency, calcium deficiency	Associated with hyperplasia of all four parathyroid glands
	Hypocalcemia, hyperphosphatemia	—
	Chronic kidney disease	—
Tertiary hyperparathyroidism	Autonomous hyperparathyroidism following prolonged secondary hyperparathyroidism	Typically with hyperplasia of all four parathyroid glands, sometimes with adenomatous change in one or more glands
Hypoparathyroidism	Congenital forms of hypoparathyroidism	Activating mutations in *CASR* (autosomal dominant)
		Failure to form adequate parathyroid tissue • DiGeorge syndrome • Deletions or mutations of genes responsible for parathyroid gland development (*GCM2*, *GATA3*) Mitochondrial DNA mutations
		PTH mutations (autosomal recessive or dominant)
	Acquired forms of Hypoparathyroidism	Surgical complication of parathyroid, thyroid, or other anterior neck surgery (most common type)
		Autoimmune • Destruction of parathyroid glands (autoimmune polyendocrinopathy-candidiasis-ectodermal dystrophy syndrome) • CaSR-activating autoantibodies Burns
		Heavy metal deposition in parathyroid glands • Copper (Wilson disease) • Iron (thalassemia, hemochromatosis)
		Hypomagnesemia or hypermagnesemia
		Radiation-induced destruction • External beam radiation • I^{131} therapy for Graves disease or thyroid cancer (rare) Tumor invasion of parathyroid glands (rare)

	PTH resistance	• Pseudohypoparathyroidism
		    • Type 1a
		    • Type 1b
		• Transient pseudohypoparathyroidism of the newborn

CaSR, calcium-sensing receptor; PTH, parathyroid hormone.

Risk Factors

Risk factors for primary, secondary, and tertiary hyperparathyroidism vary due to differing but overlapping pathophysiology. Overall risk factors include age, female gender, family history of parathyroid disease, personal or family history of an associated a genetic syndrome, and chronic low calcium intake for primary hyperparathyroidism. Risks for secondary hyperparathyroidism include severe low calcium intake, vitamin D deficiency (and other vitamin D metabolic disorders), and chronic kidney disease. Tertiary hyperparathyroidism can follow any secondary hyperparathyroidism, but is usually due to chronic kidney disease. A history of external neck irradiation is also a risk factor. Polymorphisms in the *VDR* gene (encoding the vitamin D receptor) may also be risk factors.

Pathogenesis

The most common cause of primary hyperparathyroidism is a single parathyroid adenoma. These can sometimes be ectopic in location due to the embryonic migration of the parathyroid glands down through the neck. Adenomas are usually clonal, implicating various acquired mutations or a predisposing genetic syndrome. Lithium, used to treat bipolar disorder, is associated with development of primary hyperparathyroidism, and both adenomas and multigland hyperplasia are reported, perhaps due to exacerbation of underlying disease. The mechanism is unclear, but may involve inhibition of the calcium-sensing receptor (CaSR), or suppression of inositol monophosphatase, an important enzyme in the production of inositol triphosphate, a second messenger for the CaSR system. Parathyroid cancer is a rare form of primary hyperparathyroidism. In contrast, some predisposing conditions (such as vitamin D deficiency and poor calcium intake) are associated with stimulation of parathyroid glands, which can lead to multigland hyperplasia during secondary hyperparathyroidism. Subsequent development of autonomy (tertiary hyperparathyroidism) may occur, as is commonly seen in chronic kidney disease. It may be difficult to determine whether someone has primary or tertiary hyperparathyroidism on clinical grounds.

Signs and symptoms of primary hyperparathyroidism may be severe and can include abdominal pain, altered mental status, coma, ectopic calcifications, hypertension, nephrocalcinosis, nephrolithiasis, osteoporosis and pancreatitis. Hypercalcemia can also cause increased thirst and urination due to osmotic effects in the urine. However, primary hyperparathyroidism is often mild and asymptomatic.

Therapies

Prior to the advent of routine calcium measurements, by the time people presented with primary hyperparathyroidism, most had developed severe kidney and skeletal disease. Today, most people with primary hyperparathyroidism have very few symptoms and mild disease. Mild hyperparathyroidism may be an incidental laboratory finding.

The primary treatment of symptomatic primary or tertiary hyperparathyroidism is surgery. In some cases, hyperparathyroidism is asymptomatic, but treatment may still be required, especially in younger patients and those with renal insufficiency or potential complications of hyperparathyroidism. Very mild primary hyperparathyroidism can be safely observed, as many of these patients do not progress over time. Mild hyperparathyroidism can be difficult to distinguish from familial hypocalciuric hypercalcemia, which cannot be cured with surgery, and generally does not require treatment.

Calcimimetic agents interact with the CaSR to lower PTH, and may be used for primary or tertiary hyperparathyroidism or for parathyroid carcinoma when surgery is not possible due to risks. However, surgery is still the first line treatment as it is potentially curative.

Secondary hyperparathyroidism is treated by addressing deficiencies in calcium intake or vitamin D. Those with secondary hyperparathyroidism in the context of chronic renal failure are managed with phosphate binders, limiting dietary phosphate, and analogs of $1,25(OH)_2D$. Calcimimetics have also emerged as important therapy in managing hyperparathyroidism in the setting of chronic kidney disease.

HYPOPARATHYROIDISM

Definition

Functional hypoparathyroid disorders include primary hypoparathyroidism, where there is failure to produce adequate PTH to maintain normal serum calcium concentrations, as well as disorders characterized by PTH resistance. In the primary hypoparathyroid conditions, PTH levels are either low or inappropriately normal given the degree of hypocalcemia. In addition to hypocalcemia, the functional hypoparathyroid conditions are characterized by hyperphosphatemia and low $1,25(OH)_2D$ concentrations. In PTH resistance disorders, hypocalcemia occurs despite often significantly elevated PTH levels with normal vitamin D status [a condition known as pseudohypoparathyroidism (PHP)].

Inadequate PTH impairs mobilization of calcium from bone, decreases calcium reabsorption in the kidney, and reduces stimulation of kidney 1α-hydroxylase activity, which is essential for $1,25(OH)_2D$ production. When $1,25(OH)_2D$ concentrations are low, intestinal absorption of calcium is inefficient. Low calcium levels can result in symptoms of the nerves, muscles, and heart, including tingling of the hands and feet, seizures, muscle spasms, psychiatric manifestations (including depression), heart signal conduction abnormalities, and heart failure. Neuromuscular irritability from hypocalcemia can be detected on examination. Chvostek sign is demonstrated when tapping on the facial nerve in front of the ear below the zygomatic arch elicits a twitch of the lips on the same side of the face. Trousseau sign is demonstrated by inflating a blood pressure cuff on the arm to occlude the brachial artery, causing a spasm of the hand and wrist. Patients with chronic hypoparathyroidism often develop basal ganglia calcifications, though the mechanism for this is unclear. These can be present at diagnosis and are related to the duration of hypocalcemia.

Demographics

Hypoparathyroidism is considered a rare condition, with an estimated prevalence in the United States of 80,000 to 100,000 affected persons. It can either be congenital or acquired. It has a slightly higher prevalence in women than in men due to the increased prevalence of autoimmune hypoparathyroidism in women. PHP is even rarer, with an estimated prevalence of 0.79 cases per 100,000 persons.

Risk Factors and Pathogenesis

The parathyroid glands are located in the neck posterior to the thyroid gland. The most common cause of hypoparathyroidism is surgical removal or damage during parathyroid or thyroid surgery or other neck dissection surgery.

Magnesium depletion or excess can interfere with parathyroid function and result in hypocalcemia. Magnesium is necessary for both PTH secretion and PTH receptor (PTHR) activation. When circulating magnesium concentrations are low, the parathyroid glands are unable to secrete sufficient PTH and there is a blunted response at the PTHR. Conversely, when circulating magnesium concentrations are excessive, magnesium divalent cations can activate extracellular CaSRs on the parathyroid glands, decreasing PTH release. Other etiologies of hypoparathyroidism, including parathyroid autoimmunity, parathyroid development disorders, disorders associated with parathyroid gland damage, and PTH resistance, are outlined in Table 16.3.

The term PHP refers to persons with hypocalcemia and hyperphosphatemia due to PTH resistance. Persons with PHP type 1a do not increase urinary cyclic AMP or urinary phosphate excretion when administered PTH. They may also have resistance to other hormones that use Gs-coupled receptors, such as thyroid-stimulating hormone, gonadotropins (luteinizing hormone and follicle-stimulating hormone), and growth hormone releasing hormone. Persons with

PHP type 1a also have characteristic physical features, including short stature, obesity, short fingers and toes, heterotopic subcutaneous ossification, and often varying degrees of mild mental retardation. This cluster of features is known as Albright hereditary osteodystrophy (AHO). Approximately 70% of PHP type 1a cases are caused by maternally inherited inactivating mutations of *GNAS* (encoding the stimulatory protein Gsα), which is coupled to the PTHR. However, due to imprinting resulting in differential expression of alleles in various tissues, when the same mutations are paternally inherited, the individual does not have hormonal resistances but has phenotypic features similar to AHO (termed *pseudopseudohypoparathyroidism*).

Persons with PHP type 1b have the hormone resistance but not the physical features of AHO. Like those with PHP type 1a, those with type 1b do not increase urinary cyclic AMP or urinary phosphate excretion in response to PTH. However, these people have normal Gsα activity in other tissues and do not appear to carry *GNAS* mutations. It seems instead that PHP type 1b is generally caused by the disruption of elements of the *GNAS* locus that affect long-range gene imprinting.

Therapies

Hypocalcemia can cause tetany, seizures, and arrhythmias, and in such cases therapy consists of the rapid administration of intravenous calcium until convulsions stop. Chronic treatment for hypoparathyroidism consists of oral doses of $1,25(OH)_2D$ (usually $0.5-2\,\mu g$/day for adults) given along with calcium supplementation (usually $1-3\,g$/day calcium for adults). For some patients with high serum phosphate levels, a low-phosphorus diet is prescribed. However, given the presence of phosphate in many foods in the Western diet (meats, eggs, dairy products, and colas), adherence to this is very difficult.

The goal is to treat with doses sufficient to maintain serum calcium at a level slightly lower than normal range (approximately 8.0 mg/dl), without neuromuscular symptoms. This range is targeted to minimize hypercalciuria, which is unregulated in the absence of PTH and can cause nephrocalcinosis. Careful monitoring of serum calcium levels, as well as urine calcium excretion, is necessary. Despite optimal therapy, persons with hypoparathyroidism often report reduced quality of life and have an increased risk of neuropsychiatric disease.

Studies of the use of injectable PTH for hypoparathyroidism have been published; however, the use of PTH therapy for this condition remains limited to research settings. This is a result of theoretical concerns for osteosarcoma in growing children, as occurred in animals during preclinical studies for osteoporosis. Additional concerns include calcium fluctuations due to the short half-life of injectable PTH, possible adverse effects on urinary calcium excretion and BMD, and cost.

NUTRITIONAL RICKETS/OSTEOMALACIA

Definition

Rickets refers to insufficient mineralization of osteoid at the growth plate, accompanied by an alteration in the growth plate structure. It manifests clinically as skeletal deformation. A similar process, called *osteomalacia*, involves deficient osteoid mineralization on other bone surfaces. While osteomalacia coexists with rickets in growing children, it is also possible to have osteomalacia without obvious rickets in children. Osteomalacia can occur in adults but, since the growth plates have fused, rickets does not. The process of osteomalacia should be distinguished from osteoporosis or osteopenia, which also include low bone volume but normal mineralization. Though both can present with low bone density, and therefore are indistinguishable based on DXA scans alone, the treatments are different, and most osteoporosis agents are contraindicated in the setting of osteomalacia.

Histologically, rickets and osteomalacia appear as wide osteoid seams with delay in the mineralization rate. With rickets, the growth plates become irregular, and widened, with delayed apoptosis of hypertrophic chondrocytes. Radiographic imaging of the growth plate shows expanded growth plates, along with a cupped and frayed appearance (Fig. 16.11). There may be bowing or torsional deformation of long bones, typically the legs. Additional features include frontal bossing, craniotabes (softened cranial bones), wide wrists, knees and ankles, and widening of the costochondral junctions (the so-called *rachitic rosary*). Prolonged untreated rickets can cause short stature, in addition to leg deformities. Rickets may be associated with proximal muscle weakness and delayed gross motor milestones. The clinical presentation of rickets is the same regardless of underlying etiology, though the degree of severity varies between individual patients, even among patients with the same etiology.

Rickets and osteomalacia can have acquired or inherited causes. Acquired causes are usually due to impaired nutrition or malabsorption of vitamin D and calcium, and less commonly of phosphate. Therefore, biochemical testing of 25(OH)D, calcium, creatinine, and phosphorus are useful measures to diagnose rickets or osteomalacia. Serum calcium and phosphate

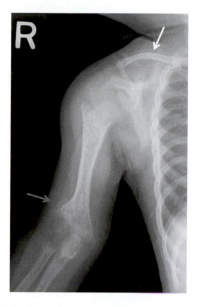

FIGURE 16.11 **X-ray of 15-month-old boy with vitamin D-deficient rickets.** Note fracture of the right clavicle (white arrow). There is also fraying and cupping of the humeral metaphyses (distal metaphysis marked with gray arrow) with some mottling of the bones consistent with rickets.

may be low-normal or low in the setting of nutritional rickets. Measurement of serum alkaline phosphatase (ALP) is also useful, as persons with rickets tend to have total (ALP) and bone-specific alkaline phosphatase (BSAP) concentrations that are elevated for age. Low ALP in the setting of clinical and radiographic rickets suggests hypophosphatasia (HPP) rather than nutritional rickets. In persons with rickets, PTH is increased in response to low vitamin D and/or calcium, resulting in low urinary calcium excretion. Measurement of the active hormone $1,25(OH)_2D$ is not useful if nutritional deficiency is expected as, due to the stimulatory effect of PTH on 1α-hydroxylase activity, the concentrations of $1,25(OH)_2D$ may be elevated, normal, or low in states of vitamin D deficiency. The ambient concentration of $1,25(OH)_2D$ is probably dependent on the amount of 25(OH)D substrate. In addition, physiologically, 25(OH)D circulates in nanomolar (10^{-9} M) concentrations, while $1,25(OH)_2D$ circulates in picomolar (10^{-12}M) concentrations. Thus, the measurement of $1,25(OH)_2D$ in rickets is limited to situations where etiologies of rickets other than nutritional deficiencies are being considered (such as genetic abnormalities affecting 1α-hydroxylase or the VDR).

Risk factors

Vitamin D deficiency (due to impaired intake or malabsorption) remains the most common cause of rickets or osteomalacia. However, vitamin D deficiency or insufficiency is not synonymous with rickets. Many individuals are vitamin D insufficient but do not have clinical evidence of rickets.

Nutritional rickets patients typically have a mixed dietary deficiency of vitamin D and calcium, though one factor usually predominates. Risk factors for vitamin D deficiency include lack of adequate skin sunlight exposure and darker skin pigmentation, which impairs the ultraviolet light-dependent step in vitamin D production. People living at higher latitudes have impairments in vitamin D production due to decreased skin exposure to sunlight and to wintertime decreases in ultraviolet B exposure caused by the angle of sunlight. Dairy avoidance, due to allergy, lactose intolerance, vegan diets, and replacement of milk with sugary drinks, all contribute to the risk of calcium and vitamin D deficiency. Breast milk has relatively low vitamin D content. This places exclusively breastfed children at a risk for vitamin D deficiency. Even though infant formula is supplemented with vitamin D, current recommendations are to provide additional vitamin D supplementation to all infants starting at birth to ensure adequate intake.

Many different clinical conditions can lead to impairments in intake or absorption of vitamin D and other fat-soluble vitamins. These include surgical resection of the intestine, celiac disease, pancreatitis, cystic fibrosis, inflammatory bowel disease, and biliary diseases. Severe liver disease may contribute to malabsorption, and in very severe liver disease there may also be impaired 25-hydroxylation of vitamin D.

Primary nutritional phosphate deficiency can occur, but is less common since phosphate is present in many foods in Western diets. Contributing factors to phosphate deficiency include generalized malnutrition or the use of phosphate-binding substances, including antacids. Additionally rickets of prematurity occurs in the setting of insufficient phosphate and calcium intake, during a time when, if in utero, there would have been the highest demand for mineral accrual (third trimester). These infants present either due to fracture in the neonatal intensive care unit, or to recognized rachitic deformities on exam or radiographs. While human breast milk is an "optimal" diet for term infants, it lacks sufficient mineral for preterm infants, and breast milk fortifiers are standard for preterm infants.

Pathogenesis

Vitamin D is either produced in the skin as cholecalciferol (vitamin D_3), or taken in through the diet as ergocalciferol (vitamin D_2) or cholecalciferol. Milk,

some juices, and other food products are fortified with vitamin D. Deficiency of vitamin D limits the absorption of calcium and phosphorus. Vitamin D metabolism and mineral absorption are discussed in Chapter 13.

One consequence of vitamin D deficiency is insufficient mineral to adequately mineralize the osteoid produced by osteoblasts. There are also direct effects of vitamin D deficiency on bone cells. At the growth plate, the hypertrophic chondrocytes fail to undergo apoptosis properly, resulting in an elongation and disruption of the hypertrophic zone. There is also impairment in the mineralization at this location and, consequent to the mineralization defect, the bone at the growth plate becomes irregular, widened, and thicker. The bone is softer than normal both at the growth plate and in the diaphysis, resulting in the clinical features of rickets.

Rickets and osteomalacia can also have inherited causes. Rare inherited disorders in vitamin D metabolism include deficiency in 1α-hydroxylase activity or in the VDR, both of which involve significant hypocalcemia as well as rickets. Inherited hypophosphatemic disorders (such as X-linked hypophosphatemic rickets, discussed below), however, are more common than other inherited rickets.

Therapies

Nutritional forms of rickets are prevented by the adequate intake of vitamin D, calcium, and phosphate. Recommended intakes, based on estimates of amounts necessary to prevent rickets and osteomalacia, are listed in Chapter 13 and vary somewhat with age. In general, though, at most ages, 600 IU/day vitamin D is recommended, while calcium intake recommendations range from 700 mg/day at 1 year of age to 1300 mg/day in adolescence. Infants require 400 IU/day vitamin D. However, these recommendations are based on the needs of healthy people and higher doses are likely to be required in the setting of malabsorption.

The purpose of treating rickets is primarily to resolve skeletal deformities, facilitate healing of fractures (these are not a common feature of rickets), and resolve bone pain if present. Repletion of deficiencies is generally highly effective for treating the skeletal features of nutritional rickets; radiographic features begin to improve rapidly. The ability to correct skeletal deformities depends in part on duration of the deficiency. Patients with prolonged nutritional rickets over several years may develop permanent abnormalities in bone shape, similar to those occurring in inherited forms of rickets.

During the initial treatment of rickets, higher doses of these nutrients are needed to begin bone healing, replenish vitamin D stores, and, in some cases, correct hypocalcemia. A variety of treatment regimens have been studied ranging from 1000 to 8000 IU/day, and in some cases much higher doses for a period of time. However, higher doses result in a higher risk of developing vitamin D toxicity. Higher doses need to be time limited, often given for only a month; if the patient fails to stop taking the higher doses, they can eventually develop complications. Key factors in the treatment of vitamin D deficiency include a limited period of higher dose vitamin D with close monitoring, followed by maintenance supplementation with age-appropriate amounts of vitamin D. Conversely, some have advocated "stoss" therapy. This is a historical term that refers to treating any condition with a single large dose of an agent, from vitamins, to anti-infective agents, to chemotherapy, as a complete treatment. Today, this term is mostly applied to giving very high doses of vitamin D over 1 or 2 days for complete repletion. Patients generally also require supplemental calcium at the onset of treatment, and may be at risk for hypocalcemia after initiating vitamin D treatment, especially with stoss therapy. Hypocalcemia is thought to be a consequence of increased calcium uptake into the osteoid as treatment allows improvements in mineralization.

Close biochemical monitoring is important and includes serial measurement of ALP, calcium, and PTH. Early in the course of treatment there will often be a transient rise in ALP but, with time and adequate replacement, ALP and PTH will normalize. Urine calcium often remains low until sufficient replenishment of bone calcium has occurred. Then hypercalciuria may develop and the doses of calcium and vitamin D must be decreased. Serum calcium and phosphate rise with treatment, and excessive levels also indicate overtreatment. Once 25(OH)D levels above 30 ng/mL have been reached, the patient should decrease to maintenance supplementation. If the cause of deficiency is malabsorption, then higher maintenance supplements may be required, with ongoing monitoring of levels. Rachitic changes are monitored with X-rays, and significant healing occurs over a few months with adequate treatment.

The primary complications of treatment relate either to either inadequate or excessive treatment. Inadequate treatment either fails to heal the bones or takes a prolonged time to heal the bones. However, excessive doses of vitamin D leads to hypercalciuria, hypercalcemia, and possibly nephrolithiasis or nephrocalcinosis and impaired renal function. An excessive calcium-phosphate product may also lead to ectopic calcification.

For inherited disorders leading to rickets, the standard treatments differ from those for nutritional rickets. With 1α-hydroxylase deficiency, 25(OH)D cannot be catalyzed to 1,25(OH)$_2$D, and bypassing this system with administration of physiologic doses of calcitriol [1,25(OH)$_2$D] can effectively overcome the defect in mineral metabolism, thus improving rickets and growth and allowing maintenance of normocalcemia. With *VDR* mutations impairing the activity of vitamin D, some patients have mild hypocalcemia, while others have severe hypocalcemia requiring intravenous calcium administration. Such patients are also treated with very large supraphysiologic doses of vitamin D or analogs, with varying degrees of responsiveness. In both these situations, the management is complicated, and there is still a risk of overtreatment resulting in hypercalciuria.

HYPOPHOSPHATEMIA

Definition

Hypophosphatemia refers to the presence of serum phosphate concentrations lower than age-appropriate normal values. Hypophosphatemic disorders are a less common cause of rickets and osteomalacia than are the nutritional disorders. However, they do represent an important class of disorders, as they have provided great insight into the mechanisms of normal physiology, and their treatment is quite different to that of vitamin D deficiency.

Hypophosphatemia may be acute or chronic. Acute hypophosphatemia causes cardiac dysfunction, arrhythmia, respiratory muscle weakness, skeletal muscle weakness, and a variety of acute neurological issues. Severe hypophosphatemia can cause hemolysis or rhabdomyolysis. Chronic hypophosphatemia results in fatigue, muscle weakness, bone pain, and the metabolic bone abnormalities of rickets and osteomalacia.

Hypophosphatemia is caused by excessive intracellular transport of phosphate (leaving the net total body phosphate unchanged), dietary deficiency/malabsorption, or inappropriate renal phosphate losses (termed *phosphate wasting*). While shifting of phosphate between extracellular and intracellular compartments occurs in many clinical situations, the net total phosphate is not changed in this process, which is short-lived, and osteomalacia does not occur. Conditions leading to impaired intestinal phosphate absorption include the use of many commonly used medications or generalized malabsorption. These conditions can cause osteomalacia from phosphate depletion. However, given the preponderance of phosphate in the diet, these are unusual causes for osteomalacia in the absence of simultaneous deficiencies in vitamin D and other nutrients.

The most common form of chronic hypophosphatemia is X-linked hypophosphatemia, with an estimated incidence of about 1 in 20,000. The mechanism of hypophosphatemia in this disorder is isolated renal phosphate wasting. Other renal phosphate wasting disorders, including both inherited and acquired disorders, are rarer. Additionally, general renal tubular disorders, which cause renal losses of multiple ions and molecules, can be genetic or drug induced. Since many renal phosphate wasting disorders are hereditary, a detailed family history is useful. However, sporadic mutations are also frequent.

Clinically, patients with chronic hypophosphatemia develop rickets and osteomalacia. Like nutritional rickets, the features become more obvious once the child starts weight-bearing with walking, although manifestations can be identified earlier. Vitamin D deficiency cannot be distinguished from hypophosphatemic rickets based on radiographic or exam features alone. Consequently, the determination must be made based on age-appropriate serum phosphate concentrations (see Chapter 13), recognizing that vitamin D deficiency can also cause hypophosphatemia due to elevations of PTH and subsequent phosphaturia. However, in primary hypophosphatemia, the 25(OH)D level is often normal and, even if 25(OH)D is low, the hypophosphatemia fails to correct even when the PTH value returns to within a normal range. Once hypophosphatemia is determined, an assessment of renal phosphate reabsorption is necessary.

Simultaneous serum and urine phosphate and creatinine measurements enable the determination of fractional excretion of phosphate, and estimation of the tubular maximum phosphate reabsorption adjusted for glomerular filtration rate (TmP/GFR). The TmP/GFR indicates the approximate serum level at which phosphate is no longer completely reabsorbed and enters the urine (Fig. 16.12). Normal ranges are similar to age-appropriate normal ranges for serum phosphate concentrations. When the serum phosphate is low (e.g. because of dietary restriction), the appropriate homeostatic response is to raise the TmP/GFR by modifying phosphate reabsorption through the actions of PTH and FGF-23. An inappropriately low TmP/GFR indicates the etiology is renal loss. Once phosphate wasting is identified, additional factors, such as wasting of other substrates (as in Fanconi syndrome) or the presence of hypercalciuria, can help in delineating non-FGF-23-mediated causes. Assessments of hypophosphatemia also include measurements of 25(OH)D, and 1,25(OH)$_2$D as factors contributing to hypophosphatemia. Measurement of the FGF-23 concentration is also beginning to be used clinically, but is not typically necessary for adequate diagnosis. In fact,

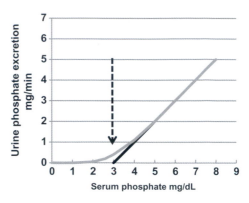

FIGURE 16.12 Tubular maximum reabsorption of phosphate adjusted for glomerular filtration (TMP/GFR). This simplified diagram indicates the normal relationship of serum phosphate concentration to urinary phosphate excretion during phosphate infusion. As serum phosphate increases, the urinary excretion of phosphate also increases (gray line). While there is some splay at the lower levels, the slope becomes steady as the phosphate concentration increases. Drawing back along this line to the intercept indicates the TmP/GFR, which indicates approximately the serum level at which one starts to lose phosphate in the urine. In conditions of abnormal renal phosphate losses, this curve is shifted to the left. Thus a low TmP/GFR means that one has urinary phosphate excretion starting at a lower serum phosphate.

multiple different disorders cause FGF-23-mediated hypophosphatemia.

X-Linked Hypophosphatemia

The X-linked dominant disorder, X-linked hypophosphatemia (XLH), is the most common form of inherited hypophosphatemia. Patients with XLH have inactivating mutations in the *PHEX* gene [which encodes phosphate-regulating neutral endopeptidase (PHEX)], with many different mutations of different types reported. Disease severity is highly variable, in terms of severity of the biochemical skeletal phenotypes. However, it is still unclear whether this gene normally has a role in phosphate metabolism. PHEX overexpression does not result in a phosphate disorder, although deficiency does.

PHEX is expressed in osteocytes/osteoblasts and odontoblasts. Patients with PHEX deficiency have increased expression and secretion of the FGF-23 peptide hormone, which has a clear role in renal phosphate metabolism. FGF-23 interaction with the FGF receptor (FGFR) and Klotho in the kidney decreases the surface expression of sodium phosphate cotransporters NPT2a (encoded by *SLC34A1*) and NPT2c (encoded by *SLC34A3*), resulting in impaired phosphate reabsorption. As phosphate is freely filtered at the glomerulus, reabsorption is critical to maintain normal phosphate homeostasis. FGF-23 also inhibits the 1α-hydroxylase, resulting in inappropriately low 1,25(OH)$_2$D concentrations. Consequently, FGF-23 excess causes the biochemical phenotype of XLH: hyperphosphaturia, hypophosphatemia, and inappropriately low or normal 1,25(OH)$_2$D concentrations. Hypophosphatemia itself decreases apoptosis of growth plate hypertrophic chondrocytes, resulting in the histologic and clinical phenotype of rickets. In addition, in PHEX (Phex)-deficient mice, reduced chondrocyte expression of the sodium-dependent phosphate transporter 1 (PiT-1) contributes to reduced cellular phosphate uptake and reduced chondrocyte apoptosis. Clinical features of XLH are similar to those of nutritional rickets. Additionally, dental abscesses are common. Odontoblast PHEX deficiency results in abnormalities in the dentin and cementum layers, making the teeth prone to infection. Patients with XLH also develop calcification of entheses, which limits their range of motion at joints, causing stiffness. Secondary hyperparathyroidism occurs commonly, probably due to effects of FGF-23 on 1,25(OH)$_2$D production and the effects of treatment with phosphate.

Other Renal Phosphate Wasting Disorders

The mechanisms of XLH and other renal phosphate wasting disorders are indicated in Fig. 16.13. Additional rare disorders cause renal phosphate wasting due to FGF-23 excess. Notably, inactivating mutations in *DMP1* [encoding dentin matrix acidic phosphoprotein 1 (DMP-1)] and *ENPP1* [encoding ectonucleotide pyrophosphatase/phosphodiesterase family member 1 (E-NPP 1)] can cause autosomal recessive hypophosphatemic rickets via excess FGF-23 production. In fibrous dysplasia, bone lesions express FGF-23, and patients with more extensive lesions are more likely to develop renal phosphate wasting and hypophosphatemia. In tumor-induced osteomalacia, FGF-23 production by mesenchymal tumors (usually small) results in the biochemical phenotype.

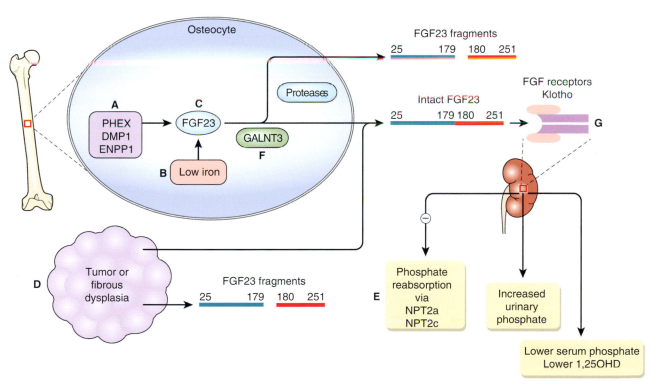

FIGURE 16.13 **Pathways involved in hypophosphatemic and hyperphosphatemic diseases.** (A) Deficiencies of PHEX (X-linked hypophosphatemia), DMP-1, or E-NPP 1 (both autosomal recessive hypophosphatemic rickets) result in increased FGF-23 production. (B) Low serum iron concentration also stimulates FGF-23 expression, which only appears to cause excess intact FGF-23 in the setting of (C) mutations in *FGF23* (in autosomal dominant hypophosphatemic rickets). (D) Some mesenchymal tumors and fibrous dysplasia lesions produce excess FGF-23. (E) Mutations affecting NPT2c impair phosphate reabsorption, resulting in hereditary hypophosphatemic rickets with hypercalciuria. Renal tubular dysfunction from a variety of causes results in a Fanconi syndrome, including wasting of phosphate as well as other ions, glucose, and amino acids. Conversely, hereditary hyperphosphatemic tumoral calcinosis occurs due to other mutations in *FGF23* (C) and *GALNT3* (F) result in increased susceptibility of FGF-23 to cleavage, with more fragments and less intact FGF-23 being secreted, causing hyperphosphatemia. Similarly chronic kidney disease and mutations in Klotho (G) impair the response to FGF-23 signaling and also cause hyperphosphatemia.

In contrast, Fanconi syndrome, caused by a variety of genetic mutations and drugs, results in loss of multiple minerals, glucose, and amino acids in the urine. Hypophosphatemia in this setting is accompanied by compensatory decreases in FGF-23. In hereditary hypophosphatemic rickets with hypercalciuria (HHRH), mutations in *SLC34A3* (encoding NPT2a) result in primary renal phosphate losses, with compensatory decrease in FGF-23, but also compensatory increases in $1,25(OH)_2D$ leading to increased intestinal calcium absorption and hypercalciuria. HHRH patients develop both rickets/osteomalacia and a propensity to nephrolithiasis and nephrocalcinosis. In contrast, mutations in *SLC34A1* (encoding NPT2a) have been reported to cause a Fanconi syndrome due to abnormal processing of the protein intracellularly.

Autosomal dominant hypophosphatemic rickets (ADHR) is caused by mutations in *FGF23* affecting a cleavage site that make the encoded protein resistant to proteolytic cleavage. Consequently, higher FGF-23 concentrations develop that lead to hypophosphatemia. However, the mutations have variable penetrance, with some subjects developing features in childhood and some only in adulthood, and with waxing and waning of FGF-23 concentrations, along with resultant changes in phosphate metabolism. Recent work in humans and mouse models has indicated that the elevated FGF-23 concentrations occur in ADHR primarily during periods of iron deficiency. Low iron concentrations result in increased production of FGF-23, but usually phosphate homeostasis is then maintained by appropriate cleaving of FGF-23 to inactive fragments. However, those with ADHR have impaired FGF-23 cleavage and develop hypophosphatemia.

Therapies

Treatment of hypophosphatemia involves treating any dietary deficiencies, or supplementing vitamin D and phosphate when malabsorption is present. In acute severe hypophosphatemia, intravenous dosing is sometimes needed, but requires close monitoring due to risk of precipitating severe hypocalcemia.

However, the treatment of renal phosphate wasting disorders is more complicated. No current treatment addresses the renal phosphate losses. FGF-23-mediated renal phosphate wasting disorders are treated with both phosphate and calcitriol, which target the results of FGF-23 excess. This requires the use of relatively high doses of these two agents to improve the rickets/osteomalacia. In addition, calcitriol increases the intestinal absorption of both calcium and phosphate. Thus, the kidneys must excrete a higher amount of calcium, as well as excreting high amounts of phosphate. This can result in one of the major complications of treatment, nephrocalcinosis, which can cause kidney failure. Thus, normalizing the serum phosphate is not the treatment target. Another important complication of therapy with phosphate is the development of hyperparathyroidism, sometimes requires surgical treatment. It has recently been demonstrated that treatment with calcitriol and phosphate also stimulates FGF-23 production in patients with XLH, though whether this has clinical consequences is currently unknown.

The primary goal of treatment of chronic hypophosphatemia is to heal the osteomalacia and improve growth and leg deformities in growing children. However, the current standard therapy is imperfect, and many children with XLH require surgical intervention to help straighten leg deformities. Likewise, there is some body disproportion that increases through childhood, with the leg length being disproportionately affected compared with the trunk length. Growth hormone treatment has been attempted, but studies with this agent remain limited.

HYPERPHOSPHATEMIA

Definition

Hyperphosphatemia is present when serum phosphorus concentrations are greater than the upper limit of normal for age. Hyperphosphatemia is most commonly due to excessive phosphate intake in the setting of chronic kidney disease, where clearance of phosphate is impaired. Chronic kidney disease is a growing problem around the world, related in part to the gradually increasing numbers of persons with diabetes. Another common cause of chronic hyperphosphatemia is hypoparathyroidism. Inherited disorders related to FGF-23 metabolism causing hyperphosphatemia are rarer. Acute hyperphosphatemia also occurs in the setting of acutely excessive phosphate intake (such as oversupplementation or overtreatment of hypophosphatemia, especially with intravenous administration) or can be caused by vitamin D intoxication. Acute hyperphosphatemia may occur during the use of phosphate-containing enemas or bowel-cleansing regimens in preparation for surgery. Hyperphosphatemia also occurs due to cell lysis, as can occur during phlebotomy or during processing of the blood sample (pseudohyperphosphatemia), but also during intravascular hemolysis, rhabdomyolysis, or tumor lysis syndrome, especially with hematologic malignancies. Typically, the cause of hyperphosphatemia is apparent from clinical history.

The biggest risk from hyperphosphatemia is the development of acute kidney failure or worsening of chronic kidney disease. In addition, ectopic soft tissue or vascular calcifications may develop due to complexing of phosphate with calcium. Mounting evidence has indicated that hyperphosphatemia, even of mild degrees, is associated with cardiovascular disease outcomes and mortality. This is most pronounced in the population with chronic kidney disease, but is also noted in the general population.

Hereditary Hyperphosphatemia

Hyperphosphatemic familial tumoral calcinosis (TC) is a rare autosomal recessive disorder of impaired phosphate excretion caused by abnormalities in FGF-23 or its action. Loss-of-function mutations in three different genes have been identified to cause TC in humans: *FGF23*, *GALNT3* (which encodes polypeptide N-acetylgalactosaminyltransferase 3 (GalNac-T3), and *KL* (which encodes Klotho). Patients with this disorder develop soft tissue calcification, especially around joints, but these may also occur in the vasculature, skin, areas of cartilage, at the dura, or in ocular structures. These calcifications can be painful and impair joint movement. After surgical resection, they tend to recur. Additionally some patients develop areas of skeletal hyperostosis, with or without TC lesions. Biochemically, these patients have an "opposite" phenotype to that of the FGF-23-mediated renal phosphate wasting disorders (e.g. XLH). Patients with TC develop hyperphosphatemia and impaired phosphate excretion (increased tubular reabsorption of phosphate), and have inappropriately high or high-normal $1,25(OH)_2D$ concentrations. The high $1,25(OH)_2D$ plus elevations in phosphate and calcium absorption, coupled with impaired phosphate excretion, result in deposition of calcium-phosphate crystals in soft tissue structures. A similar clinical phenotype can sometimes occur due to chronic kidney disease, though advanced chronic kidney disease patients typically have low $1,25(OH)_2D$ concentrations.

Mutations in *GALNT3* cause hyperphosphatemic TC due to lack of O-linked glycosylation of FGF-23. This makes FGF-23 more susceptible to cleavage and impairs secretion of intact full-length FGF-23. Consequently, these patients develop increased production of FGF-23 fragments (which can be detected with a C-terminal FGF-23 assay), and low concentrations of intact FGF-23 (detected using a separate assay). The result is that, despite greatly increased *FGF23* gene expression, as observed in *Galnt3*-knockout mice, the circulating intact FGF-23 concentration is low, and hyperphosphatemia develops.

Mutations in *FGF23* itself that affect the N-terminal portion of FGF-23 can also lead to increased intracellular proteolytic degradation and impaired secretion of intact FGF-23, causing TC. A similar biochemical phenotype of increased FGF-23 fragments, low intact FGF-23, and hyperphosphatemia also occurs in these individuals. The phenotype is similar to that seen in *Fgf23*-null mice (except for the absence of FGF-23 in these mice), though the mouse phenotype is considerably more severe with a much shortened lifespan.

Finally a mutation in *KL*, which encodes a critical FGF-23 coreceptor with FGFR1, 2 or 3, results in impaired FGF-23 signaling. The *Kl*-null mouse has a phenotype strikingly similar to that of the *Fgf23*-null mouse, with biochemical abnormalities of hyperphosphatemia and high $1,25(OH)_2D$. Reduced FGF-23-mediated signaling through FGF receptors can impair end-organ effects of FGF-23, causing a compensatory increase in FGF-23 production that results in high levels of intact FGF-23 (in contrast to those TC patients with excessive cleaving of FGF-23).

Therapeutic Agents

Hyperphosphatemia is difficult to effectively manage, especially chronically. Management includes adequate hydration (in acute forms), as well as limiting dietary phosphate intake or absorption. Acetazolamide is a carbonic anhydrase inhibitor that can in some situations increase renal phosphate excretion rate, though the effect may be somewhat limited, especially for inherited disorders. Dialysis does effectively remove phosphate. However, in most situations, even with standard dialysis methods, it is necessary to engage in limiting dietary phosphate intake, but the high amounts of phosphate in Western food sources limits the effectiveness of this strategy. Consequently, a variety of phosphate-binding agents are used, including calcium-, magnesium-, or aluminum-based antacids. In the setting of hyperphosphatemia, there is concern that also increasing the amount of absorbed calcium could increase calcium-phosphate deposition. Specific binding resins have also been developed such as sevelamer, which avoid the problems of excess calcium, magnesium, and aluminum intake. Phosphate binders are used in the management of hyperphosphatemic TC, but the evidence is limited to case reports and the reported effectiveness is quite variable.

HYPOPHOSPHATASIA

Definition

HPP describes a spectrum of disorders of impaired mineralization of bone due to mutations in the *ALPL* gene (formerly *TNSALP*) encoding tissue-nonspecific alkaline phosphatase (TNSALP). These missense mutations are inherited in either an autosomal recessive or dominant pattern, and result in loss or decreased activity of the TNSALP enzyme. The more severe perinatal and infantile forms are inherited in an autosomal recessive pattern, while dominant forms are generally milder. The clinical spectrum of the disease varies widely, with most manifestations being the result of undermineralized bone and consequences on skeletal structure. Younger presentation is typically associated with more severe forms and the perinatal form is usually lethal.

Perinatal severe HPP can be identified in utero or at birth, with bony deformities and extreme hypomineralization of the skeleton, sometimes to the degree that many bones are not visible on radiographs. Where mineralized bone is visible on radiographs, severe rickets is usually evident. Weakness of the rib cage and hypoplastic lungs means that patients may die due to respiratory failure or infectious complications. However, there are reports of a "benign" perinatal form, with intrauterine skeletal abnormalities detected but without a lethal course.

Infantile HPP is also severe and about half of patients die during infancy. Moderate to severe

mineralization defects are present (Fig. 16.14). Presentation is usually before 6 months of age, with poor feeding, failure to thrive (poor weight gain), low muscle tone and rachitic features in long bones. Fontanelles may appear wide due to hypomineralization, but premature fusion may occur, and some patients require surgery for craniosynostosis. Similar to perinatal HPP, respiratory weakness occurs due to both skeletal and muscular insufficiency. Seizures may occur in perinatal or infantile forms. Hypercalciuria and nephrocalcinosis are also important complications in the more severe forms.

Childhood HPP usually presents at under 5 years of age. Patients may have delayed motor milestones and short stature or dolichocephaly. Odontohypophosphatasia is often the presenting finding, with early loss of deciduous teeth. Normal shedding of deciduous teeth at appropriate ages involves resorption of the dental root, but in odontohypophosphatasia the full tooth including the root is shed. This is related to abnormalities in the cementum layer, which alters the periodontal ligament attachment to the tooth. Permanent teeth may not manifest difficulties. Proximal muscle weakness may occur, and some patients develop intermittent joint pain and swelling. Radiographs often demonstrate tongues of lucency extending from the growth plate, representing hypomineralized bone (Fig. 16.14).

Adult HPP is generally discovered by the presence of fractures starting in middle age. Recurrent stress fractures in the feet or legs are typical. Adult HPP may be the initial manifestation of the disorder, or patients may have a history consistent with having had childhood HPP.

Additional features of HPP include nephrocalcinosis, which is a complication of hypercalciuria. Patients may sometimes have hypercalcemia and hyperphosphatemia. More severely affected infants may develop seizures. Generalized muscle weakness also occurs, which can delay motor milestones and contribute to respiratory failure. Prolonged healing time for fractures is noted. Joint pain and swelling may result from pseudogout (calcium pyrophosphate dehydrate crystal arthropathy), which relates to elevations of inorganic pyrophosphate.

Demographics

In European populations, prevalence estimates of moderate HPP are about 1 in 6400 and of severe HPP are about 1 in 300,000. HPP is less common in patients of African descent. Some populations with high consanguinity have higher prevalence. Presentation may be at any age, with greater severity of disease in those that present at the youngest ages.

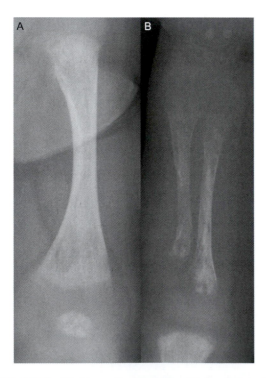

FIGURE 16.14   Eight-month-old child with hypophosphatasia. The patient has generalized poorly mineralized bone, irregular metaphyseal ossification, some metaphyseal lucencies, and flared and irregular metaphyseal borders, shown in (A) left femur and (B) radius and ulna.

Pathogenesis

TNSALP functions extracellularly. Mutations result in reduced catalytic function, intracellular retention, and decreased gene expression or mRNA stability. In the absence of TNSALP function, buildup of mineralization inhibitors, such as inorganic pyrophosphate, occurs. Additional substrates are also increased including phosphoethanolamine and pyridoxal 5'-phosphate, due to inability to dephosphorylate these compounds.

Impaired degradation of inorganic pyrophosphate, a mineralization inhibitor, limits hydroxyapatite deposition on osteoid matrix. This results in the histologic and radiographic features of rickets. Histologically, the findings cannot be distinguished from other causes of osteomalacia/rickets, although there may be some benefit to using ALP activity staining. An additional consequence of mutations can be deposition of calcium pyrophosphate dihydrate crystals in articular cartilage, causing chondrocalcinosis, pseudogout, or periarthritis. Clinically, testing for serum total ALP activity and/or BSAP can identify the disorder due to low levels. Additionally elevations of pyridoxal 5'-phosphate in serum and of phosphoethanolamine in the urine are supportive; pyridoxal 5'-phosphate is more specific than phosphoethanolamine.

Therapies

Currently, no approved therapy exists for HPP, though many have been attempted. Treatment has been largely palliative for the severe complications of the disorder. Unlike nutritional rickets, there is no benefit to supplementation with calcium and vitamin D. In fact such treatment is likely to result in harm through increased hypercalciuria or hypercalcemia, and a higher risk of nephrocalcinosis in HPP. If hypercalcemia is developing, fluid hydration and calcium restriction may be needed. Case reports have indicated some improvements in fracture or bone density in adult HPP patients treated with PTH (1−34) or PTH(1−84). Other case reports or series identified positive responses to marrow cell transplantation.

Previous attempts were made to treat severe HPP with serum from patients with Paget disease of bone (PDB; also called osteitis deformans), who have very high ALP levels. However, this strategy had limited and inconsistent success in improving radiographic features. Retrospectively, this was probably due to the enzyme failing to enter the bone matrix in sufficient amounts for the target clinical effect. Recently, the most promising development in HPP involved enzyme replacement with a fusion protein of TNSALP with the human immunoglobulin G1 (IgG1) Fc domain and a deca-aspartate, which successfully targeted the enzyme to the bone for action. Phase I clinical trial data published in 2012 demonstrated dramatic improvement in skeletal mineralization and respiratory and motor function in severely affected infants and children. Thus, successfully targeting the mechanism of disease may improve multiple aspects of HPP.

PAGET DISEASE OF BONE

Definition

While some metabolic bone diseases are caused by generalized abnormalities in bone and mineral metabolism, some are associated with more focal areas of abnormal metabolism within one or more bones. PDB is one example of this process.

PDB typically presents in adults over age 50 years, with elevated markers of bone turnover and focal areas of bone disease affecting one or more bones. PDB lesions manifest as variably sized expansile skeletal lesions with both lytic and sclerotic components, and cortical thickening. Bone pain and deformation may occur, as well as compression of neurovascular structures. Lesions involve focal areas of coupled increased osteoclastic and osteoblastic activity. Commonly affected bones include the skull, spine, pelvis, femur, and tibia, although any bone may be involved. Most patients are found incidentally, with asymptomatic disease diagnosed from radiographs performed for another purpose, or during evaluation for an elevated serum ALP activity.

When symptoms are present, localized bone pain is the usual complaint. Due to high vascularity of the bone lesions, there may be warmth to the overlying skin compared to other areas. Lesions near joints may cause joint deformation and arthritic pain, and severe deformity of long bones may cause gait abnormalities, which contribute to joint pain. In addition, bony lesions in the vertebrae may cause pain or neurologic symptoms due to compression on the spinal cord or on nerve roots. Pathologic fractures may occur through PDB lesions, especially in the femur.

Involvement of the skull is important and may manifest as headaches, increased skull size, frontal bossing, or deformation of cranial or maxillofacial bones, as well as dental abnormalities. Hearing loss and cranial nerve palsies sometimes occur. In addition, softening of the skull base due to PDB may cause basilar invagination and brainstem compression or obstructive hydrocephalus. Angioid streaks may occur in the retina.

Some rare complications of PDB include congestive heart failure and hypercalcemia. Congestive heart failure may be exacerbated by severe PDB, due to high

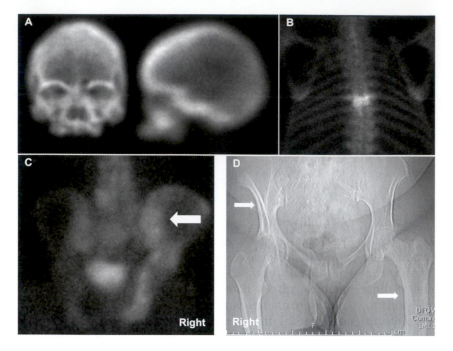

FIGURE 16.15 **Paget disease of bone.** This 63-year-old male patient with Paget disease has increased uptake on Technetium-99 scanning (A) diffusely in the skull, (B) also at the T7 vertebra, and (C) in the right hemipelvis (viewed posteriorly), as indicated by the arrow. (D) A radiograph of a 71-year-old woman with Paget disease in the left femur and right hemipelvis demonstrates areas of sclerosis with cortical thickening, coarse trabeculae, and abnormal shape of the femur (arrows).

vascularity of lesions. Hypercalcemia usually occurs in the setting of immobilization in PDB, although primary hyperparathyroidism may also coexist. Malignant degeneration is also rare (<1%). Sarcomas (chondrosarcoma, fibrosarcoma, or osteosarcoma) may develop and have poor prognosis.

Technetium bone scanning of the whole skeleton facilitates identification of lesions (Fig. 16.15). Areas of increased uptake on bone scans are highly sensitive, but nonspecific. A variety of other lesions including bone metastases will appear similar to PDB with this technique.

Plain radiographs reveal characteristic areas of bony expansion with thick cortices, coarse trabecular markings, and a mixture of lytic and sclerotic bone areas (Fig. 16.15D). Fractures, cortical fissures, and deformation may be identified. Occasionally metastases to bone may look similar to PDB on plain radiographs as well. New symptoms in a previously identified lesion may indicate fracture in a lesion or sarcomatous change and should be evaluated by repeating radiographs of the PDB lesion. Bone formation markers, especially ALP, and resorption markers, are elevated during active disease.

Demographics

After osteoporosis, PDB is the second most common metabolic bone disease. It affects men slightly more often than women. The disease is uncommon before the age of 40, and juvenile forms are even rarer. PDB occurs in 1–2% of white adults over 55 years, but the prevalence increases with each decade after age 50. An estimated 1.5 million people in the United States have PDB. PDB is most common in people of European, and specifically Anglo-Saxon, background. PDB is rare in in Scandinavian, African, and Asian populations. Family history may be present in up to 15–30%.

Pathogenesis

PDB is primarily an osteoclast problem characterized by localized lesions of coupled increased osteoclast activity and osteoblastic bone formation. Lesions begin as areas of increased numbers of osteoclasts, with larger numbers of nuclei per cell and elevated resorptive capacity. Pagetic osteoclast precursors are more sensitive to the stimulatory effects of 1,25 $(OH)_2D$, due to expression of a coactivator of the VDR TATA box-binding protein associated factor 17 (TAFII-17). Subsequent to the increased osteoclast activity, there is accelerated formation of new bone by osteoblasts. The high rate of turnover leads to the production of woven bone as the collagen fibers are deposited in a less organized fashion, although the mineralization rate is typically normal or increased, depending on the availability of calcium and phosphorus. Extensive vascularization occurs, along with development of fibrous connective tissue within the marrow. In more quiescent lesions, bone turnover is no longer elevated, but the abnormal bone remains.

The etiology of PDB remains somewhat unclear, and both genetic and environmental factors have been proposed, including a possible contribution of paramyxoviral infections. A genetic component is indicated by the significant percentage of patients with a family history. An autosomal dominant pattern is described, although penetrance may be incomplete.

Genome-wide association studies (GWAS) have identified associations with genetic loci containing CSF1 (colony stimulating factor 1), OPTN (optineurin) and TNFRSF11A [encoding tumor necrosis factor receptor superfamily member 11 A (RANK)]. Additional genes have also been identified in GWAS or candidate gene approaches. A rare mutation in TNFRSF11A is reported in PDB, and the rare recessive juvenile PDB can be caused by mutation in the gene encoding osteoprotegerin (TNFRSF11B). Mutations affecting RANK or osteoprotegerin cause PDB by disrupting the normal regulation of RANK-RANKL signaling and osteoclast generation.

However, linkage analysis identified the most commonly identified mutation, which occurs in SQSTM1, which encodes sequestasome-1, an ubiquitin binding protein. Roughly 30−40% of familial PDB patients are explained by this mutation. Up to 10% of sporadic cases also harbor SQSTM1 mutations. Although patients with these mutations may present at an earlier age and have more severe disease, some have no phenotype suggestive of PDB.

The SQSTM1 mutation may only be a predisposing factor, requiring some additional exposure or mutation, because the mutation does not result in all features of PDB. This mutation increases the response of osteoclast precursors to RANKL and TNF-α, and the size of osteoclasts, but not the number of osteoclast nuclei. Nor does it make these cells more sensitive to $1,25(OH)_2D$. Genetically modified mice harboring the SQSTM1 P392L mutation do not have PDB lesions.

Studies suggest that the presence of an important environmental factor leads to a decreased incidence of PDB. Paramyxoviral nucleocapsid proteins from canine distemper virus and measles virus nucleocapsid protein (MVNP) and transcripts have been identified in Pagetic osteoclasts. Transfection of the gene encoding MVNP into osteoclast precursors results in some of the phenotypes of Pagetic osteoclasts. These osteoclasts have increased nuclei and an increased response to $1,25(OH)_2D$. Corroborating evidence comes from animal studies, in which transgenic expression of the MVNP in osteoclasts generates PDB lesions.

Therefore, both genetic and potential environmental factors are likely to play a role in the development of PDB. However, the relative contribution of specific environmental factors remains controversial and many questions remain regarding the integration of genetic and/or environmental contributors to PDB.

Therapies

The goals of treatment are to relieve pain due to lesions and to prevent other complications of PDB. Existing bone deformity is unlikely to improve with treatment. Although not yet proven by clinical trials, suppression of bone turnover in PDB may prevent development of new complications due to deformity over time. Thus even asymptomatic patients with elevated ALP and PDB affecting the skull, vertebrae, or weight-bearing bones may benefit from treatment.

Treatment of PDB involves suppressing osteoclastic activity. Calcitonin was used in the past, but issues included limitations in the degree of suppression of bone turnover, along with limited duration of effect following cessation of therapy. Calcitonin has been supplanted by the more potent antiresorptives, bisphosphonates, which have demonstrated effectiveness in PDB. Risks of bisphosphonates are similar to those when treating for osteoporosis (hypocalcemia, acute renal failure, and acute phase reaction; and for oral agents, gastrointestinal symptoms or esophageal erosions and rarely atypical fractures or osteonecrosis of the jaw). However, for oral bisphosphonates, the PDB treatment regimen is different than in osteoporosis (see Chapter 17), requiring daily dosing for 2−6 months. Bone turnover markers decline to normal in most patients by 6 months after beginning treatment. Zoledronate is given intravenously as a single 5 mg dose, and in one clinical trial normalization of ALP was faster and occurred in more subjects than in those treated with oral risedronate for 2 months. The older agents, etidronate and tiludronate, are used less frequently than the later nitrogen-containing bisphosphonates because of their greater propensity to inhibit skeletal mineralization. Remission with bisphosphonates assessed by bone turnover markers can be sustained for months to years, especially with intravenous zoledronate, and is related to the degree of suppression of bone turnover following initial treatment. With bisphosphonate therapy, there is some filling in of lytic lesions, although deformity remains.

OSTEOPETROSES

Definition

The osteopetroses (marble bone diseases) are a series of bony disorders characterized by high bone mass. They are caused by genetic defects in osteoclast function that lead to decreases in bone resorption relative to bone formation. Persons with osteopetrosis have diffuse, symmetric excessive bone accumulation. The bone has pathognomonic primary spongiosa

TABLE 16.4 Osteopetrosis

Type	Mode of Inheritance	Gene	Protein Name	Protein Function
Infantile	Autosomal Recessive	CLCN7	H⁺/Cl⁻exchange transporter 7	Cl⁻/H⁺ antiporter
	Autosomal Recessive	OSTM1	Osteopetrosis-associated transmembrane protein 1	ß-subunit of CLCN7
	Autosomal Recessive	TCIRG1	V-type proton ATPase 116 kDa subunit a isoform 3	Subunit of v-ATPase proton pump that acidifies the resorption lacunae
	Autosomal Recessive	TNFRSFIIA	Tumor necrosis factor ligand superfamily member 11/receptor activator of the NF-κB ligand	Osteoclast formation, function, survival
Infantile/intermediate with renal tubular acidosis and cerebral calcification	Autosomal Recessive	CA2	Carbonic anhydrase 2	H⁺ generation for acidification
Intermediate	Autosomal Recessive	CLCN7	H⁺/Cl⁻ exchange transporter 7	Cl⁻/H⁺ antiporter
	Autosomal Recessive	PLEKHM1	Pleckstrin homology domain-containing family M member 1	Vesicular trafficking
Adult onset type 2	Autosomal Dominant (Dominant-negative effect - with incomplete penetrance	CLCN7	H⁺/Cl⁻ exchange transporter 7	Cl⁻/H⁺ antiporter
Mild Osteopetrosis with Ectodermal Dysplasia and Immune Defect	X-linked	IKBKG	NF-κB essential modulator	NF-κB activation

remnants that can be observed by histology. The defect in bone turnover results in skeletal fragility despite increased bone mass because the bone is highly mineralized. There is also a decrease in the caliber of cranial nerve and vascular canals and decreased marrow space in the long bones. Characteristic laboratory findings include increased concentrations of tartrate-resistant acid phosphatase (TRAP) and the BB-isoenzyme fraction of creatine kinase.

Demographics

Osteopetrosis is a rare inherited bone disease with an overall estimated incidence of between 5.5 in 100,000 for the dominant forms and 1 in 500,000 for the recessive forms.

Pathogenesis

Osteopetrosis is a heritable disease, and therefore risk is conferred by inheritance of one or more mutated gene(s). Gene defects are known to account for 70% of infantile and intermediate cases (Table 16.4). Mutations in the CLCN7 gene [which encodes H⁺/Cl⁻exchange transporter 7 (CLC-7)] can underlie mild, moderate, or severe forms.

The osteopetroses are highly heterogeneous, with severity ranging from a perinatal lethal form to forms that are asymptomatic even in adulthood. The infantile form presents shortly after birth: nasal stuffiness due to sinus malformation is often the presenting feature, and affected babies have hypocalcemia, fracture, and poor growth and weight gain. Defective bone tissue tends to replace bone marrow hematopoietic space, resulting in anemia, large livers and spleens, and recurrent infections from low white blood cell counts. The bony skull deformities can lead to deafness and visual loss from nerve compressions. Untreated, these babies die within a few years from infection or hemorrhage.

The autosomal dominant form was once divided into types I and II. Type I, due to mutations in LRP5, which encodes low-density lipoprotein receptor-related protein 5 (LRP5), is now known not to be a

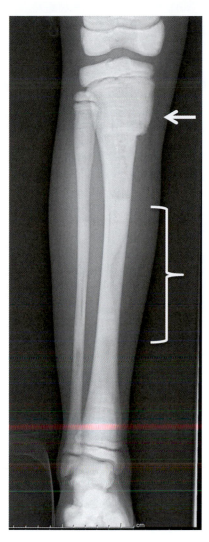

FIGURE 16.16   **Osteopetrosis.** Right tibia and fibula of 11-year-old child with osteopetrosis. Note increased bony mineralization with sclerotic bands throughout the metaphyses of the long bones. There is also an endobone (bone within bone) visible within the tibia (to left of bracket), and cortical irregularity and Erlenmeyer flask deformity from a remodeling defect of the proximal tibia metaphysis (arrow).

true osteopetrosis, as fractures are not increased. Type II has 75% gene penetrance. Bones may be uniformly sclerotic, or sclerotic areas may alternate with lucent bands. Sometimes *endobones* (i.e. bones within bones) are seen. Affected individuals are characterized by a "rugger jersey" appearance (similar to the striping on a rugby player's jersey) of the spine and endobones in the pelvis by X-ray. Erlenmeyer flask deformities can be seen at the distal ends of the long bones (Fig. 16.16). However, there is great phenotypic heterogeneity, and a tendency of the disease to worsen with increasing age. Complications of the disease can include pathologic long bone fractures, bone pain, bone infections (particularly of the jaw), and entrapment of the optic nerve leading to vision loss.

Osteomyelitis of the maxilla or mandible typically occurs in older adults. Patients can also have bone marrow failure.

Therapies

There is no established therapy for osteopetrosis. Bone marrow transplant can be curative and is often performed in children with the severe autosomal recessive form of the disease. A variety of other treatments have been tried in children with the recessive forms or more severe dominant forms, with the goal of stimulating bone turnover or reducing bone formation. These include very high dose $1,25(OH)_2D$ along with restricted calcium intake, glucocorticoids, human interferon gamma (IFN-γ), and even thyroid hormone. $1,25(OH)_2D$ appears to help by stimulating osteoclasts and thus stimulating bone resorption. However, the clinical response is limited and not sustained after therapy is discontinued. Treatment with IFN-γ has been reported to produce long-term benefits; in addition to beneficial effects on bone, it improves white blood cell function, thereby decreasing the incidence of new infections. Glucocorticoids have been used to stimulate bone resorption and treat anemia, although this therapy is not a preferred option. Bone marrow transplantation markedly improves some cases of infantile osteopetrosis. Furthermore any effective treatment is likely to increase urinary calcium excretion due to increased bone resorption, thus placing the patient at risk of nephrocalcinosis and kidney failure. No specific medical treatment exists for the adult type, although complications of the disease may require intervention.

In summary, metabolic bone disease affects a large number of patients, resulting collectively in significant health care costs. The most common diseases are osteoporosis and PDB, although hyperparathyroidism also affects a large number of adults. Although we have discussed the details of only a limited number of diseases, the principles involved relate to many others. Studying the mechanisms of pathophysiology of various rare or common metabolic bone diseases has provided the source of much of our understanding of normal skeletal and mineral homeostatic processes.

STUDY QUESTIONS

1. Discuss the pathogenesis of glucocorticoid-induced osteoporosis.
2. Define osteoporosis. How is it diagnosed? What are the modifiable and nonmodifiable risk factors for osteoporosis?
3. Discuss the pathogenesis and treatment options in osteogenesis imperfecta.

4. What is rickets? Discuss its causes and clinical features. How does it relate to osteomalacia?

5. What is hypophosphatasia? Discuss its clinical features and laboratory abnormalities.

6. What are the clinical features of Paget disease of bone? How is it diagnosed? What are the available treatment options?

Suggested Readings

Carpenter, T.O., et al., 2011. A clinician's guide to X-linked hypophosphatemia. JBMR. 26 (7), 1381–1388.

Forlino, A., et al., 2011. New perspectives on osteogenesis imperfecta. Nat. Rev. Endocrinol. 7 (9), 540–557.

Mantovani, G., 2011. Pseudohypoparathyroidism: diagnosis and treatment. JCEM. 96, 3020–3030.

Marcus, R., Feldman, D., Kelsey, J., 2012. Osteoporosis, fourth ed. Academic Press, San Diego (multiple relevant chapters).

Naot, D., 2011. Paget's disease of bone: an update. Curr. Opin. Endocrinol. Diabetes Obes. 18 (6), 352–358.

Orwoll, E.S., Bilezikian, J.P., Vanderschueren, D., 2010. Osteoporosis in Men, second ed. Academic Press, Amsterdam (multiple relevant chapters).

Roodman, G.D., 2010. Insights into the pathogenesis of Paget's disease. Ann. N. Y. Acad. Sci. 1192 (1), 176–180.

Rosen, C.J., Glowacki, J., Bilezikian, J.P., 1999. The Aging Skeleton. Academic Press, San Diego (multiple relevant chapters).

Rosen, C.J., 2008. Primer on the Metabolic Bone Diseases and Disorders of Mineral Metabolism, seventh ed. American Society for Bone and Mineral Research, Washington, DC (multiple relevant chapters).

Rubin, M.R., Levine, M.A., 2008. Chapter 75. Hypoparathyroidism and Pseudohypoparathyroidism. In Primer on the Metabolic Bone Diseases and Disorders of Mineral Metabolism, seventh ed. Washington, DC, pp. 354–361.

Shoback, D., 2008. Clinical practice. Hypoparathyroidism, N. Engl. J. Med. 359 (4), 391–403.

Weinstein, R.S., 2011. Glucocorticoid-induced bone disease. N. Engl. J. Med. 365 (1), 62–70.

Pharmaceutical Treatments of Osteoporosis

Bruce H. Mitlak[1], David B. Burr[2] and Matthew R. Allen[2]

[1]Distinguished Medical Fellow, Lilly Research Laboratories, Indianapolis, Indiana, USA [2]Department of Anatomy and Cell Biology, Indiana University School of Medicine, Indianapolis, Indiana, USA

Osteoporosis is a common disease and the societal burden of osteoporosis-related fractures is substantial (Fig. 17.1). One in three women and one in five men is predicted to suffer an osteoporosis-related fracture in their lifetime. Fractures of the hip cause the most morbidity and are associated with an increase in mortality compared to other fractures. Worldwide, approximately 1.6 million hip fractures occur each year and, with many individuals living longer, by 2050 this number could reach 4.5 million. Interestingly, even 20 years ago there were few pharmacologic treatments for patients with osteoporosis. Over the past few years, progress in the development of new therapies has come from our growing understanding of bone biology. Recall from Chapter 16 that the pathophysiology

of osteoporosis is related to an imbalance in the usually coordinated activity of osteoclasts and osteoblasts. Postmenopausal osteoporosis is associated with increased proliferation, differentiation, and activity of osteoclasts resulting in a higher number of active remodeling sites. Bone loss occurs because osteoblasts do not fully balance the effect of osteoclasts in these patients. In contrast, a type of secondary osteoporosis associated with the use of glucocorticoids is primarily characterized by diminished osteoblast function, which contributes to bone loss.

Strategies for treatment of osteoporosis include nutritional and pharmacological, the latter of which are generally characterized based on their primary mode of action (Fig. 17.2). Anticatabolic agents target the osteoclast to reduce the amount of bone resorption and remodeling, thus slowing bone loss. Conversely, anabolic agents target osteoblasts to increase the amount of net bone formation and bone mass. In 2013, there are five classes of antiremodeling agents approved in the United States for the treatment of osteoporosis. There is only one approved anabolic agent: teriparatide [rhPTH(1−34)].

CALCIUM AND VITAMIN D

Because of the importance of calcium in the skeleton (Chapters 13 and 14) and the relatively low intake of dietary calcium by most individuals, it is not a surprise that calcium supplementation has been considered an essential part of treatment to reduce fracture risk. Insufficient dietary calcium increases the rate of remodeling and creates an imbalance between resorption and formation that results in bone loss. Supplementing daily diet with calcium can return the remodeling rate to normal and restore balance at the individual basic

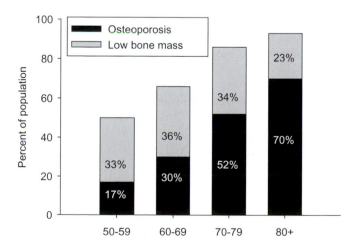

FIGURE 17.1 **Prevalence of low bone mass and osteoporosis in women over the age of 50 in the USA.** Over 14 million women are estimated to have osteoporosis [bone mineral density BMD <−2.5 standard deviation], while an additional 12 million have low bone mass (BMD between −1.0 and −2.5 standard deviations). Both osteoporosis and low bone mass are also prevalent in men over the age of 50.

Basic and Applied Bone Biology.
DOI: http://dx.doi.org/10.1016/B978-0-12-416015-6.00017-4

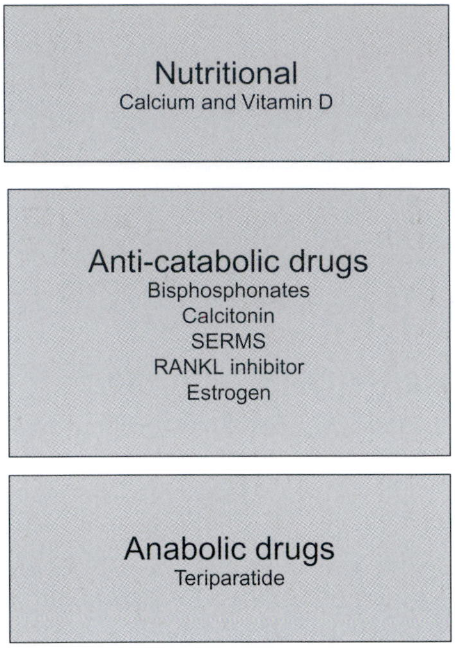

Nutritional
Calcium and Vitamin D

Anti-catabolic drugs
Bisphosphonates
Calcitonin
SERMS
RANKL inhibitor
Estrogen

Anabolic drugs
Teriparatide

FIGURE 17.2 Treatment of osteoporosis can occur through nutritional or pharmacological intervention. Pharmaceutical agents used to reduce fracture risk are generally characterized based on their primary mode of action. Anticatabolic agents target osteoclasts to either inhibit their formation or activity. There are five different anticatabolic drug classes that have US Food and Drug Administration approval for treatment of osteoporosis; some of the classes [such as bisphosphonates and selective estrogen receptor modulators (SERMs)] have several different drugs approved. Anabolic agents, for which there is only one approved in the USA, target osteoblasts and stimulate increase in the production of bone. RANKL, tumor necrosis factor ligand superfamily member 11/receptor activator of the NF-κB ligand.

multicellular unit (BMU) level. One current view is that osteoporosis treatments should be supplemented with a total intake of 1000−1500 mg of calcium/day. However, postmenopausal osteoporosis is not just the result of low calcium, and can occur even in the presence of normal dietary calcium. Supplementing in people who are already calcium replete might not be expected to have much of an effect on remodeling rate and bone mass. Interestingly, recent studies have raised questions about the use of calcium supplements. In the Women's Health Initiative (WHI) study, administration of calcium and a vitamin D supplement to healthy postmenopausal women was associated with a small improvement in bone mineral density but no change in fracture risk (Fig 17.3) and an increase in the risk of kidney stones. Other recent studies at least suggest a potential association between calcium supplements, but not dietary calcium, and cardiovascular disease. While these studies are not viewed as definitive, they may result in a preference for calcium intake to come more from dietary sources. Interestingly, a

very recent pooled analysis of major vitamin D trials found that mortality was reduced with vitamin D and calcium supplementation but not with vitamin D alone.

Vitamin D is very important for increasing the efficiency of calcium absorption by the gastrointestinal (GI) tract (see Chapter 13). Low serum concentrations of vitamin D negatively affect calcium absorption, which causes the body to compensate by stimulating the parathyroid glands to produce parathyroid hormone (PTH). PTH acts at several levels to normalize the level of calcium in the blood. It acts at the kidney to reduce calcium excretion and stimulates renal hydroxylase increasing the concentration of 1,25-dihydroxyvitamin D [1,25(OH)$_2$D], which in turn promotes calcium absorption from the GI tract. PTH also increases bone remodeling to mobilize calcium from skeleton. Vitamin D insufficiency can increase the risk of fracture through this effect on calcium metabolism. Vitamin D also has direct effects on muscles. As a result vitamin D deficiency can be associated with loss of muscle strength resulting from a selective loss of the rapid type II muscle fibers, and increased the occurrence of falls.

Several studies of vitamin D have examined the relationship between vitamin D treatment and falls and fractures (Fig. 17.3). A series of clinical trials have come to conflicting conclusion, with some trials concluding a benefit of vitamin D supplementation, and others concluding that there is no additional benefit. One possible explanation for the heterogeneity of these results is that the studies gave a wide range of vitamin D doses. Studies using relatively low doses of vitamin D (200−600 IU/day) showed no evidence of fracture benefit. Patients who received higher doses of vitamin D sustained significantly fewer falls, which would probably translate into fewer fractures. A recent meta-analysis examined some potential reasons and found that in persons of 65 years or older only a high intake of vitamin D was associated with a reduction in the risk of nonvertebral fractures, including fractures of the hip. Nevertheless, the Institute of Medicine recommends that persons of 65 years or older receive only 800 IU vitamin D/day.

ANTICATABOLIC THERAPIES

Estrogen

In 1947, Fuller Albright described the relationship between estrogen deficiency and osteoporosis. He showed that estrogen therapy could reverse the negative calcium balance observed in postmenopausal women and later in a small placebo-controlled trial showed that estrogen could prevent bone loss. We now

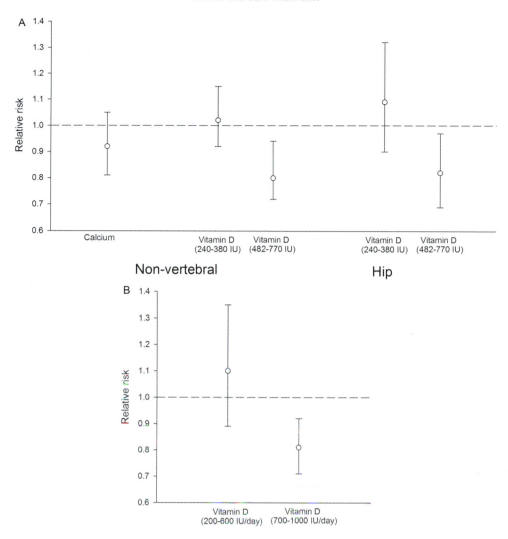

FIGURE 17.3 Meta-analyses, in which several clinical studies are combined statistically, provide a way to observe an overall pattern of a particular intervention. The relative risk of an event, i.e. the probability of that event occurring in a treated group compared to a control, can be useful for assessing the overall efficacy of a treatment: if the 95% confidence intervals cross 1.0, then there is no difference between groups. Meta-analyses have shown that (A) calcium supplementation has no effect on fracture risk reduction in populations that are not severely calcium deficient. Vitamin D supplementation reduces (A) fracture risk and (B) falls but only at higher doses. Interestingly, even the high doses studied are still below the recommendations for people over the age of 65 (800 IU). *Data from Bischoff-Ferrari et al. BMJ 2009:b3692–b3692, Bischoff-Ferrari et al. Am J Clin Nutr 2007;86:1780–1790, and Bischoff-Ferrari et al. Arch Intern Med 2009;169:551–561.*

know that estrogen receptors are present on both osteoblasts and osteoclasts (see Chapters 2 and 15). Estrogen replacement therapy with or without a progestin reduces bone turnover by about 50% and improves bone balance (formation versus resorption) at the BMU level in postmenopausal women. While estrogen had been used as therapy to treat osteoporosis for many years, the evidence for clinical efficacy and safety was limited to observational studies in which women who took estrogen were observed to fracture less frequently.

In 1991, a set of clinical trials, collectively called the WHI study, were initiated by the National Institutes of Health (Fig. 17.4). One focus of these studies was to determine the role of estrogen on bone health, but other aims included evaluating whether estrogen had

positive effects on the health of other systems, such as the cardiovascular system. One of the trials treated 16,000 relatively healthy postmenopausal women for 8 years with what at the time was the standard regimen for management of osteoporosis—conjugated equine estrogen and medroxyprogesterone acetate. The combination of estrogen and progestin was given because estrogen alone is known to cause thickening of the endometrial lining of the uterus and to increase the risk of endometrial cancer, but using it in combination with progesterone reduces the risk of this adverse effect. The study included several different end points, including cardiovascular outcomes such as stroke and heart attacks, breast cancer, and overall mortality, as well as fractures. The study was stopped by the Data

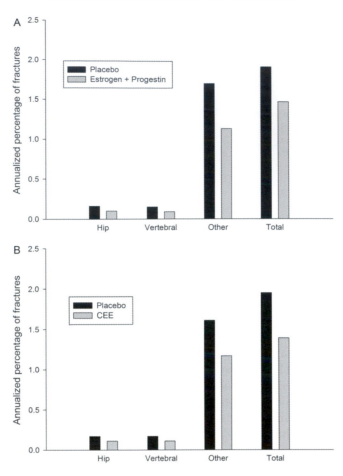

FIGURE 17.4 The Women's Health Initiative (WHI) study assessed the effects of estrogen plus progestin on several health parameters including fracture risk. These data showed a clear and significant reduction in all fracture classifications compared to untreated individuals. A separate arm of the WHI studied similar parameters in women treated with conjugated equine estrogen (CEE) following hysterectomy and found similar efficacy. *Data from Rossouw JE et al JAMA 2002;288:321–333 and Cauley et al JAMA 2003; 290:1729–1738.*

and Safety Monitoring Board, with some controversy, after 5.5 years because they concluded that estrogen plus progestin significantly increased the risk of cardiovascular events in women. The studies also found a significantly lower risk of fractures, including fractures of the hip, in patients assigned to estrogen and progestin as well as those assigned to estrogen alone compared to the placebo-treated controls. This study was the first to definitively demonstrate fracture reduction with hormone therapy. Follow-up analyses from this study cohort demonstrated that there was a greater reduction in fracture risk for women with baseline calcium intake above 1200 mg/day compared with those who had lower calcium intake. Fracture risk was also reduced more in those women with a body mass index (BMI) below 30 compared to those with a higher BMI.

Another arm of the WHI included women who had a hysterectomy and were thus given estrogen only, as there was no longer a possibility of causing endometrial cancer (Fig. 17.4). This arm was stopped after 6.5 years due to an increased risk of stroke in women taking estrogen. As with the combination treatment study, patients treated with estrogen alone had a significantly lower incidence of fracture compared to placebo-treated controls.

The WHI studies are important as they were the first large clinical trials of estrogen therapy. Results from the WHI have altered the use of estrogen therapy such that most women who now receive estrogen usually take it acutely for short-term relief of menopausal symptoms rather than for chronic disease management. However, questions remain about whether the results of the WHI apply to all women or if there are subgroups of women who still might benefit from estrogen or estrogen plus progestin. This is an area still being explored.

Selective Estrogen Receptor Modulators

While the skeletal benefits of estrogen have been clearly demonstrated, the overall balance of benefits and risks has limited its use for osteoporosis. The development of selective estrogen receptor modulators (SERMs) was undertaken to provide new therapies that have a more favorable balance of benefits and risks. This drug class includes nonsteroidal molecules that mimic some of the favorable effects of estrogen and decreases some of the potential negative side effects. These agents interact with the estrogen receptor but not in exactly the same way as estrogen, resulting in different conformations in the estrogen receptor, which ultimately translates to tissue-specific effects.

Several SERMs are used in clinical practice. Clomiphene, based on its effect as an anti-estrogen at the level of the pituitary gland causing increased levels of gonadotropins, is used to stimulate ovulation in patients who are having problems with regular ovulation. Tamoxifen is used primarily to treat women with breast cancer or to lower the risk of breast cancer in women at high risk because it blocks the effects of estrogen on breast tissue. Raloxifene is the only SERM that is indicated for the prevention and treatment of osteoporosis in the United States. It is also indicated for reducing the risk of breast cancer in women at increased risk for breast cancer. Another SERM, bazedoxifene, is approved in a few European countries and in Japan for treatment of osteoporosis.

The primary study to examine the clinical utility of raloxifene on fractures included women with and without prior vertebral fractures. The Multiple Outcomes of Raloxifene Evaluation (MORE) trial treated patients with raloxifene or a placebo for 3 years, with a primary outcome of morphometric spine fractures evaluated by radiographs (Fig. 17.5). In any fracture trial, fracture data are often reported as both absolute and relative risk reduction compared to the placebo-treatment and/or comparative drug. Using the MORE trial as an example, 4.3% of the patients without preexisting vertebral fractures experienced a vertebral fracture over 3 years, while only 1.9% of those treated with raloxifene experienced a fracture. This represents an absolute fracture risk reduction of 2.4% (i.e. 4.3% minus 1.9%) and a relative risk reduction of 55% (i.e. 1.9% divided by 4.3%). In those patients with a preexisting vertebral fracture, raloxifene treatment imparted a 6.1% absolute risk reduction and a 30% relative risk reduction compared to placebo. Another method to express the results from this trial is a calculation of the number of women who need to be treated to prevent a fracture, or the number needed to treat (NNT). In the MORE trial, the NNT for 3 years was 42 for women without a preexisting fracture and 16 for women with a prior fracture. The MORE study also showed that bone mineral density (BMD) was modestly increased (2%) at the femoral neck and spine, but was unable to demonstrate fracture risk reduction at sites other than the spine. Nonbone-related findings

from the MORE trial showed that raloxifene reduces the risk of breast cancer and did not increase endometrial hyperplasia, cancer, or coronary events. There was an increased risk of venous thrombotic events similar in magnitude to that seen in prior studies with estrogen. Hot flashes and leg cramps were reported more commonly in women treated with raloxifene compared with placebo.

Raloxifene does not appear to primarily reduce fracture risk by increasing bone mass. Whereas the bisphosphonates and denosumab may increase BMD by 7% or more during the first 3 years of treatment, raloxifene has a modest effect on BMD, increasing it by only 1–2%. It has been estimated that only 4% of the fracture risk reduction with raloxifene can be attributed to changes in BMD. The mechanism by which raloxifene is associated with a reduction in the fracture risk is thus far unknown. Evidence now suggests that raloxifene may work by altering the properties of the bone matrix itself, through mechanisms that increase the fraction of bound water at the interface between the mineral and the collagen. Whether this additional mechanism can account for the larger than expected fracture risk reduction remains to be confirmed.

Calcitonin

This 32-amino-acid peptide hormone is secreted from the thyroid gland in response to an increase in serum calcium, thus acting essentially as a counter-regulatory hormone to PTH. It interacts with specific G protein-coupled receptors (G_s and G_q) on osteoclasts to reduce bone resorptive activity. Calcitonin was first used clinically in the 1970s for the treatment of Paget

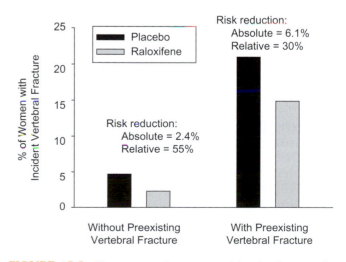

FIGURE 17.5 Three years of treatment with raloxifene, a selective estrogen receptor modulator, significantly reduces vertebral fracture risk compared to placebo-treated postmenopausal women.

disease and in the 1980s was registered for the treatment of osteoporosis. Clinical studies of calcitonin have demonstrated a significant but modest increase in BMD but an inconsistent effect on the reduction in fractures compared to placebo. For a number of years it was used as a mainstay in the management of patients with osteoporosis although it is currently used infrequently in the USA because of its relatively weak antiresorptive effects compared to other agents. Its use is also limited due to the routes of administration, either nasal spray or injection, which are less attractive than oral administration to many patients. Emerging information in 2013 about the risk-benefit of calcitonin therapy has led to a proposal that it no longer be used for long-term management of osteoporosis.

Bisphosphonates

The majority of antiremodeling agents approved in the United States are bisphosphonates, which are analogs of naturally occurring pyrophosphate. Very simply, these agents reduce bone remodeling, often by as much as 70% or more, and thereby stem the loss of bone, thus allowing BMD to be maintained or even increased by a small amount. Bisphosphonates can be classified as those that contain nitrogen, and those that don't (Fig. 17.6). The earliest generations of bisphosphonates used for osteoporosis were non-nitrogen containing (e.g. clodronate and etidronate), and these are less potent in affecting osteoclasts than the nitrogen-containing bisphosphonates (alendronate, ibandronate, pamidronate, risedronate, and zoledronate (also known as zoledronic acid)).

A series of clinical trials have shown that the nitrogen-containing bisphosphonates all reduce bone remodeling compared to placebo-treated patients (Fig. 17.7) and have clinically important effects on reducing vertebral and nonvertebral fractures, although there are some subtle differences among the individual agents. These include differences in the rate of onset of effect and on the rate at which bone metabolism will return to pretreatment levels when the drugs are withdrawn.

The core structure of bisphosphonates is a phosphate-carbon-phosphate with two side chains (R1 and R2) (Fig. 17.6). The structure is important because it allows these molecules to avoid hydrolysis in the GI tract, thus allowing them to be given orally. In contrast, orally administered phosphate is degraded in the GI tract, which prevents it from being absorbed in a biologically active form. Careful work has shown that substitutions that occur at R1 and R2 are important for increasing the binding affinity of the drug to the bone mineral, as well as altering potency toward osteoclast activity (Fig. 17.6). Increased binding affinity is associated with longer retention in the bone, providing the

possibility of longer intervals between dosing. The cellular and molecular effects (i.e. potency) also depend on the substitutions at the R1 and R2 positions. Both of these effects—potency and binding affinity—are important and vary substantially among the different bisphosphonates. Because the absorption of bisphosphonates when taken orally is low, only about 1%, several bisphosphonates are given by injection (ibandronate and pamidronate), or intravenous infusion (zoledronate). This permits less frequent dosing (ranging from monthly to yearly) compared to the oral forms, which are given daily or weekly.

The non-nitrogen-containing bisphosphonates are taken up by osteoclasts upon bone resorption and create nonhydrolyzable ATP analogs, which disrupt cell function and induce apoptosis because they cannot be activated as a source of energy. The nitrogen-containing bisphosphonates may act in two ways: one physicochemical and the other through inhibition of metabolic pathways. The physicochemical action is through the binding of nitrogen-containing bisphosphonates to bone (Fig. 17.8), which helps stabilize the bone matrix and prevents it from being resorbed. These molecules have high affinity for bone surfaces, particularly where bone matrix is exposed in areas where resorption is occurring. It has also now been shown that bisphosphonates enter the bone through canaliculi and can be found on many internal surfaces (Fig. 17.8). Further, bisphosphonates stabilize the bone crystal (a physicochemical property) and prevent bone resorption. The nitrogen-containing bisphosphonates also have important metabolic effects that decrease the efficacy of osteoclasts or stop them from resorbing bone by disrupting the cholesterol synthetic pathway from which the prenylated proteins required for osteoclast function and survival are derived. A key enzyme in this pathway, farnesyl pyrophosphate synthase (FPS) is inhibited by nitrogen-containing bisphosphonates, thus reducing the availability of the proteins. Therefore, in this case, the osteoclasts do not die; they are just "crippled." (Fig. 17.9) Consequently, there may not be fewer osteoclasts present; they just function poorly. In addition, because their lifespan is prolonged, some of the osteoclasts become quite large with many nuclei. However, they are unable to effectively resorb bone; consequently, bone remodeling is reduced and bone mass retained.

Bisphosphonates are very stable and are not degraded once they are incorporated into bone. They can be released again once bone resorption takes place; once they are released, they can potentially reattach to bone. Recycling bisphosphonates from one bone surface to another may contribute to the very long duration of their effect on bone turnover, which is reported to be about 10 years for alendronate. The more potent

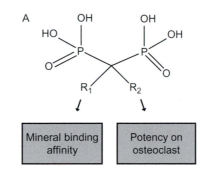

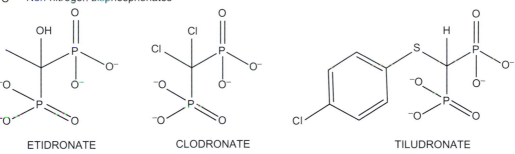

FIGURE 17.6 The bisphosphonate drug class is comprised of several different drugs all with a similar core structure. (A) The P-C-P backbone is accompanied by two important side chains, one that dictates the mineral binding affinity and the other that dictates the potency against osteoclasts. The bisphosphonates are classified as (B) nitrogen and (C) non-nitrogen containing, the former being most prevalent in the clinic today.

bisphosphonates, such as zoledronate, may reside in the skeleton for even longer. A recent study showed that a single infusion of zoledronate can suppress bone resorption and maintain increased BMD for 5 years in postmenopausal women and 6.5 years in patients with Paget disease (Fig. 17.10).

Several clinical trials have been conducted that have begun to address the important clinical question about how long patients should be treated with bisphosphonates. One clinical study following patients after their participation in a fracture trial where they were treated with alendronate showed that serum biomarkers do not

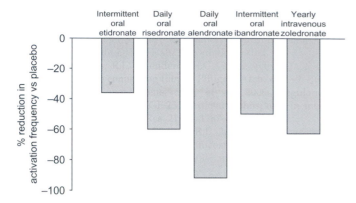

FIGURE 17.7 **The main mechanism of action for bisphosphonates is remodeling suppression.** Iliac crest biopsies have been assessed for the five different bisphosphonates currently approved for treatment of osteoporosis in the USA. They all significantly reduce activation frequency, an index of bone remodeling, albeit to various degrees. *Data from Hodsman et al. Bone and Mineral 1989; 5:201–212; Recker et al. Osteoporosis Int 2004;15:231–237; Chavassieux et al. J Clin Invest 1997;100:1475–1480, Recker et al. J Bone Miner Res, 2008;23:6–16, and Eriksen et al. Bone 2002;31:620–625.*

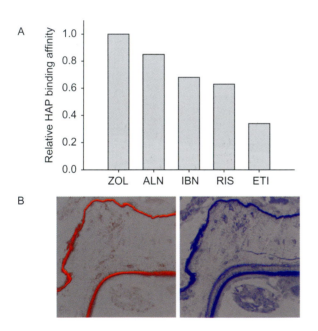

FIGURE 17.8 (A) Bisphosphonates have significant differences in mineral binding affinity which dictates where they first become attached to the matrix and how easily they come off. Those with highest affinity are thought to concentrate at the bone surfaces, while lower affinity ones penetrated deeper into the matrix through the canaliculi. (B) Proof of this concept has been generated by using fluorescently tagged bisphosphonates. The photomicrograph shows that the agent with lower affinity (blue) penetrates deeper into the bone matrix compared to the agent with higher affinity (red). ALN, alendronate; CLO, clodronate; ETI, etidronate; IBN, ibandronate; RIS, risedronate; ZOL, zoledronate. *Data in (A) from Nancollas, G. H., et al. Bone 2006;38:617–627.*

return to control levels even 5 years after withdrawal of alendronate treatment. In this study, vertebral BMD was maintained over that period even without treatment, although hip BMD began to decline slightly by the end of the 5-year period. More importantly, the total rate of vertebral and nonvertebral fractures was not different in those who received alendronate for 10 years (17.7%) compared to those who received alendronate for 5 years and were then switched to a placebo for 5 years (16.9%). When data from three extension trials were analyzed, no consistent effect on vertebral and nonvertebral fracture risk could be identified with continued treatment, beyond the risk reduction seen in the respective original trials, though some subgroups of patients at high risk did appear to have some additional effect on fracture risk. This has resulted in the FDA recommending caution when considering continuing treatment with a bisphosphonate for more than 5 years.

The bisphosphonates have been associated with several side effects. Oral bisphosphonates can cause irritation of the GI tract. About one-third of patients who take intravenous bisphosphonates will experience an *acute phase reaction*, which is accompanied by flu-like symptoms. It has been suggested that this is caused by inhibition of the mevalonate pathway, which is accompanied by accumulation of isoprenoid lipids. These lipids are somehow released and bind to γδ T-cell receptors, causing activation and proliferation of these T cells and an acute phase response. A more serious but uncommon side effect that has been associated with high dose administration of bisphosphonates for cancer is the development of exposed bone in the jaw, or osteonecrosis of the jaw. Although the suppression of remodeling caused by high doses of the bisphosphonates is probably a part of the etiology for this condition, poor dental hygiene and bacterial infiltration causing an osteomyelitis-like condition also probably play some role. More recently, unusual fractures of the femoral shaft (atypical femoral fractures) have been observed in a very small number of patients taking bisphosphonates for osteoporosis (for an average of 5–7 years), although these fractures are also seen in patients who have not taken bisphosphonates. These fractures typically occur without a fall and can be preceded by pain and a visible periosteal reaction (formation of woven bone) on the lateral cortex of the femur on radiographs. They may occur bilaterally and radiographically are similar to stress fractures in appearance. Further, they may be impacted by the suppression of remodeling, although a causal relationship has not yet been established.

Denosumab

Recently, advances in the understanding of the central importance of RANK ligand [tumor necrosis factor

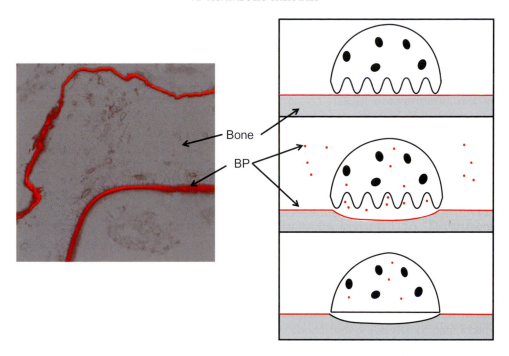

FIGURE 17.9 **Bisphosphonates reduce remodeling by frustrating osteoclasts.** Upon dosing, the drugs bind to bone surfaces. When osteoclasts fuse with the surface and begin to resorb the matrix, the surface-bound bisphosphonate is liberated and endocytosed into the osteoclasts where they either accumulate as ATP analogs (non-nitrogen-containing bisphosphonates) or disrupt protein prenylation necessary for osteoclast function (nitrogen-containing bisphosphonates). This prevents the osteoclasts from resorbing additional bone, although it may not necessarily cause the cell to die. The initial resorption of bisphosphonate-containing matrix may also liberate drug into the general circulation, where it will then rebind with exposed mineral. BP, bisphosphonate.

ligand superfamily member 11/receptor activator of the NF-κB ligand (RANKL)], its receptor [tumor necrosis factor receptor superfamily member 11A (RANK)], and osteoprotegerin/tumor necrosis factor receptor superfamily member 11B (OPG) for osteoclast recruitment and function have resulted in a fundamentally new approach to the treatment of osteoporosis: the use of an antibody (denosumab) specifically developed to target RANK ligand. As outlined in Chapter 2, RANKL is produced by osteoblasts and stromal cells and binds to the RANK receptor to stimulate the development and activity of osteoclasts. The antibody mimics the natural inhibitor of RANKL, OPG.

Denosumab acts at several levels to prevent bone resorption by inhibiting osteoclast formation, function, and survival. RANKL is important in the differentiation of osteoclasts. The antibody prevents RANKL from binding to RANK on the osteoclast precursor, which inhibits the development of osteoclasts. It also prevents the activation of fully differentiated osteoclasts, and interferes with their eventual survival. Although both the antibody to RANKL and bisphosphonates are primarily inhibitors of bone resorption, they have different effects on osteoclasts (Fig. 17.11). By blocking differentiation and reducing survival, denosumab reduces the number of osteoclasts. Bisphosphonates, on the other hand, disable the

osteoclast by blocking the FPS pathway and causing prenylation of GTPases, but the osteoclasts themselves, although inactive, may persist in the tissue. Thus, viewed histologically, one might not see a reduction of osteoclast numbers following treatment with the bisphosphonates, but one would see fewer osteoclasts following treatment with denosumab.

Because denosumab works through a receptor-ligand complex and is not retained in bone, once antibody levels decline, bone remodeling increases and BMD tends to decline toward pretreatment levels within about 12 months (Fig. 17.12). This contrasts with the effects of bisphosphonates, which are retained once dosing is stopped. Denosumab withdrawal may cause an "overshoot" in which BMD temporarily declines below pretreatment levels, but then recovers to pretreatment levels.

The Fracture Reduction Evaluation of Denosumab in Osteoporosis (FREEDOM) trial provides the basis for understanding the efficacy and safety of denosumab. In the trial, 7868 postmenopausal women were treated with either 60 mg denosumab by subcutaneous injection twice a year, or placebo. Compared with placebo treatment, denosumab resulted in a relative risk reduction in the incidence of new vertebral fractures (RR = 0.32) as well as nonvertebral fractures (RR = 0.80) and hip fractures (RR = 0.6). Bone turnover, assessed

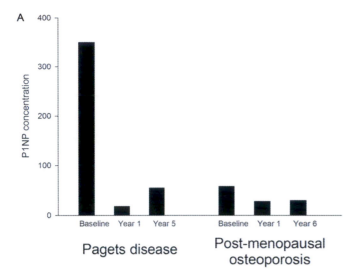

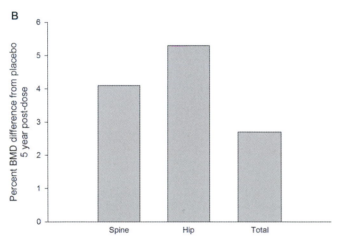

FIGURE 17.10 The most potent and high affinity bisphosphonate, zoledronate, has been shown to have long-lasting efficacy following a single dose. (A) Five years following a single dose in
postmenopausal osteoporotic women, P1NP levels (a biomarker of
bone remodeling activity) remain significantly lower than baseline
while bone mineral density (BMD) gains in the spine, hip, and total
bone remain present. (B) Sustained efficacy in remodeling suppression has also been shown 6.5 years following a single dose in Paget
disease patients, who have very high baseline turnover rates (A).
*Data from Reid et al. Journal of Bone and Mineral Research 2011;
26:2261−2270 and Grey et al. Bone 2012;50:1389−1393.*

with biomarkers, was dramatically reduced by 86%.
Common adverse events include back and musculoskeletal pain, hypercholesterolemia, and cystitis.
Warnings include hypocalcemia, serious infections,
osteonecrosis of the jaw, and suppression of bone turnover. Recently, two cases of an atypical femoral fracture
in patients treated with denosumab have been reported,
both in extension trials. One patient had been on denosumab for 3 years, following 3 years on placebo. The
other patient had been on denosumab for 7 years. This
suggests that this type of fracture is associated with

long-term suppression of remodeling (although some
fractures have occurred with a shorter treatment, or
with no treatment at all), and that it is not a bisphosphonate class effect.

While both denosumab and the bisphosphonates
reduce fracture risk primarily by increasing bone
mass, denosumab is slightly more potent than the
bisphosphonates and may almost completely inhibit
bone remodeling, as evidenced by the absence in
many cases of double fluorochrome labels in iliac
crest biopsies. This level of suppression can create
potential hypocalcemia in a patient whose bone turnover is already low. It may also cause the accumulation of unrepaired microdamage in bone. However,
the drug is only administered (by injection) once
every 6 months, and biomarkers of remodeling suggest that remodeling begins to increase over the couple of months prior to the next injection in some but
not all patients. There may be sufficient remodeling
during this time to prevent the long-term accrual of
damage, although remodeling suppression 6 months
following a dose is still about 70%, i.e. about the
same as with oral alendronate.

ANABOLIC THERAPY

The 84-amino-acid PTH is the primary regulator of
calcium and phosphate metabolism in bone and kidney (see Chapters 13 and 15.). Physiologic actions of
PTH include regulation of bone metabolism, renal
tubular reabsorption of calcium and phosphate, and
intestinal calcium absorption.

Teriparatide is identical to the 34 N-terminal amino
acids of endogenous PTH. Both full length PTH and
teriparatide bind to both PTH receptors with the same
affinity and have the same physiologic actions on bone
and kidney. The skeletal effects of teriparatide result
from the pattern of systemic exposure resulting from a
subcutaneous injection, which is considered intermittent because serum levels rise quickly after a dose and
then decline over a few hours. Once-daily administration of teriparatide stimulates new bone formation on
trabecular and cortical (periosteal and/or endosteal)
bone surfaces by the preferential stimulation of osteoblastic activity over osteoclastic activity. In contrast,
continuous excess of endogenous PTH, as occurs in
hyperparathyroidism, stimulates bone resorption more
than bone formation with a characteristic loss of bone
(Fig. 17.13).

At the level of bone tissue, daily administration of
teriparatide has several actions. Its first effect is to
stimulate the apposition of bone directly to trabecular
surfaces, probably through modeling processes that do

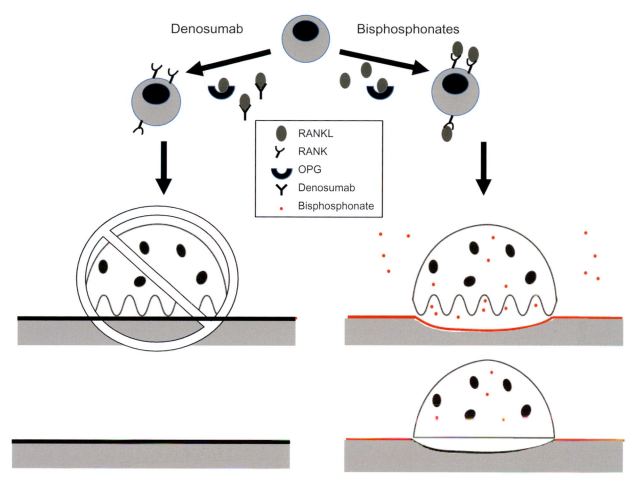

FIGURE 17.11   Denosumab, an antibody that targets the RANK ligand [tumor necrosis factor ligand superfamily member 11/receptor activator of the NF-κB ligand (RANKL)], suppresses remodeling but does so in a fundamentally different way than do bisphosphonates. Denosumab inhibits osteoclast formation, function, and survival and thus affects the number of mature osteoclasts. Conversely, bisphosphonates inhibit only the activity of osteoclasts and thus the mature cells are often found in normal numbers on bone surfaces when examined histologically. OPG, osteoprotegerin/tumor necrosis factor receptor superfamily member 11B; RANK, tumor necrosis factor receptor superfamily member 11A.

not require prior bone resorption. Teriparatide subsequently stimulates bone remodeling, like endogenous PTH would, but with a greater effect on osteoblasts compared to osteoclasts. The first initial effect, i.e. formation without an increase in prior resorption, creates an "anabolic window" in which bone volume increases rapidly. The subsequent increase in bone remodeling increases bone volume by allowing osteoblasts to "overfill" resorption spaces, resulting in a positive BMU balance (see Chapter 4, Fig. 4.11). In other words, although resorption is increased, each of the erosion spaces that is created is filled with more bone than is removed, thus increasing bone mass further. This results in an increase in bone mass, improved microarchitecture (as trabecular thickness and connectivity increase), and reduced fracture risk. The precise mechanism for the differential effects of intermittently administered PTH and endogenous PTH is unknown,

although the brief elevation of rhPTH(1−34) levels may modulate the cell cycle in positive ways to create some of the effect (See Chapter 15 for more details).

The relationship between the duration of exposure and the anabolic effect on bone was explored in an animal study in which rats were given a subcutaneous injection of PTH as an infusion for 1 h/day, 2 h/day, or continuously. Treatment for longer periods was associated with more osteoclasts and fewer osteoblasts, as would occur with constitutively active endogenously released PTH. How this happens is incompletely understood, but it may occur through differential effects on osteoclast signaling molecules.

While a favorable skeletal effect of intermittent PTH administration was first suggested by animal studies in the 1920s it was not until the past few decades that this was evaluated in humans. Very small clinical trials

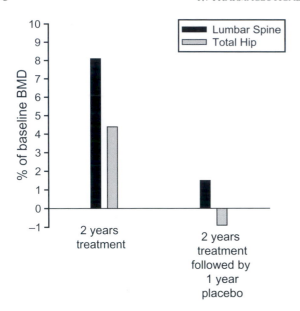

FIGURE 17.12 Women with postmenopausal osteoporosis exhibit robust bone mineral density (BMD) responses to denosumab treatment at both the hip and spine. In contrast to bisphosphonates, which bind to the bone and have long-lasting remodeling suppression effects after treatment withdrawal, the withdrawal of denosumab treatment leads to reversal of effect. Just 1 year after cessation of treatment, the gains in BMD over the previous 2 years are lost at both the hip and spine. *Data from Miller et al. Bone 2008;43:222–229 and Miler et al. J Clin Endocrinol Metab 2011;96:394–402.*

were initiated in the late 1970s and early 1980s, but not until the mid-1990s was a large fracture outcome trial undertaken. In the pivotal clinical trial, teriparatide at a dose of 20 μg/day increased bone density at the spine and hip, and decreased the risk of vertebral fractures by 65% and nonvertebral fractures by 35% (Fig. 17.14). The most common adverse events reported by people taking teriparatide include arthralgias, pain, and nausea. In long-term toxicology testing, teriparatide was given to rats throughout their lifetime. Some of the rats developed osteosarcomas, a malignant type of bone tumor. In the rodent study, the incidence of osteosarcoma was related to the dose and duration of treatment. Subsequently, no bone lesions were found in two long-term studies in nonhuman primates. Because of the uncertain relevance of the findings to humans, administration of teriparatide is limited to adults at high risk for fracture and should be given for a period of only 2 years. After 10 years of clinical use, no significant increase over the background incidence for osteosarcoma has been detected.

One important factor to consider in making treatment decisions is that once teriparatide is stopped, bone formation decreases more quickly than bone resorption and the new bone that was added to the skeleton during treatment tends to decrease. In part because of this, experiments using combinations of teriparatide and bisphosphonates have been undertaken.

COMBINATION TREATMENTS

With the availability of several therapies that have different mechanisms of action, investigators have begun to explore whether using more than one treatment for a patient might provide greater benefit than using a single treatment (Table 17.1). From a mechanistic perspective, the most interesting combinations involve teriparatide, the only approved bone anabolic agent, with one of the antiresorptives. The hypothesis in this case is that the anabolic agent will increase bone mass by stimulating formation, and the antiremodeling agent will reduce loss caused by increased remodeling rate and the imbalance of resorption over formation, thus giving benefits to both the formation and resorption sides of the remodeling process. There are three basic ways that these treatments could be combined: concurrent treatment; antiresorptive followed by anabolic; and anabolic followed by antiresorptive. A smaller number of studies have also examined the efficacy of combining two antiresorptive agents. None of these studies have been designed to test the effect of combining treatments on fracture risk. This information will be essential for informed clinical decision making.

Concurrent Treatment with Antiresorptive and Anabolic Agents

The rationale for concurrent treatment with these two different classes of agents is to determine whether a more favorable balance of bone formation and resorption can be established compared with treatment with either agent alone.

Most studies of concurrent treatment with alendronate and PTH have not provided evidence for a substantial clinical advantage for increasing BMD. A recent study using a single dose of zoledronic acid combined with teriparatide showed that spine and hip BMD was higher in the combination group compared to either treatment alone at 13 and 26 weeks. However, after 1 year spine BMD was similar in the combination and the teriparatide groups and greater than zoledronate alone. After 1 year, hip BMD was similar in both the combination and zoledronate groups and greater than with teriparatide alone.

Although the use of hormone replacement therapy (HRT) for osteoporosis has significantly declined since the results of the WHI were reported, prior studies investigated combination HRT and teriparatide

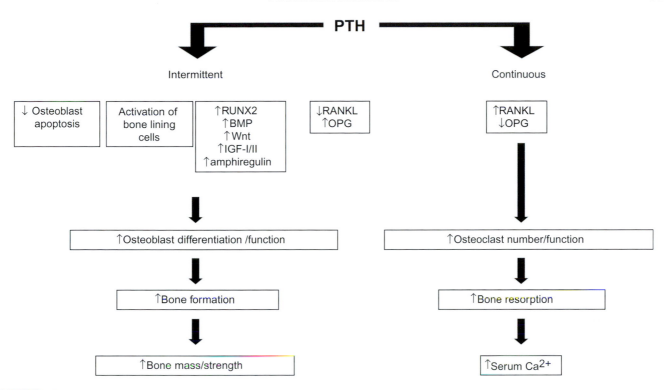

FIGURE 17.13 Parathyroid hormone (PTH) exerts paradoxical effects depending on whether it is chronically elevated (such as with hyperparathyroidism) or intermittently elevated (such as with pharmacological treatment). Continuous elevation leads to increased resorption, while intermittent elevation is associated with increased formation. Ca^{2+}, calcium; IGF, insulin-like growth factor; OPG, osteoprotegerin/tumor necrosis factor receptor superfamily member 11B; RANKL, tumor necrosis factor ligand superfamily member 11/receptor activator of the NF-κB ligand.

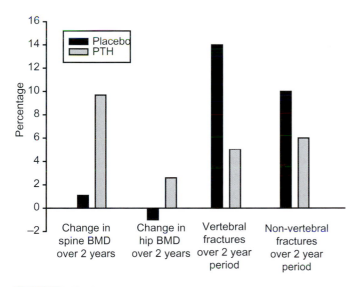

FIGURE 17.14 Postmenopausal women treated for 2 years with teriparatide (PTH) experience significant increases in hip and spine bone mineral density (BMD) compared to placebo-treated controls. This is associated with a reduction in both vertebral and nonvertebral fractures. *Data from Neer et al. N Engl J Med 2001;334:1434–1441.*

treatment. Women who were treated with HRT for more than 2 years experienced significant increases in spine and hip BMD when teriparatide was concurrently administered compared to those given HRT alone. Related work on the combination of raloxifene and teriparatide is consistent with the HRT results, showing robust increases in femoral neck BMD that exceed those of either raloxifene or teriparatide alone.

Sequential Treatment with an Antiresorptive Followed by Anabolic Agents

Antiresorptive therapy represents the first line of treatment for those with osteoporosis. Despite the efficacy of these agents, some patients continue to fracture and have low BMD during or after treatment with an antiresorptive agent. Therefore, several studies have examined whether switching to an anabolic agent could be a good strategy.

Sequential treatment with teriparatide after an antiremodeling agent has positive effects that vary between the antiremodeling agents and the region of the skeleton. A delay or blunting of the expected BMD increase with teriparatide in these patients, if present, can usually be overcome. For example, 1 year of teriparatide treatment has a greater effect on bone formation, measured by serum biomarkers, and BMD of the spine in patients treated previously for 2 years with risedronate than in those treated previously with alendronate. These data contrast with those of other

TABLE 17.1   Summary of Combination Treatments

Antiresorptive followed by Anabolic	BP followed by PTH	PTH less effective
	SERM followed by PTH	Less attenuation of PTH effect than with BP
Antiresorptive + anabolic	BP + PTH	PTH less effective with ALN; equal efficacy as monotherapy with ZOL
	SERM + PTH	Less attenuation of PTH effect than with BP
	HRT + PTH	Less attenuation of PTH effect than with BP
Antiresorptive + cyclic anabolic	ALN with PTH on/off for 3 months each	BMD increase similar to PTH alone
Anabolic followed by antiresorptive	PTH followed by BP	BP maintains PTH benefit
Antiresorptive + antiresorptive	HRT + BP	Small additive effects on BMD
	ALN + SERM	Small additive effects on BMD

ALN, alendronate; BMD, bone mineral density; BP, bisphosphonate; HRT, hormone replacement therapy; PTH, parathyroid hormone, SERM, selective estrogen receptor modulator; ZOL, zoledronate.

studies that observed progressive increases in both spine and hip BMD regardless of whether patients previously received alendronate or risedronate. Importantly, this later study included a group that was treatment naive prior to teriparatide treatment and showed that, while the response to teriparatide was similar in these patients to bisphosphonate-treated patients at the spine, the response at the hip was significantly blunted.

Pretreatment with HRT did not affect the BMD response at the spine, total hip, or femoral neck compared to those who were treatment naive prior to teriparatide therapy. Head-to-head studies looking at the conversion to teriparatide from either alendronate or raloxifene treatment demonstrated a 10% increase in spine BMD in patients previously treated with raloxifene compared to only 4% in those treated with alendronate. Hip BMD was increased by teriparatide only in those previously treated with raloxifene.

Sequential Treatment with Anabolic Agents Followed by an Antiresorptive

Because there is a 2-year limit on teriparatide therapy and clinical data demonstrates loss of BMD after teriparatide is discontinued, it is also logical to evaluate whether antiresorptive therapy can maintain the gains in BMD following cessation of teriparatide.

Patients treated with PTH(1−84) for 1 year followed by treatment with alendronate maintained BMD at the spine in year 2, whereas those who switched to placebo in year 2 lost significant BMD. Several other clinical studies have examined the effects of giving teriparatide for 2 years and then switching patients to either alendronate or ceasing treatment. Those switched to alendronate gained additional BMD at both the spine and hip, while those who ceased all treatment lost significant BMD over the following 2.5 years. These results illustrate the effect of subsequent bisphosphonate treatment to stabilize or further increase BMD after PTH. They also suggest, although a bit less clearly, that sequential therapy, specifically anabolic followed by antiresorptive, can have significant benefits on BMD at the hip beyond those of antiremodeling agents given alone.

Switching to raloxifene after teriparatide results in site and time-specific effects. At the spine, those who were treated with teriparatide for 2 years continued to experience increases in BMD throughout the treatment period, while those switched to raloxifene after a year simply maintained the gains from the first year. This is contrasted by results at the hip, in which at the end of 2 years, those taking teriparatide for a year and raloxifene for a year had similar BMD values as those who took teriparatide for the whole duration.

Concurrent Treatment with Two Antiresorptive Agents

Despite the absence of a strong rationale for this combination treatment, there is a fair amount of data suggesting that it is effective. Most of this stems from studies that were undertaken soon after the emergence of bisphosphonates into the clinical arena—when the use of HRT was still common. These studies showed consistently greater effects from combining HRT and bisphosphonate compared to either monotherapy alone. This is true whether the treatments were given together or sequentially (HRT followed by bisphosphonate) and is independent of the type of bisphosphonate (both alendronate and risedronate were studied). More recent data combining raloxifene with alendronate show that BMD tends to increase to a greater extent than with either monotherapy alone. This effect on BMD has not been associated with a greater effect on fracture risk reduction and of course needs to be balanced with any concerns that would arise from a greater suppression of bone turnover.

EMERGING THERAPIES

There are several emerging therapies that show promise for treating osteoporosis. One of these, anti-sclerostin antibody, may become the second anabolic agent for treatment of osteoporosis, after teriparatide. The most recent research suggests that it may uncouple formation and resorption, thus increasing formation but reducing resorption slightly. This is a distinctly unique mode of action compared to teriparatide, which promotes net bone formation, but mechanistically at the metabolic expense of increased local resorption. The other emerging agent, anti-cathepsin K, attempts to specifically target bone resorption by interfering with the osteoclast's ability to enzymatically digest and remove collagen in the sealed zone between the osteoclast and the bone surface. Unlike the bisphosphonates, which disrupt the internal cellular processes of osteoclasts and may promote osteoclast apoptosis, inhibiting cathepsin K (Cat K) does not inactivate or kill the cell itself; it just prevent its products from performing their normal enzymatic functions. These two novel agents could each fill a unique niche in the treatment of osteoporosis.

Anti-Sclerostin Antibody

Sclerosteosis and van Buchem disease are two rare sclerosing bone disorders that were first described in the 1950s. The skeletal manifestations in affected individuals include high bone mass and are the result of endosteal hyperostosis that is most pronounced in the mandible and skull. Patients with sclerosteosis have a more severe phenotype than those with van Buchem disease that includes syndactyly (webbed or conjoined fingers). Sclerosteosis is associated with mutations in the SOST gene (encodes the sclerostin protein), whereas patients with van Buchem disease have deletions in a downstream region of the gene that contains regulatory elements for SOST transcription.

In adults, sclerostin is secreted by osteocytes and regulates bone mass by actions on Wnt signaling, leading to changes in gene transcription. Specifically, sclerostin has been shown to antagonize canonical Wnt signaling by binding to LRP4/5/6, thereby decreasing Wnt-β-catenin signaling through frizzled receptors and downregulating bone formation. The antibody to sclerostin blocks binding to LRP4/5/6, which would otherwise inhibit Wnt signaling and allow accumulation of the intracellular signaling molecule β-catenin, which can translocate to the nucleus and induce gene transcription that stimulates bone formation. Histomorphometric analyses show that blocking sclerostin with the antibody increases osteoblast number but does not change osteoclast number. Therefore, an antibody to sclerostin may represent a future anabolic approach to osteoporosis therapy.

Recent evidence suggests that anti-sclerostin antibody may also have a slight effect on reducing overall bone resorption, in addition to its anabolic effect. This is because the antibody releases the inhibition on osteoblasts, thus allowing them to produce more OPG, which reduces the activation of osteoclasts by binding to RANKL (see Chapter 2). Recent reports using Sost knockout mice also suggest that the absence of sclerostin induces apoptosis in B cells and downregulates the immune system. This has not been reported with the use of anti-sclerostin antibody, and the role of sclerostin in modulating immune function remains to be proven conclusively.

Interestingly, intermittent PTH, the only approved anabolic therapy for osteoporosis, stimulates bone formation in part by inhibiting the expression of sclerostin. Postmenopausal women treated with teriparatide have lower serum sclerostin levels, and serum sclerostin levels are negatively correlated to PTH levels in healthy women. Animals in which the Sost gene has been knocked out are unable to respond to intermittent PTH. In contrast, glucocorticoids stimulate sclerostin expression and thereby inhibit bone formation.

A Phase I (dose-ranging) study to assess the effects of anti-sclerostin antibody in postmenopausal women with low BMD showed significant increases in BMD at the lumbar spine and hip, with a similar number of adverse events compared with the control group (Fig. 17.15). The rise in BMD occurred within 3 months following a single injection and was the result of both stimulation of bone formation and inhibition of bone resorption (Fig. 17.15). Anti-sclerostin antibody is currently being studied in Phase III clinical trials that have fracture incidence as the primary end point.

Cathepsin K Inhibitors

Cat K is a lysosomal cysteine protease. Although found in several cell types, its expression in osteoclasts is at least several hundred-fold higher compared to other cells. Upon osteoclast attachment to the bone surface, Cat K is secreted into the resorption lacuna

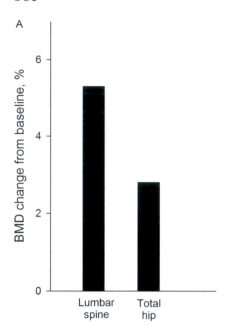

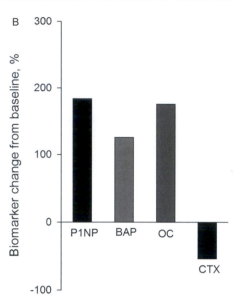

FIGURE 17.15 Three months following a single dose of anti-sclerostin antibody bone mineral density (BMD) of the hip and spine were both significantly increased over baseline in postmenopausal women. These results were manifest through both an increase in bone formation biomarkers (P1NP, BAP, OC) and reduction in bone resorption biomarkers (CTX). These data suggest the processes of formation and resorption are uncoupled and thus the potential exists for prolonged treatment to have a larger effect on BMD than the anticatabolic or anabolic agents currently in use. *Data from Padhi et al. J Bone Min Res 2011;26:19–26.*

where it functions in the degradation of the organic matrix (see Chapter 2). Specifically, it cleaves the N-telopeptide and C-telopeptide fragments of type I collagen. The genetic disorder pycnodysostosis, an osteochondral dysplasia resulting in low bone resorption and high bone mass, is due to the perturbation of Cat K function. This phenotype has been reproduced in the Cat K-null mouse. Beyond this genetic mouse model, most preclinical work has been done in nonhuman primates and rabbits due to the significant differences in the Cat K amino acid sequence in humans and rodents making treatment with human Cat K inhibitors ineffective.

Inhibition of Cat K shows promise as a pharmacologic target for the treatment of bone disease. Several Cat K inhibitors (balicatib, odanacatib, and relacatib) have been investigated as pharmacologic agents, but only odanacatib is still under consideration. Phase I clinical dose-ranging studies with odanacatib have shown that both daily and weekly oral dosing is well tolerated and produces the expected reductions in urine resorption biomarkers across various doses. Odanacatib imparts progressive dose-related increases in lumbar spine and total hip BMD over 2 years (Fig. 17.16). In a small cohort of individuals treated for 3 years with once-weekly dosing (50 mg), BMD increased at the spine (7.9%), hip (5.8%), and femoral neck (5.0%) during the treatment period. Phase II trials show continued efficacy of odanacatib up to 5 years for spine and hip BMD. Large Phase III fracture trials with once-weekly odanacatib are in progress.

Cat K inhibition has several unique features compared to the other antiresorptive agents used as treatments for metabolic bone disease. Odanacatib treatment specifically targets osteoclast activity, with no apparent effect on cell number. Both indirect assessment of cell number by serum assessments of the tartrate-resistant acid phosphatase, TRAP5b, as well as direct measures of cell number by histology, have shown that osteoclast number is unchanged with Cat K inhibition. This means that, although resorptive activity of the cells is disrupted, other functions of these cells, such as signaling in the local microenvironment, are retained.

The effects on bone formation also appear to differ from those of other antiremodeling agents. Despite significant reductions in bone resorption markers, odanacatib either transiently suppresses or has no effect on bone formation biomarkers [bone-specific alkaline phosphatase (BSAP) and osteocalcin]. This is supported by iliac crest biopsy analyses, in which no difference is documented in osteoid surface or bone formation rate between odanacatib-treated and placebo-treated patients. Histologic analyses of vertebrae and long bones in ovariectomized rabbits and nonhuman primates also show no effect on bone formation. Some preclinical studies have also shown stimulation of periosteal bone formation (Fig. 17.16). The primate and rabbit studies have clearly shown benefits of odanacatib on bone strength, both in the vertebrae and in the long bones.

Similar to denosumab and different from the bisphosphonates, discontinuation of odanacatib results in a reversal of effect, as assessed by both biomarkers and BMD (Fig. 17.16). Following 2 years of treatment with

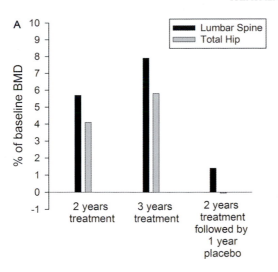
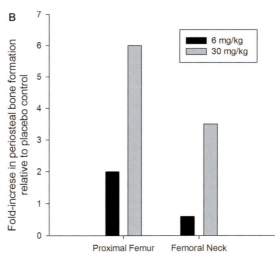

FIGURE 17.16 Odanacatib, a cathepsin K inhibitor, significantly increases both hip and spine bone mineral density (BMD). Similar to denosumab, withdrawal of treatment results in rapid loss of the BMD gain. One potential unique characteristic of cathepsin K inhibitors among the anticatabolic agents is their effect on periosteal bone. Several preclinical studies have documented that cathepsin K stimulates periosteal bone formation. Given the importance of periosteal expansion for biomechanical properties this could have important effects on fracture risk reduction at sites with large amounts of cortical bone (e.g. femoral neck). *Data from Bone et al. J Bone Min Res 2010;25:937−947; Eisman, et al. J Bone Min Res 2011;26:424−251; and Cusick et al. J Bone Min Res 2012;27:524−537.*

weekly odanacatib, withdrawal of treatment leads to a rapid loss of BMD over the first 6 months, reaching pre-treatment values by 1 year posttreatment. Biomarkers were substantially increased upon withdrawal but were comparable to baseline values after 1 year.

TREATMENT DECISIONS

The choice of therapy for an individual patient should include consideration of the patient's clinical history and preferences, as well as the risks and benefits of a particular therapy. A comparison of the efficacy of different treatments would be helpful in decision making, however direct head-to-head comparisons between treatments can require large and lengthy studies requiring 2000−10,000 patients or more. In the absence of prospective trials directly comparing the effect of different treatments on fracture incidence, researchers have used statistical approaches to compare results across studies (Fig. 17.17).

The patient's clinical history and the epidemiology of osteoporotic fractures may influence the choice of treatment. Younger patients are at highest risk for spine fracture and older patients at higher risk for hip fractures. If a relatively young woman is considering treatment, then it may be most important that the treatment should have a significant effect in reducing the risk of spine fractures. The tolerability or side effects of a treatment must also be considered. Although significant progress

has been made since the 1990s in treating osteoporosis and preventing fractures, no agent can prevent all fractures. Another important question is how long to treat a patient. The answer to this is not known with certainty. A few years ago, antiresorptives in bone were considered in the same way that glucose-lowering agents were used for diabetes, or cholesterol lowering agents were used for heart disease, i.e. as a long-term treatment that will continue to benefit the patient. In 2011, an FDA review of clinical studies measuring the effectiveness of long-term bisphosphonate use showed that some patients may be able to stop using bisphosphonates after 3−5 years and still continue to benefit from their use.

Remarkable progress has been made since the 1990s in developing new treatments for osteoporosis. Not that long ago patients at high risk for fracture could only receive estrogen, calcitonin, or supplements of calcium plus vitamin D; now there is a range of agents that have demonstrated substantial efficacy based on large scale controlled trials, and which can be used under different patient circumstances. Several newer classes of agents that may further improve our ability to manage the disease are on the horizon.

STUDY QUESTIONS

1. Compare estrogen and SERMs as treatments for postmenopausal osteoporosis.
2. Describe the mechanism of action of nitrogen-containing bisphosphonates.

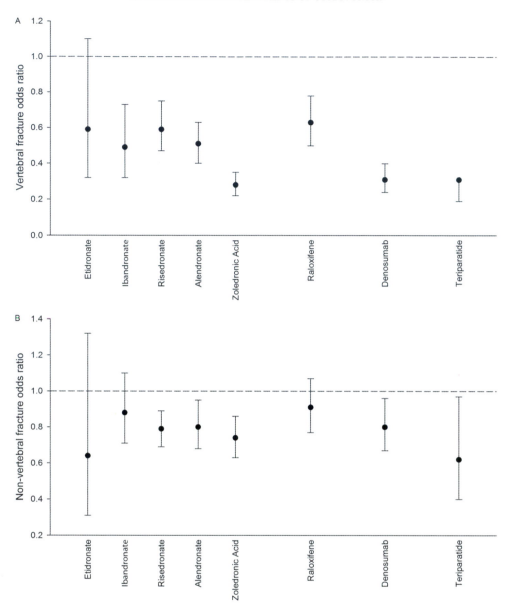

FIGURE 17.17   Direct head-to-head comparisons of fracture risk reduction between treatments do not exist and thus statistical approaches must be used to compare results across the individual fracture studies. Odds ratios can be interpreted as the percent decrease of an event relative to a control population. An odds ratio of 0.8 means that the treatment showed a 20% reduction in the risk of fracture for a particular trial. *Data from Hopkins et al. BMC musculoskeletal disorders 2011;12:209.*

3. What is the potential advantage of combination therapy over monotherapy for treating osteoporosis? What are the current limitations of combination therapy?

4. Your patient has osteoporosis but is unable to take bisphosphonates. What are the two treatment alternatives for this patient? Explain briefly how these treatments work.

5. Describe an emerging bone drug and explain its mechanism of action.

Suggested Readings

Austin M., Yang Y.-C., Vittinghoff E., Adami S., Boonen S., et al., for the FREEDOM Trial, 2011. Relationship between bone mineral density changes with denosumab treatment and risk reduction for vertebral and nonvertebral fractures. J. Bone Miner. Res. 27, 687–693.

Bischoff-Ferrari, H.A., Dawson-Hughes, B., Willett, W.C., Staehelin, H.B., Bazemore, M.G., Zee, R.Y., et al., 2004. Effect of vitamin D on falls: a meta-analysis. JAMA. 291, 1999–2006.

Bischoff-Ferrari, H.A., Willet, W.C., Oray, E.J., Lips, P., Meunier, P.J., Lyons, R.A., et al., 2012. A pooled analysis of vitamin D dose requirements for fracture prevention. N. Engl. J. Med. 367, 40–49.

Black, D.M., Delmas, P.D., Eastell, R., Reid, I.R., Boonen, S., Cauley, J.A., et al., 2007. Once yearly zoledronic acid for treatment of postmenopausal osteoporosis. N. Engl. J. Med. 356, 1809–1822.

Body, J.-J., Bergmann, P., Boonen, S., Devogelaer, J.-P., Gielen, E., et al., 2012. Extraskeletal benefits and risks of calcium, vitamin D and anti-osteoporosis medications. Osteoporos Int. 23 (Suppl. 1), S1–S23.

Burr, D.B., Russell, R.G.G., (Eds.), 2011. BONE: Special Issue on Bisphosphonates. 49, 1 – 146.

Costa, A.G., Cusano, N.E., Silva, B.C., Cremers, S., Bilezikian, J.P., Cathepsin, K., 2011. Its skeletal actions and role as a therapeutic target in osteoporosis. Nat. Rev. Rheumatol. 7, 447–456.

Cummings, S.R., San Martin, J., McClung, M.R., Siris, E.S., Eastell, R., et al., for the FREEDOM Trial, 2009. Denosumab for prevention of fractures in postmenopausal women with osteoporosis. N. Engl. J. Med. 361, 756–765.

Ettinger, B., Black, D.M., Mitlak, B.H., Knickerbocker, R.K., Nickelsen, T., et al., 1999. Reduction of vertebral fracture risk in postmenopausal women with osteoporosis treated with raloxifene: results from a 3-year randomized clinical trial. Multiple Outcomes of Raloxifene Evaluation (MORE) Investigators. JAMA. 282, 637–645.

Jackson, R.D., LaCroix, A.Z., Gass, M., Wallace, R.B., Robbins, J., Lewis, C.E., et al., 2006. Calcium plus vitamin D supplementation and the risk of fractures. N. Engl. J. Med. 354, 669–683.

Papapoulos, S., Chapurlat, R., Libanati, C., Brandi, M.L., Brown, J. P., et al., 2012. Five years of denosumab exposure in women with postmenopausal osteoporosis: results from the first two years of the FREEDOM extension. J. Bone Miner. Res. 27, 694–701.

Rejnmark, L., Avenell, A., Masud, T., Anderson, F., Meyer, H.E., et al., 2012. Vitamin D with calcium reduces mortality: patient level pooled analysis of 70,528 patients from eight major vitamin D trials. J. Clini. Endocrinol. Metab. 97, 2670–2681.

Rossouw, J.E., Anderson, G.L., Prentice, R.L., LaCroix, A.Z., Kooperberg, C., Stefanick, M.L., et al., 2002. Risks and benefits of estrogen plus progestin in healthy postmenopausal women: principal results from the women's health initiative randomized controlled trial. JAMA. 288, 321–333.

Russell, R.G., Watts, N.B., Ebetino, F.H., Rogers, M.J., 2008. Mechanisms of action of bisphosphonates: similarities and differences and their potential influence on clinical efficacy. Osteoporos Int. 19, 733–759.

The Womens Health Initiative Steering Committee, 2004. Effects of cojnugated equine estrogen in posstmenopausal women with hysterectomy: the womens health initiative randomized controlled trial. JAMA. 291, 1701–1712.

Index

Note: Page numbers followed by "*f*" and "*t*" refers to figures and tables, respectively.